T0138933

NANOTECHNOLOGY
AND
GLOBAL EQUALITY

DONALD MACLURCAN

NANOTECHNOLOGY
AND
GLOBAL EQUALITY

PAN STANFORD PUBLISHING

Published by

Pan Stanford Publishing Pte. Ltd.
Penthouse Level, Suntec Tower 3
8 Temasek Boulevard
Singapore 038988

Email: editorial@panstanford.com
Web: www.panstanford.com

British Library Cataloguing-in-Publication Data
A catalogue record for this book is available from the British Library.

ISBN 978-981-4303-39-2 (Hardcover)
ISBN 978-981-4303-40-8 (eBook)

Printed in the USA

*"Anyone who believes exponential growth
can go on forever in a finite world
is either a madman or an economist."*

—Kenneth Boulding

Acknowledgements

My deepest thanks go to Associate Professor James Goodman for venturing with me into the unknown and guiding me so wisely. To Associate Professor Mike Ford, and Professor Michael Cortie, thank you for your confidence in sponsoring me as a researcher from outside the sciences and for the valuable direction, insights and opportunities you provided. To my colleagues from the Institute for Nanoscale Technology: Dr Carl Masens, Dr Benjamin Soule de Bas, Dr Burak Cankurtaran and Dr Dakrong Pissuwan, thank you for welcoming me to your ranks, and to my office-companion and mentor Dr David Xu, my heartfelt appreciation. To the staff from the Faculty of Arts and Social Sciences at the University of Technology, Sydney, particularly Associate Professor Paul Ashton, Carolyn Carter and Juleigh Slater, thank you for your guidance and administrative assistance. Special thanks to Dr Patrick Tooth for your assistance with Endnote™ and Dr Leigh Wood for your help with NVivo. To all the interviewees from Thailand and Australia and Emeritus Professor Tony Moon and Dr Patarapong Intarakumnerd for assisting in the facilitation of this research, thank you for making my 'field' experience so valuable. Thanks to Drs Noela Invernizzi and Guilermo Foladori for adding immense depth to my inquiries and special thanks to Dr Sue-Anne Wallace, Dr Natalia Radywyl, Joe Fitzgerald and Kate and James Maclurcan for copyediting, proofreading and advising on matters of structure for various parts of my work. To the Bracken, Coote, Kremer, Inana, Logan, Norton-Knight, and Zehtner families, thank you for caring for me so warmly on my many 'last' writing retreats. To my family and friends, particularly Andrew McMillan, Lauren Simpson, Geoff Moore, Wafa Chafic, Suyin Hor, Tori Saint, Mickey Martin, Anna Louis, Jess and Jon Watkins and Jake and Fiona Logos who long endured my struggles with this work, thank you for your ongoing care and concern. Finally, to my partner, Julie Dardel, thank you for your assistance, patience and ever-loving support.

Preface

This book's journey began with an upturned brochure in a park. I found it walking home one afternoon, spread open to a page about courses at the University of Technology, Sydney on nanotechnology — an area 'set to revolutionise industrial processes'. Whilst the brochure's content made for an interesting read, I was far from a scientist and, once binned, I didn't give it another thought... until two weeks later. I was having a conversation with a staff member at that same university, when the word 'nanotechnology' popped into my head. I casually asked this person if they knew anything about the subject. "No, but I think there are people on level 16 doing that stuff", came the reply. Sure enough, I easily found the newly formed 'Institute for Nanoscale Technology' and, upon entering, was greeted by the Associate Director, Mike Ford. I deferentially explained that my background was not in science and that I was merely looking for more information, given my surprise at finding a brochure about the Institute in a park ten kilometres away. I had been in the room less than five minutes when Dr Ford proposed that, given the Institute was looking to 'branch out', I undertake research looking at nanotechnology's social implications. I sat stunned. Mike and I had never met, I had little idea what 'research' entailed and I still did not have a clue as to what nanotechnology was, let alone the nature of its social implications! Revisiting the basic research proposal I submitted a week later still brings me a laugh, but the novelty of having an 'outsider' in a scientific institute must have blinded everyone to my lack of research experience because three months later I sat down in an office vacated by a visiting professor and started to think about nanotechnology.

At this time, I had been working with The Fred Hollows Foundation, a non-governmental organization whose work in reducing avoidable blindness in the global South had pioneered new technologies and approaches to capacity building. I thus leapt forward with my research, having decided to explore nanotechnology's potential implications for the South and global inequality.

Through this journey I have been able to look in a broad, exploratory manner, at a largely uncharted area that remains surprisingly understudied. Given my research began nine years ago, much of the work in this book is best viewed as a snapshot from and reflection upon an early period of nanotechnology's development — although, arguably, the relevance of many of the insights remain.

Of real privilege has been the chance to ride at 'the boundaries', engaging with the very different approaches of a sociologist and a physicist who oversaw my research. Working across faculties, whilst exploring the interdisciplinary field of nanotechnology, has only added to my interest.

As my understanding and critique of development has deepened, it has become clearer to me that nanotechnology is, at its sociological best, a medium to assess the processes and possible trajectories accompanying technological futures. In an unjust world, where struggles to avoid the co-option and mainstreaming of ideals are ever-present, there would seem to be value in bold creativity, grounded in existing wisdom.

As this book goes to print, I realise that its writing has helped uncover in me a passion for exploring alternatives to the 'growth' paradigm and a particular interest in collaborative junctures between feminist, indigenous, peasant, Marxist and ecological thought. I now see 'the boundaries' as exciting spaces for new reflexivity, and finally recognise that the greatest resilience to avoiding co-option lies, as it always has, at the periphery.

Donnie Maclurcan

Contents

List of Figures

List of Tables

List of Appendices

Commonly Used Acronyms

AFM: atomic force microscope
AMO: atomically modified organism
APEC: Asia Pacific Economic Co-operation
BIOTEC: National Centre for Genetic Engineering and Biotechnology (Thailand)
DNA: deoxyribonucleic acid
ELSI: ethical, legal and social implications
EPO: European Patent Office
ETC Group: Action Group on Erosion, Technology and Concentration
EPO: European Patent Office
E.U.: European Union
FDA: Food and Drug Administration
GDP: gross domestic product
GM: genetically modified
GMO: genetically modified organism
IP: intellectual property
IPRs: intellectual property rights
JPO: Japanese Patent Office
LDC: least developed country
MDG: millennium development goal
MM: molecular manufacturing
MTEC: National Metal and Materials Technology Centre (Thailand)
NANOTEC: National Nanotechnology Centre (Thailand)
NECTEC: National Electronics and Computer Technology Centre (Thailand)
NGO: non-governmental organisation
NNI: National Nanotechnology Initiative (United States of America)
NSF: National Science Foundation (United States of America)
NSTDA: National Science and Technology Development Agency (Thailand)
OECD: Organisation for Economic Cooperation and Development
OTOP: One Tambon One Product
RS&RAE: Royal Society and Royal Academy of Engineering

R&D: research and development
STM: scanning tunnelling microscope
TRIPs: Agreement on Trade Related Aspects of Intellectual Property Rights
U.K.: United Kingdom
U.N.: United Nations
UNCTAD: United Nations Conference on Trade and Development
UNDP: United Nations Development Program
UNESCO: United Nations Educational, Scientific and Cultural Organization
U.S.: United States of America
USPTO: United States Patent and Trademark Office
UTJCB: University of Toronto Joint Centre for Bioethics
WHO: World Health Organisation
WTO: World Trade Organisation

Scientific Glossary

Molecular Manufacturing: An anticipated technology based on Richard Feynman's vision of factories using nanoscale machines to build complex products, including additional nanoscale machines.

Nanometre: One billionth of a metre or 10^{-9} metres.

Nanoparticle: A particle having one or more dimensions of the order of 100 nanometres or less.

Nanoscale: A length scale between 1–100 nanometres and the level of most atoms and some molecules.

Nanotechnology: The understanding and control of matter at dimensions between 1 and 100 nanometres, where unique phenomena enable novel applications.

Nanotube: A structure comprising atoms that form a hollow, nanoscale cylinder.

Quantum Dot: Semiconducting nanocrystals that differ in their ability to absorb and emit energy, based on the size of the crystal.

Quantum Mechanics: A set of scientific principles describing the known behavior of energy and matter that predominate at the atomic and subatomic scales.

Quantum Physics: The branch of physics which studies matter and energy at the level of atoms and other elementary particles, and substitutes probabilistic mechanisms for classical Newtonian ones.

Self-assembly: A method by which atoms or molecules arrange themselves into ordered nanoscale structures by physical or chemical interactions between the units.

Parts of this book have originally appeared in the following publications:

Invernizzi, N., Folladori, G., and Maclurcan, D.C. (2007), "The Role of Nanotechnologies in Development and Poverty Alleviation: A Matter of Controversy", *Journal of Nanotechnology Online*, Available: www.azonano.com/Details.asp?ArticleID=2041.

Invernizzi, N., Foladori, G., and Maclurcan, D.C. (2008), "Nanotechnology's Controversial Role for the South", *Science, Technology and Society*, vol. 13, no. 1, pp. 123–148.

Maclurcan, D.C. (2005), "Nanotechnology and Developing Countries: Part 1 — What Possibilities", *AzoNano Online Journal of Nanotechnology*, Available: http://www.azonano.com/Details.asp?ArticleID=1428.

Maclurcan, D.C. (2005), "Nanotechnology and Developing Countries: Part 2 — What Realities", *AzoNano Online Journal of Nanotechnology*, Available: http://www.azonano.com/Details.asp?ArticleID=1429.

Maclurcan, D.C. (2009), 'Nanotechnology and the Global South: Exploratory Views on Characteristics, Perceptions and Paradigms', in S. Arnaldi, A. Lorenzet, F. Russo (Eds), *Technoscience in Progress: Managing the Uncertainty of Nanotechnology*, IOS Press, Amsterdam, pp. 97–112.

Maclurcan, D. (2009), "Southern Roles in Global Nanotechnology Innovation: Perspectives from Thailand and Australia", *Journal of Nanoethics*, vol. 3, no. 2, pp. 137–156.

Chapter 1

Introduction

Although the term 'nanotechnology' was not coined until 1974,[1] its conceptual foundation is most commonly attributed to Richard Feynman in his 1959 address: 'There's Plenty of Room at the Bottom' (see Feynman, 1960). In this address, Feynman considered the concept of individual, atomic manipulation, stating: "the principles of physics, as far as I can see, do not speak against the possibility of manoeuvring things atom by atom" (*ibid.*, p. 35). In 1981, Eric Drexler, a graduate from the Massachusetts Institute of Technology, built upon Feynman's ideas through work in which he proposed, based on protein engineering, "...a path to the fabrication of devices to complex atomic specifications..." (p. 5275). Discussed in greater detail in subsequent publications (see 1986, 1992), Drexler's aspirations focussed on the construction of self-replicating molecular machinery that could perform production tasks at the nanoscale — one billionth of a metre and the level of most atoms and some molecules. Through popular science, Drexler's work not only proved inspirational but simultaneously raised public fears, such as the 'grey goo': an apocalyptic scenario in which self-replicating, omnivorous nanoscale robots consume the global ecosystem. However, as the end of the second millennium approached, Drexler's views were increasingly criticised by the general scientific community, given a belief that:

> A number of very serious technical challenges would have to be overcome before it would be possible to create nanoscale machines

Nanotechnology and Global Equality
Donald Maclurcan
Copyright © 2012 by Pan Stanford Publishing Pte. Ltd.
www.panstanford.com

that could reproduce themselves in the natural environment. Some of these challenges appear to be insurmountable with respect to chemistry and physical principles, and it may be technically impossible to create self-reproducing mechanical nanoscale robots of the sort that some visionaries have imagined (Roco and Bainbridge, 2001, p. 11).

Increasingly marginalised and ostracised to the point that his proposals were "...regarded as obsolete" (Gordijn in UNESCO, 2005, p. 3), Drexler recast his understanding and aspirations for nanotechnology as 'molecular manufacturing' (see Drexler, 2004). Molecular manufacturing is explored in this book only so much as it is privileged by the literature on nanotechnology and development and by the people I interview. My views on the relevance of molecular manufacturing to the matters of global inequity are discussed in other work (see Maclurcan, 2006).[2]

However, catalysed by the hype and momentum built via molecular manufacturing, a form of nanotechnology, somewhat independent of Drexler's visions, entered the mainstream at the dawn of the new millennium (Malsch, 2002b).[3] Whilst similarly involving work on the nanoscale, the new understanding of nanotechnology avoided the idea of self-replicating machines, and is more commonly defined as:

> ...the understanding[4] and control of matter at dimensions between approximately 1 and 100 nanometers, where unique phenomena enable novel applications....nanotechnology involves imaging, measuring, modeling, and manipulating matter at this length scale (National Nanotechnology Initiative, 2003).

Scientifically, this new field emerged as a result of certain developments, particularly the materialisation of "...tools to see, measure, and manipulate matter at the nanoscale..." (Ratner and Ratner, 2002, p. 39). Critical here has been the more widespread use from the 1960s onwards of the 'scanning electron microscope'[5] and the discovery of scanning probe microscopes, such as the 'scanning tunnelling microscope'[6] (STM) in 1981 and 'atomic force microscope'[7] (AFM) in 1986. Drawing on various forms of surface interaction, such instruments have enabled imaging of a sample's topography, composition and scientific properties at the atomic level. Furthermore, the STM's ability to move single atoms on surfaces has given humans the ability to engineer with atomic precision (Harper, 2003b).[8]

Enabling techniques have also included quantum mechanical computer simulation,[9] soft X-ray lithography[10] and new synthesis methods, such as chemical vapour deposition,[11] all stimulating an ever-accelerating understanding of scientific endeavour at the nanoscale. The final, major piece in nanotechnology's scientific evolution has been the discovery of materials such as quantum dots,[12] around 1983 (see Brus, 1984), fullerenes[13] — including the spherical forms known as buckyballs[14] — in 1985 (see Kroto, Heath, O'Brien, Curl and Smalley, 1985), and nanotubes[15] — particularly carbon-based — in 1991 (see Iijima, 1991).

Buckyballs and carbon nanotubes are prominent examples of 'self-assembly', a method by which "...atoms or molecules arrange themselves into ordered nanoscale structures by physical or chemical interactions between the units" (The Royal Society and Royal Academy of Engineering, 2004, p. 27). This 'bottom-up' approach is seen as offering a new paradigm for science (see, for example, Tegart, 2001; El Naschie, 2006), with the nanoscale "...not just another step toward miniaturization, but a qualitatively new scale" (Roco and Bainbridge, 2001, p. 4).

However, the construction of nanomaterials still incorporates 'top-down' techniques in which "...very small structures [are produced] from larger pieces of material, for example by etching to create circuits on the surface of a silicon microchip" (The Royal Society and Royal Academy of Engineering, 2004, p. viii). This diversity of approaches highlights the unifying, interdisciplinary nature of nanotechnology (Haberzettl, 2002; Ratner and Ratner, 2002; Welland, 2003; Wood, Geldart and Jones, 2003), with the nanoscale forming the point of integration for the exploitation of biological principles, physical laws and chemical properties (see Fig.1.1).

This convergence is particularly relevant given that, at the nanoscale,[16] the laws of quantum physics supersede those of traditional physics, resulting in certain size-dependent phenomena, such as changes in the elastic, electrical, magnetic, optical, and tensile behaviour of materials (The Royal Society and Royal Academy of Engineering, 2004; Juma and Yee-Cheong, 2005). Titanium dioxide and zinc oxide, for example, appear opaque at the macroscale but transparent at the nanoscale (Maynard, 2007).

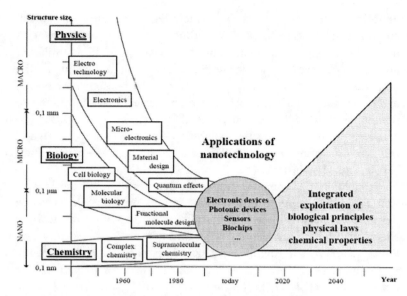

Figure 1.1 Physics, Biology and Chemistry Meet in Nanotechnology (VDI-Technology Centre, Future Technologies Division in Bachmann, 2000, p. 75).

In terms of potential benefits, given atoms are the basic building blocks of all physical things (National Science and Technology Council, 2000), the nanoscale is seen as the most efficient length scale for manufacturing and addressing problems such as disease (Ratner and Ratner, 2002; Gross, 2003; Roco, 2004). Building upon this knowledge of size-based phenomena is the belief that nanotechnology will enable cheaper, lighter, faster, stronger, 'smarter', more energy-efficient and environmentally friendly, safer and more precise solutions (Morrison, 2003; Etkind, 2006; Waruingi and Njoroge, 2008). For example, nanotechnology's cost-effectiveness is said to stem, in part, from low energy requirements for both production and maintenance (Roco and Bainbridge, 2005b), as control on the nanoscale offers opportunities to use materials and energy more efficiently whilst simultaneously reducing waste (Lloyd, Lave and Matthews, 2005). In this way, processes such as self-assembly are also said to exemplify nanotechnology's potential for energy efficiency and environmental friendliness (The Royal Society and Royal Academy of Engineering, 2004). In the case of nanotubes, nanotechnology is believed to offer alternatives that can be stronger than steel, yet lighter than aluminium (Moniruzzaman and Winey, 2006). In healthcare, nanotechnology's faster solutions are said to stem from size-dependent phenomena such as increased reaction rate co-efficients (Heines, 2003),[17] whilst design is seen as smarter in the case of drug delivery given the potential for slow-release, quick-release, specific-release, temperature-release, pH-release, and pressure-release methods (de Villiers, Aramwit and Kwon, 2008). Continuing with examples in healthcare, drug delivery is envisaged as being more precise (Gillis, 2002; Ratner and Ratner, 2002; Graham, 2003; Saxl, 2003), subsequently providing greater safety via reductions in unwanted side-effects (Haberzettl, 2002; Ratner and Ratner, 2002).

Nanotechnology's 'new' materials have heralded particularly grand aspirations. Said to offer exciting prospects as substitutes to copper wiring and silicon chips for areas such as information and communications technology and electronics, nanotubes "...can be either semiconductors or insulators, depending on how their carbon sheets are rolled up" (ETC Group, 2005c, p. 12). As buckyballs are hollow, they are said to make ideal nanosized vessels (*ibid.*), offering the ability to overcome challenges and concerns in healthcare such as permeability, solubility and toxicity (Haberzettl, 2002; Malsch, 2002a; Saxl, 2003). Quantum dots also offer great potential in areas such as the treatment and monitoring of disease, as dots of differing sizes are able to be attached, like barcodes, to biological materials (ETC Group, 2005c).

What can be seen from the above discussion and the broad application of nanotechnology is that nanotechnology's transformative potential stems from its capacity as an enabling 'platform technology' (Harper, 2003b), with applications crossing several sectors at once (Meridian Institute, 2007; ETC Group, 2008). According to Bowman and Hodge (2006), these sectors include: agriculture, chemicals and cosmetics, electronics, the environment and energy, food science, materials, medicine, military and security and scientific tools (applications within these sectors are outlined more fully in Fig. 1.2).[18] Reciprocally, therefore, advances made across multiple sectors can also be advances for nanotechnology (Foster, 2002).

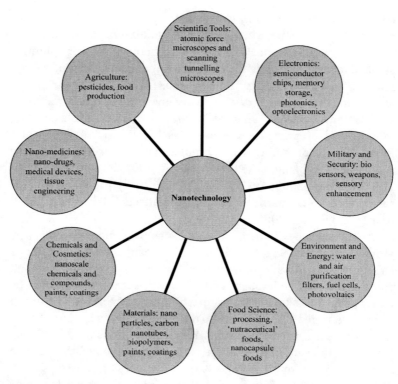

Figure 1.2 Short- and Medium-Term Nanotechnology Applications by Category (Adapted from Bowman and Hodge, 2006, p. 1062).

In stark contrast to biotechnology's early days, nanotechnology's applications have already been vast, producing circumstances in which there are "...products in the marketplace and almost half of the start-up nano-nichers are selling their wares" (ETC Group, 2003a, p. 42). Nanotechnology's products are regarded as having been available for commercial use since 1997[19] and, as of 2009, the Woodrow Wilson database of nanotechnology products[20] had registered a total of 1015 products across eight categories (shown in Fig. 1.3).

Examples from this product inventory include wound dressings coated with silver nanoparticles to prevent infection, nanoscale titanium-dioxide-based sunscreens and clear plastic food wraps for protection from ultraviolet light (ETC Group, 2005c).

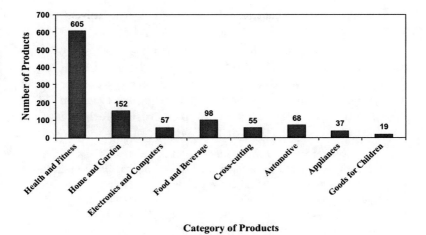

Figure 1.3 Products Incorporating Nanotechnology that are Currently in the Market (Adapted from Woodrow Wilson International Centre for Scholars, 2009).

In terms of its scale, as of 2003, 40 countries were said to be engaged in nanotechnology research and development (R&D) (Huang, Chen, Chen and Roco, 2004), involving the employment of an estimated 20,000 researchers (National Nanotechnology Initiative, 2003). In 2006, worldwide investment in nanotechnology was over $US4 billion,[21] with 29% coming from the United States of America (U.S.), 25% from the European Union (E.U.), 21% from Japan and the remaining percentage from other countries (Roco, 2007). Projected by the U.S. National Science Foundation (NSF) to have a global market value of $1 trillion by 2011 (Roco, 2001), it is estimated that the value of nano-incorporated goods will rise to $2.6 trillion by 2014 — ten times that of biotechnology (Lux Research cited in ETC Group, 2008).

In this light, nanotechnology is often heralded as 'the next industrial revolution' (see, for example, Interagency Working Group on Nano Science, 2000; Ratner and Ratner, 2002; Hood, 2004) and set to have dramatic implications:

> The effect of nanotechnology on the health, wealth, and lives of people could be at least as significant as the combined influences of microelectronics, medical imaging, computer-aided engineering, and man-made polymers developed in this century (National

Science and Technology Council, Committee on Technology and Subcommittee on Nanoscale Science, 2002, p. 13).

Nanotechnology's dramatic implications are said to extend to the global South,[22] with the claim that: "over the next two decades, the impacts of nano-scale convergence on farmers and food will exceed that of farm mechanisation or of the green revolution" (ETC Group, 2004b, p. 1). As the Action Group on Erosion, Technology and Concentration — the ETC Group (2003a) note:

> The convergence of technologies at the nano-scale may seem a long way from rural communities in Africa, Asia or Latin America. It is not...Though its impact will be felt first in the North, Atomtech[23]—like biotech before it — will have early economic and environmental consequences for developing countries (p. 5 and p. 9).

With regard to the nature of nanotechnology's implications, some, such as this senior official within the U.S. Department of Commerce, optimistically claim:

> ...these technologies could eventually achieve the truly miraculous: enabling the blind to see, the lame to walk, and the deaf to hear; curing AIDS, cancer, diabetes and other afflictions; ending hunger; and even supplementing the power of our minds...Nano[technology] also holds extraordinary potential for the global environment through waste-free, energy-efficient production processes that cause no harm to the environment or human health (Bond in Hanson, 2006, p. 3).

Yet others, such as the ETC Group (2008), see things very differently, claiming: "...nanotech[nology] threatens to widen the gap between rich and poor and further consolidate economic power in the hands of multinational corporations" (p. 11).

Thus, as will be demonstrated more fully in Chapter 4, key perspectives in the literature are significantly polarised between those who claim nanotechnology is part of the 'development solution' and others who claim it is part of the 'development problem' (Wood *et al.*, 2003; Munshi, Kurian, Bartlett and Lakhtakia, 2007; Invernizzi *et al.*, 2008). As will be shown in Chapter 3, this polarity can be said to stem from contestations of 'development' itself, as well as technology more broadly.

1.1 Central Questions

I locate my central investigation at exactly this rift between these two opposing views, thereby assessing their respective claims about nanotechnology's foreseeable consequences. I therefore ask: to what extent does nanotechnology offer hope for a more equitable world?

Whilst my research commences with the use of a broad understanding of nanotechnology, I focus on 'human-engineered' nanotechnology and, given the literature rarely refers to nanotechnology in such forms, exclude consideration for what has been termed 'incidental' nanotechnology, such as diesel exhaust, and 'naturally occurring' nanotechnology, such as forest fire combustion (see Colvin cited in Goldman and Coussens, 2005). In this light, nanotechnology is understood to refer to "…an umbrella term for a range of enabling technologies" (Rip, 2006, p. 270) that are bound merely by their common engagement with the nanoscale (The Royal Society and Royal Academy of Engineering, 2004). Nanotechnology's broad nature, and the subsequently distinct differences between products, are said to create difficulties for assessing associated social implications (Rip, 2006). Some, such as The Royal Society and Royal Academy of Engineering — RS&RAE (2004), thus prefer to pluralise the field and use the term 'nanotechnologies'.

However, I hold what Grimshaw (2008) calls "…an alternative view of technology" (p. 18), in which I see the social processes accompanying an object's emergence as on a par, in terms of importance, with the object itself. As Everts (1998) explains:

> An object in itself, however well designed with the purpose of enhancing human capabilities, cannot be called a technology but is merely an artefact, an object or a thing. What *is* to be called 'technology' is the total package of this artefact plus the organizational, informational and human contexts that are required for its functioning (pp. 5–6).

Following this perspective, I take a holistic view of nanotechnology.[24] This view drives my investigation into the reflexivity around nanotechnology's emergence with respect to key development concerns that are identified in my review of the literature. Thus, to comprehensively understand nanotechnology's implications for Southern development, I follow Grimshaw's (2008) suggestion and consider the historical and contemporary contexts in which it is emerging.

From this constructivist viewpoint, whilst the issues raised by nanotechnology are predominately not unique (Wood *et al.*, 2003; The Royal Society and Royal Academy of Engineering, 2004), "... this does not make them any less important or relevant to the social sciences" (Wood *et al.*, 2003, p. 41). On the contrary, I see nanotechnology as a useful site for the analysis of the technology-development nexus. As Wood, Geldart and Jones (2003) note:

> Given nanotechnology's apparent importance, and that it is evolving at a time when there are other pressures on issues such as ageing, intellectual property and risk management, it provides social scientists with an opportunity to study the effects of technology on these issues. Similarly, it provides them with an opportunity to study issues associated with any emerging technology (p. 42).

The question of whether nanotechnology can 'offer hope' addresses its perceived potential and foreseen implications simultaneously. Hope, in relation to emerging technology and world futures, is a well explored theme (see, for example, Berry, 2002; Arvanitakis, 2007), with Pieterse (2001) highlighting:

> The dilemmas of development parallel the dilemmas of modernity on one question at least: what of a politics of hope?...The challenge facing development is to retrieve hope from the collapse of progress (p. 163).

Finally, in focussing on hope relating to 'a more equitable world', I make the important distinction between inequity and inequality, a distinction that is aptly explained by the Pan American Health Organisation (1999):

> ...inequity refers to differences which are unnecessary and avoidable but, in addition, are also considered unfair and unjust. Not all inequalities are unjust, but all inequities are the product of unjust inequalities (p. 11).

I therefore position my argument in relation to a range of inequities, such as those marked by income, access to technologies and innovative capacity. This is a crucial aspect of my question, given "there are moral and practical arguments for an equitable society" (Lowe, 2009, p. 27). Juxtaposed with the dangers of increased inequity,[25] a shift to a more equitable world is said to equate to improved social and environmental sustainability (Jackson, 2009; Lowe, 2009; Wilkinson and Pickett, 2009), with approaches to technological change envisaged as a critical aspect of any such transition (Lowe, 2009).

To answer my central research question and in order to supersede the simplistic, polarised debate that is emerging, "...the study of nanotechnology should be organized around a set of key issues and manageable sub-questions" (Wood *et al.*, 2003, p. 40). As shall soon be discussed, my analysis of the literature (Chapters 3 and 4) highlights four themes for contemporary technological assessment, from which emerge the four sub-questions underpinning my research framework. These questions are deliberately less open-ended, as this limits their exploration in terms of scope, enabling greater pragmatic engagement with the key issues.

It is evident in the literature on nanotechnology, development and inequity that authors often assume universal understandings about what nanotechnology is and is not, despite evidence of definitional differences. Hence, in order to clarify the legitimacy and limitations of the discussions about nanotechnology's foreseen implications for global inequity, in my first sub-question I ask: is nanotechnology understood in ways that allow common discussion about its implications for global inequity?

Moving to the more substantive issue of how nanotechnology fits within an historical and contemporary context, my review of the literature on technology, development and inequity shows that, throughout the history of development, centralised control over *innovative capacity* has continually created greater inequity. Hence, in order to assess nanotechnology against this trend, in my second sub-question I ask: will nano-innovation and innovative capacity be globally and locally decentralised and autonomous?

Alongside the debate about innovative capacity, the issue of *technological appropriateness* has been both directly and indirectly central to critiques of the green revolution, biotechnologies and other technologies emerging within the development context. Hence, in order to evaluate the implications arising from nanotechnologies, in my third sub-question I ask: do nanotechnologies offer appropriate technologies for the South?

Intricately tied to both issues of technological appropriateness and innovative capacity are the various *approaches to technological governance* employed when it comes to engaging with emerging technology. In this respect, the democratisation of science remains a critical struggle, with centralised control over guided and restricted technological trajectories continuing as the dominant form of engagement with emerging technology in both the South and the North. Hence, in order to investigate nanotechnology in light of these claims, in my final sub-question I ask: do the present and foreseen approaches to nanotechnology's governance in the South enable empowering, democratic processes?

In order to more reliably assess the extent to which nanotechnology offers hope for a more equitable world, its promises and early trends will benefit from technological comparison (see Mehta, 2004; Wolfson, 2004). Whilst any comparison is limited by differing contexts, across this book I will also be carefully considering the extent to which nanotechnology's promises, issues and early trends differ from those experienced with the emergence of previous technology, such as biotechnology.

In addition to answering my central research questions, the principal aims of this book are to: place nanotechnology's emergence within a broad historical and contemporary global context and develop and test an interpretive framework through which to assess relevant claims; establish greater clarity about the nature of global engagement with nanotechnology R&D; and explore a range of perspectives, from within both the South and North, regarding nanotechnology's foreseen implications for global inequity.

1.2 Overview of My Approach

My approach to addressing these aims utilises an exploratory, mixed methods design in three phases (to be explored more fully in Chapter 2). In the first phase of my work I seek *contextualisation*, given that any contemporary assessment of nanotechnology, development and inequity must be grounded in an assessment of historical and theoretical underpinnings (Foladori, Rushton and Zayago Lau, 2008). By reviewing the literature on technology, development and inequity, I am able to establish a framework that includes three themes — 'innovative capacity', 'technological appropriateness' and 'approaches to technological governance'

— and associated criteria by which to evaluate the extent to which nanotechnology offers hope for a more equitable world. The importance of such an approach is highlighted by Kearnes, Macnaghten and Wynne (2005), who state:

> We need to develop new frameworks of theoretical reflection to understand the emergence of nanotechnologies...approaches that move beyond conceptualising the future in terms of prediction and control...[that consider] the complexities of multiple 'futures' (p. 285).

To finish this contextualisation phase, I test and subsequently refine my framework using the literature on nanotechnology, development and inequity, adding a fourth theme — understandings — and simultaneously exploring what gaps remain in the research.

Given an identified lack of quantitative research which comprehensively outlines the global 'state of play', in the second phase of my work I seek *quantification* of global nanotechnology data. To do this I assess search engine data in relation to national engagement, conference and event hosting and participation and global-health-related patenting. The results provide the justification and some directions for my qualitative research methodology and establish a broad base on which to analyse nanotechnology's foreseen implications for the global South.

Given an identified lack of qualitative research, in the third phase of my work I seek *interpretation* of nanotechnology's foreseen implications for the South. Here I take into particular consideration two claims that emerge from my review of the literature on development, technology and inequity. Firstly, 'development' is a global problem, requiring that the impact of technological change be considered across North-South divides. Secondly, a reflexive response to the legitimate requirements of equitable development can emerge via shared insights that, in turn, lead to pragmatic self-reflection. In this respect, my research design closely reflects my commitment to North-South reflexivity by seeking data across the North-South divide through interviews that explore the perspectives of 31 'key informants'[26] from Thailand and Australia, supplemented by surveys of 24 Thai nanotechnology practitioners.

For the interviews, I sought diverse perspectives that I felt would be the most revealing (I will outline interviewee characteristics more fully in Chapter 2), with my detailed reasoning behind the selection of perspectives from Thailand and Australia explained in *Appendix*

A. In brief, as of 2004, these countries offered an adequate balance between distinctive and similar features to meaningfully discuss the issues at hand. Whilst there is a significant divide between the two countries on certain measures of 'development', the national governments of both seek strong engagement with nanotechnology and both countries have a common history of endogenous critiques relating to emerging technology that now extends to existing or anticipated critiques of nanotechnology.

Despite being unorthodox, there are precedents for approaches to qualitative research on nanotechnology that cut across the North-South divide. Based in Zimbabwe, Grimshaw, Stilgoe and Gudza's (2006) 'nanodialogues' engaged local community groups and scientists from both the South and North in an assessment of nanotechnology's appropriateness for development needs. Similarly, participants in the Meridian Institute's (2006) 'Nanotechnology, Water and Development' workshop moved beyond local case studies to discuss broader challenges people in the South may face as a result of nanotechnology's emergence.

Whilst the development of my work followed a typically nonlinear, iterative process (as described by Neuman, 1997; Johnson and Onwuegbuzie, 2004; Sandelowski, Voils and Barroso, 2006), my investigation has been specifically organised so as to flow from a broad, macroview in Phase 1, to an 'industry' perspective in Phase 2, to the primary qualitative perspectives from individuals 'on the ground' in Phase 3. Such an approach is useful because it allows me to consider my research questions from a number of perspectives, each of which builds on the former.

1.3 General Limitations

Research into development issues has been characterised by a considerably high number of studies having been conducted by researchers from the North (Kumar and Siddharthan, 1997), with the relevance of these studies persistently questioned by academics within the South (Hettne, 1995). This has subsequently caused difficulties regarding the 'position' of researchers when undertaking such studies. In this light, throughout the process of my research I felt increasingly uncomfortable about my position as a relatively affluent, Australian male. The problem, as Spivak (cited in Kapoor,

2004) summarises clearly, is that "our discursive constructions are intimately linked to our positioning (socioeconomic, gendered, cultural, geographic, historical, institutional)..." (p. 628). Despite my unorthodox approach in seeking reflexivity across the South and North,[27] I encountered this discomfort most strongly when considering the nature of my research topic and the chance that, in focussing on inequity and the North-South divide, my work could be construed as tacitly reinforcing the mainstream belief that the development 'problem' lies in the South. I also had difficulty identifying and using appropriate language on a consistent basis. Such problems are well documented in development literature, where Fukuda-Parr (2002), for example, declares that the concepts and language are filled with hierarchy and inequality. Whilst Fukuda-Parr points to exemplary terms such as 'aid', 'developing', 'donors' and 'recipients', Esteva (1997) believes 'development' is the term needing the most scrutiny:

> Development cannot delink itself from the words with which it was formed — growth, evolution, maturation...No matter the context in which it is used, or the precise connotation that the person using it wants to give it, the expression becomes qualified and coloured by meanings perhaps unwanted. The word always implies a favourable change, a step from the simple to the complex, from the inferior to the superior, from worse to better... but for two-thirds of the people on earth, this positive meaning of the word 'development'...is a reminder of *what they are not...* (p. 10).

1.3.1   Self-Reflexivity

To respond to these dilemmas, Spivak (cited in Kapoor, 2004) argues for, "...a heightened self-reflexivity" (p. 628). As Kapoor (2004) notes, there is value in acknowledging one's complicities and 'contamination' which, in turn:

> ...helps temper and contextualise one's claims, reduces the risk of personal arrogance or geoinstitutional imperialism, and moves one toward a non-hierarchical encounter with the Third World/ subaltern (p. 641).

With these limitations in mind, I return to my use of language and approach to the research. In terms of language, instead of referring to the 'First/Developed World' and 'Third/Developing World', I instead employ[28] the social and partly spacial concept[29] of 'global North' and 'global South' — seen, by some (see, for example, Rye Olsen, 1995; James, 1997; Schummer, 2007; Slater, 2004), as less constrained with embedded meanings, although the term remains contested.[30] However, whilst Schummer (2007) notes that many Southern countries have a number of common characteristics,[31] my use of the North-South dichotomy has its limitations — as with any attempt to homogenise essentially non-homogeneous groups (Hettne, 1995). Here it is important to acknowledge that there may be variations amongst regions, nations or groups[32] (2003) and that 'third worlds' exist within the 'first world' and vice versa (Hoogvelt, 1997). Subsequently, there is "...very little generic development knowledge..." (Fukuda-Parr *et al.*, 2002, p. 18), meaning that my qualitative research should be viewed as illustrative, rather than representative. Thus, when I use expressions such as: "development settings" and "...a country such as Thailand", I am endeavouring to say that the points being made may be applicable for various circumstances, but that I am unable to specify what these circumstances are and do not want to suggest their universality.

1.4 Chapter Outlines

In Chapter 2 I will further develop my approach by investigating the use and limitations of the mixed methods methodology before detailing the methods I used in the quantitative and qualitative phases of my research. I will then commence my study in Chapter 3 by reviewing the literature surrounding development, technology and inequity in order to establish a context and create an analytical framework for assessing the extent to which nanotechnology offers hope for a more equitable world. I will test this framework in Chapter 4 through an analysis of the literature about nanotechnology, development and inequity. This will also allow me to continue developing an appropriate context for assessment as well as leading to the identification of residual gaps in the literature. Having confirmed and refined my theoretical framework, in Chapter 5 I will launch into a quantitative mapping of the contemporary state of play relating to global engagement with nano-innovation. In addition to establishing critical data, this quantitative phase will serve to justify and provide direction for the qualitative phase of my research, establishing a broad base on which to analyse nanotechnology's foreseen implications for the South. This analysis will commence in Chapter 6 by examining how nanotechnology is understood by Thai and Australian key informants. Having then established both the legitimacy and limitations of engaging differing perspectives in my assessment of nanotechnology, in Chapter 7 I will move to exploring interviewee perspectives about Southern innovative capacity with respect to nanotechnology R&D. In Chapter 8 I will continue exploring the themes that emerged in my literature review by assessing interviewee perspectives relating to the appropriateness of nanotechnologies to the South. In Chapter 9, I will complete my qualitative study by investigating interviewee perspectives on Southern approaches to nanotechnology's governance. In Chapter 10 I will conclude this book by re-examining the central questions and aims associated with my study. I will then respond to my central questions, whilst reflecting on the field in relation to my findings. I will, at this point, note the major limitations of my work and suggest some avenues for further research. I will finish by discussing the possible implications of my work and making some broader recommendations for the field and its research agendas.

Endnotes

1 The term 'nanotechnology' was first used by Japanese scientist Norio Taniguchi in an address to the International Conference of Production Engineering (see Taniguchi, 1974).

2 In essence, I argue that, despite its claims of a revolution of an entirely different nature, the key debates facing molecular manufacturing are the same as those facing most other technological revolutions and that, therefore, it is wiser and more tangible to focus on assessing the immediate and foreseeable consequences of what is now understood as 'nanotechnology'.

3 Although, as shown in Chapter 4, nanotechnology has been around a lot longer than the start of the third millennium, the highly visible launch of the U.S. National Nanotechnology Initiative in 2001 is said to have led to a 'cascade response' by governments in Europe, Japan and other nations (see Malsch, 2002b).

4 Inherent in this definition is the inclusion of 'nanoscience' - the study of matter at dimensions between 1 and 100 nanometres.

5 The scanning electron microscope is an instrument that uses feedback from the interactions between an electron beam and the surface at which it is aimed in order to scan the surface's topography.

6 The scanning tunnelling microscope is an instrument that uses the difference in voltage between a conducting tip and a surface to scan the surface's topography.

7 The atomic force microscope is an instrument that uses the difference in atomic force between a cantilevered tip and a surface to map the surface's topography.

8 The ability to manoeuvre atoms was made famous by the 1990 manipulation of 35 xenon atoms into the letters: 'I.B.M.' (see Eigler and Schweizer, 1990)

9 Quantum mechanical computer simulation is a technique which facilitates the theoretical modelling of atoms or small molecules for the purpose of predicting the scientific characteristics of such matter.

10 Soft X-ray lithography is a technique by which a pattern is etched onto a surface via X-rays.

11 Chemical vapour deposition is a process by which matter, once exposed to volatile agents, will leave a material residue on a surface.

12 Quantum dots are semiconducting nanocrystals that differ in their ability to absorb and emit energy, based on the size of the crystal.

13 Fullerenes are a class of carbon molecule that can be arranged in spherical, ellipsoidal, or cylindrical formations.

14 A buckyball is a spherical fullerene.

15 Nanotubes can be further disaggregated into those with single walls and those with multiple walls.

16 Matter below approximately 50 nanometres (ETC Group, 2003a).

17 An increased surface area per unit accompanies materials on the nanoscale (Oberdörster, Oberdörster and Oberdörster, 2005). This leads to the situation in which "a single gram of catalyst material that is made of 10 nanometre particles is about 100 times more reactive than the same amount of the same material made of particles one micrometre in diameter" (Hume cited in ETC Group, 2003a, p. 14).

18 There are many different interpretations of the 'sectors' across which nanotechnology can be applied. For another categorisation of nanotechnology, see the Meridian Institute (2004).

19 Developed in Australia, the AMBRI biosensor has been claimed as the first commercial 'nanomachine' (see Adams, 1998).

20 See: http://nanotechproject.org/inventories/consumer/analysis_ draft.

21 All cited financial figures throughout this book are in U.S. dollars, unless otherwise stated.

22 In this book I argue that nanotechnology's hopes for global equity can be alternatively measured by considering its implications for the global South. The 'global South' is a definitional term I employ to refer to what is otherwise described as the 'non-industrialised-', 'developing-' or 'third-world'. My use of this term is explained in greater detail later in Chapter 1.

23 'Atomtech' is a term used by the ETC Group (2003a) to describe "...a spectrum of techniques involving the manipulation of molecules, atoms and sub-atomic particles to produce materials. Atomtech also involves the merging and manipulation of living and non-living matter to create new and/or hybrid elements and organisms" (p. 7).

24 Others see, for example, Mee, Lovel, Solomon, Kearnes and Cameron, 2004) also use the term 'nanotechnology' as an 'umbrella term' within research about the technology's implications.

25 Inequity is said to: exacerbate almost every modern social and environmental problem (Wilkinson and Pickett, 2009); create social instability (Lowe, 2009); increase "tolerance for inequality" (Birdsall cited in UNDP, 2001, p. 17); and deny the possibilities for peace (Max-Neef in Simms, Johnson and Edwards, 2009).

26 As I shall explore more fully (Chapter 2), key informants are defined as "...those who can provide relevant input to the process, have the highest authority possible and are committed and interested" (Gutierrez, 1989, p. 33).

27 As my research involved analysing non-engaged data, that is, one-on-one interviews, the reflexivity involved was less between interviewee perspectives and more between my comparative analyses.

28 To disaggregate the data I often use the Organisation for Economic Co-operation and Development (OECD) country classification terms: 'least developed', and 'transitional'.

29 The terms 'North' and 'South' are more of a concept, given a literal reading would support the geographical absurdity that India be considered in the Southern hemisphere whilst Australia be considered in the Northern hemisphere.

30 For Monceri (2004), the 'North-South' dichotomy remains politically-charged, resurrecting notions of colonialism and imperialism.

Moreover, according to Dudley (1995), the term "...tends to subtly reinforce parochialist attitudes...[and is] patently false in its most literal sense and misleading in others" (p. 1332). In light of a 'fast moving world', Slater (2004) questions whether the term 'North-South' is becoming increasingly obsolete, suggesting as more appropriate a focus on the relational, ahead of spatial, nature of power and knowledge distribution.

31 Schummer (2007) states: "...for instance, historically, they were former colonies and frequently still have some special ties (economical, political, or military) to their former colonial powers...Large parts of their populations suffer from very basic needs, like malnutrition and the lack of safe drinking water, sanitation, education, and healthcare, despite devastating epidemics like AIDS and malaria. Rural exodus has even increased these needs through exploding slums around big cities. They have only poor infrastructures of public and private research and development, including small public research budgets and virtually no venture capital...they have little experience in technology governance, including the launch and conduct of research programs, safety and environmental regulations, marketing and patenting strategies, and so on" (p. 292).

32 Thailand, for example, does not have a colonial history.

Chapter 2

Methodology

My work is best viewed as a 'snapshot' investigation of nanotechnology at a critical stage in its early development. Given the emergent nature of nanotechnology as a field of analysis, my work is situated within what Neuman (1997) describes as an 'exploratory' research framework. Here, the focus is on "...gaining insights and familiarity with the subject area for more rigorous investigation at a later stage" (*ibid.*, p. 10). Exploratory research is said to be particularly appropriate for areas in which measures are not available, variables are unknown or there is no guiding framework (Cresswell and Plano Clark, 2007). Furthermore, when an area holds a diversity of views and complex issues and problems demand an exploratory study, an open, wide-ranging mixed methods approach is seen as particularly appropriate (Seidman, 1998; Cresswell and Plano Clark, 2007). Such an approach can be seen in assessments of biotechnology's implications for the South (see, for example, Acharya, Rab, Singer and Daar, 2005).

As will be shown in Chapter 4, diverse views are held regarding nanotechnology's global implications (Invernizzi *et al.*, 2008). Moreover, there is a need for both quantitative, descriptive data to map the worldwide 'state of play' in terms of where and by whom nanotechnology R&D is being conducted (Carroll, 2001; Crow and Sarewitz, 2001; Roco and Bainbridge, 2001; Kearnes *et al.*, 2005) and qualitative data in terms of assessing societal attitudes and phenomena surrounding nanotechnology's emergence (Carroll,

2001; Roco and Bainbridge, 2001; Kearnes *et al.*, 2005). Hence, my approach to assessing nanotechnology's implications for equitable development encompasses multiple data sources within a single study, using the increasingly accepted, distinct research design and methodology known as 'mixed methods' (Johnson and Onwuegbuzie, 2004; Cresswell and Plano Clark, 2007). Amongst other advantages that will be explored shortly, mixed methods research is seen by some (see, for example, Greene, Caracelli and Graham, 1989; Johnson and Onwuegbuzie, 2004) as a useful means to help bridge the schism between quantitative and qualitative research. However, to make such a claim it is firstly useful to explore the different research approaches inherent to quantitative and qualitative methodology (Cresswell and Plano Clark, 2007).

Dominating the 20th Century, the quantitative approach is a predominantly deductive methodology that describes and accounts for social phenomena that can be quantified (Payne and Payne, 2004), thereby allowing the researcher to test theories and extrapolate data (Cresswell and Plano Clark, 2007). A quantitative methodology is often associated with objectivity, assuming that social processes can be extracted and studied in isolation given that they "...exist outside of individual actors' comprehension" (Payne and Payne, 2004, p. 182). Here, the researcher remains in the background and takes steps to remove bias (Cresswell and Plano Clark, 2007).

On the other hand, emerging as a scientific methodology since the 1980s, the qualitative approach is a predominantly inductive methodology focussing on what materialises in a study, rather than using a predetermined set of hypotheses to frame outcomes (Seidman, 1998). In contrast to the quantitative approach, this methodology "...provides understanding and description of people's personal experiences of phenomena...as they are situated and embedded in local contexts" (Johnson and Onwuegbuzie, 2004, p. 20). It assumes that actions are "...part of a holistic social process and context" (Payne and Payne, 2004, p. 176) and that there are "...multiple definitions of reality embedded in various respondents' experiences" (Sale, Lohfeld and Brazil, 2002, p. 49). Here, there is no distinction between the researcher and the researched, which combine to form a subjective, mutually evolving reality (Phillips, 1988; Johnson and Onwuegbuzie, 2004), although the researcher identifies their personal stance and bias (Cresswell and Plano Clark,

2007). In this sense, a qualitative methodology enables authoritative accounts of both an objective and subjective nature.

Alone, each methodology is limited in what it can offer an exploratory research study assessing the global emergence of a technology. Quantitative research does not incorporate the voices of participants or provide in-depth knowledge on a subject matter (Cresswell and Plano Clark, 2007), whilst qualitative research is more easily influenced by personal biases and idiosyncrasies (Britten, 1995; Johnson and Onwuegbuzie, 2004) and is often unable to support generalisations (Johnson and Onwuegbuzie, 2004; Payne and Payne, 2004; Cresswell and Plano Clark, 2007).

Mixed methods research seeks to overcome these limitations by using the strengths of both approaches to offset inherent biases and weaknesses within the other (Greene *et al.*, 1989; Tashakkori and Teddlie, 2003; Johnson and Onwuegbuzie, 2004; Sandelowski *et al.*, 2006; Cresswell and Plano Clark, 2007). Qualitative research is said to fill the gaps of quantitative research in terms of context and representation, whilst quantitative research fills the gaps of bias or interpretation and generalisability (Cresswell and Plano Clark, 2007). A mixed methods methodology is therefore seen as providing "...a better understanding of research problems than either approach alone..." (*ibid.*, p. 5). This increases "...the interpretability, meaningfulness and validity of constructs and inquiry results" (Greene *et al.*, 1989, p. 259) and makes data more acceptable to audiences that bias one methodology over another (Cresswell and Plano Clark, 2007). Furthermore, selection of a mixed methods methodology is highly appropriate for my study given my academic position bridges both science and social science, and that these faculties provide tacit pressure for collection of both 'hard' and 'soft' data, respectively. In this sense, by using mixed methods methodology, I engage with an interdisciplinary, pragmatic approach, viewing knowledge as "...both constructed and based on the reality of the world we experience and live in" (Johnson and Onwuegbuzie, 2004, p. 18). Moreover, as Carroll (2001) notes: "if social scientists can be introduced earlier into partnerships and collaborations with nanoscientists and nanotechnologists, there is a better chance to learn..." (p. 191).

In terms of my particular use of the mixed methods approach, I connect the data by arranging complementary findings into a line of argument (Greene *et al.*, 1989; Sandelowski *et al.*, 2006). Here one data set is said to provide a supportive, secondary role in the study, helping to develop or inform the other, primary data set (Johnson and Onwuegbuzie, 2004; Sandelowski *et al.*, 2006; Cresswell and Plano Clark, 2007), thereby emphasising one of the two paradigms (Morse, 1991; Morgan, 1998). As is typically the case in exploratory research (Cresswell and Plano Clark, 2007), I emphasise the qualitative methodology, with quantitative research playing an introductory role by informing the focus of my qualitative study. This further reinforces my theoretical position: that technological development must be interpreted as a social process if its implications for equitable development are to be understood. However, I also use quantitative methods (surveys) within my qualitative research as part of a 'within-stage mixed-model design' (Johnson and Onwuegbuzie, 2004), an approach said to improve generalisability (*ibid.*).

Exploratory research is limited in a number of ways. Firstly, it infrequently yields definitive answers (Neuman, 1997; Johnson and Onwuegbuzie, 2004), as social dynamics ensure "...predictions will be uncertain and causal explanations will be difficult to validate" (Carroll, 2001, p. 192). As is the case for any nascent, fast-moving and extremely broad field, nanotechnology faces such challenges (Roco and Bainbridge, 2001; Strand, 2001). Hence, my interest is in looking for nanotechnology's overall trajectory. Secondly, cross-sectional research like mine that captures a snapshot of various phenomena, cannot capture social processes or change (Neuman, 1997).[1] It is therefore suggested that any reading of social science research into nanotechnology keep these restrictions in mind (Carroll, 2001).

Mixed methods methodology is also said to present a number of other limitations. In addition to the sequential nature of my research, by unequally weighting methodologies and investigating different phenomena across my research phases, I rule out the possibility for triangulation (Greene et al., 1989; Cresswell and Plano Clark, 2007).[2] Moreover, Sale, Lohfeld and Brazil (2002) claim that the mixed methods methodology is often adopted uncritically and that, as currently practised, it can diminish the value of both methods by failing to acknowledge basic differences in the fundamental premises of each. This, they believe, leads to a situation in which:

> In order to synthesize results obtained via multiple methods research, people often simplify the situation under study, highlighting and packaging results to reflect what they think is happening (*ibid.*, p. 47).

In this light, many have claimed over the last century that "... qualitative and quantitative research paradigms, including their associated methods, cannot and should not be mixed" (Howe cited in Johnson and Onwuegbuzie, 2004, p. 14). The two paradigms are seen as unable to be assimilated, given they study different aspects of phenomena or see the same phenomena in different lights (see, for example, Sale *et al.*, 2002; Sandelowski *et al.*, 2006). However, there is acceptance that, as with my work, mixed methods can be combined in a single study for complementary purposes if each method studies different phenomena or addresses different questions (Sale *et al.*, 2002).

2.1   Methods

I will begin this section by introducing my unorthodox approach to my review of the literature. I will then describe the quantitative methods I used to gather and analyse descriptive data relating to global engagement with nanotechnology R&D, as well as the orientation of early research. I will conclude by examining the qualitative methods I used to gather and analyse a range of perspectives within both the South and the North regarding nanotechnology's foreseen implications for global inequity, as well as the supplementary quantitative methods I used to further strengthen the qualitative study.

2.1.1   Literature Review

In this book I conduct a broad review of the literature split into two parts. In the first part (Chapter 3) I chart the mainstream and alternative schools of thought that have emerged surrounding development, technology and inequity, given exploratory research is typically grounded in an historical context (Seidman, 1998). By establishing a context for development, I am then able to look more broadly at emerging and consistent themes and build an interpretive framework through which to assess relevant claims. I separate this part of my review of the literature from that which places nanotechnology in a development context given that development debates have significant and complex histories predating nanotechnology's emergence. I draw on global perspectives throughout this part of my review, predominantly using journals, books and then reports. However, by purposefully covering a great deal of theoretical ground, I am limited in my ability to deeply explore various aspects of certain debates, such as counter-critiques of appropriate technology.

In order to place nanotechnology in a development context and test the framework I construct at the end of Chapter 3, in the second part of my review (Chapter 4) I look at the literature relating to nanotechnology, development and inequity across instrumentalist and contextualist perspectives,[3] categorised by my three themes. As the literature in this field is diverse and fast-moving, in addition to considering books and journals I refer to a great number of electronic news articles and reports. By deductively applying my own framework, it is important to consider that I may have excluded significant literature, although the flexibility within my model is demonstrated by the inductive inclusion of a fourth theme, following my review of the literature on nanotechnology, development and inequity.

2.1.2 Quantitative

In order to establish greater clarity about the nature of global engagement with nanotechnology R&D, in the quantitative phase I use search engine data to make preliminary assessments about national levels of nanotechnology R&D activity and international research participation as well as a more comprehensive assessment of health-related nanotechnology patenting. Search engines have been noted for their usefulness in studies assessing modern realities via data collection (see, for example, Kim, Eng, Deering and Maxfield, 1999). Moreover, website data has been identified as an important resource for assessing nanotechnology's social impacts (Carroll, 2001), with Court *et al.*'s (2004) study of nanotechnology activity across selected Southern countries employing "Internet searches... to identify developing countries with NT [nanotechnology] activity".

2.1.2.1 National levels of R&D engagement

Various studies have been conducted to assess national nanotechnology activity,[4] including comparative assessments of international data (see Siegel, Hu and Roco, 1999b; Court *et al.*, 2004; Huang *et al.*, 2004). Such studies highlight the importance of comparative, international nanotechnology assessment. Yet, as of 2004, I found no indication of research which had assessed national engagement with nanotechnology R&D across every single country despite such research having occurred for biotechnology, accompanied by the implicit claim that such an approach was useful for assessing the technology's foreseen global trajectory (see Runge and Ryan, 2004).

Thus, in the first part of my quantitative research, I collected search engine data relating to national engagement with nanotechnology R&D across every country in the world.[5] Using Google,[6] the world's most foremost search engine,[7] on 16 January 2004 and then updated on 1 September 2004, I conducted a Boolean word and phrase search,[8] individually combining the truncated term 'nano*'[9] with the title of each of the 208 economies[10] recognised by the World Bank in 2004 (see *Appendix E*). The search parameters for data collection were based on a typology of engagement I created that included three sets of data (as outlined in Table 2.1). By using the sets: 'countries in which there exists national activities or national funding in nanotechnology', 'countries in which individual or group research is underway' and 'countries demonstrating an interest in nanotechnology', I was able to present engagement in a graded fashion, similar to Court *et al.*'s (2004) study of nanotechnology activity across selected Southern countries.[11]

Table 2.1 Categorisation of National Nanotechnology Activity.

Category	Evidential Requirements
National Activities or Funding	Either: A national strategy for nanotechnology Nationally coordinated nanotechnology activities Government funding for nanotechnology research
Individual or Group Research Project	At least one individual or group currently conducting work identified as 'nanotechnology research'
Country Interest	An expression of interest from a country's government, representatives or international delegates

To assess the distribution of engagement across recognised global groupings, countries registering activity were then categorised based on development classifications presented by the 2003 Organisation for Economic Development (OECD) Development Assistance Committee (see *Appendix F*).

In terms of the limitations of this method, not all nanotechnology activity can be expected to be reported via the World Wide Web. Furthermore, there is a reported national bias within the major search engines (Vaughan and Thelwall, 2004), compounded by my exclusive use of English language searches. If anything, these factors suggest my classifications for certain countries could be understated. Moreover, the new classification of research as 'nanotechnology' may mean that some nanotechnology activity is yet to be reported using this terminology, although the converse could be said to be true in terms of hype driving an overstatement of research activities in this field. Moreover, my classifications are uni-dimensional, lacking deeper clarification, such as distinguishing the strength and/or research directions of each country's engagement with nanotechnology activity — although I present some aspects of this analysis in my related review of the literature..

2.1.2.2 International research participation

Scientific conferences are an important venue for announcing and communicating research findings (Martens and Saretzki, 1993) and are increasingly being given attention and focus on a national level (Rutherford and O'Fallon, 2007). Consequently, Martens and Saretzki (1993) show the importance of reviewing conference data, noting, for example, which countries are engaged as hosts so as to detect early trends in science and technology. Such analysis, they claim, can provide important information ahead of delayed indicators such as patenting. Furthermore, a number of studies have shown how assessing delegate numbers at key science and technology conferences can be a useful indicator of the levels of international research participation in a certain field (see Mooney, 1999; Thorsteinsdóttir, Quach, Martin, Daar and Singer, 2004).

Hence, in the second stage of my search engine research, I assessed the location of all 2004 nanotechnology conferences or events by host country, as found on the premier nanotechnology conference database at the time: 'Nanotechnology Now'.[12] In order to get an indication of the breakdown of national representation at these conferences I then assessed participation at three key international nanotechnology meetings. In promotional materials, each had been explicitly noted as global in nature and focus: the 2004 International

Nanotechnology Congress[13] (INC) was one of the first events to claim to provide a forum for people, worldwide, from government, industry, academia and the nongovernmental organisation (NGO) sector to discuss nanotechnology; the 2004 International Dialogue on Responsible Research and Development of Nanotechnology[14] (IDRRDN) was the first intergovernmental dialogue of its kind; and the 2005 North–South Dialogue on Nanotechnology[15] (NSDN) was the first United Nations (U.N.)-sponsored meeting to specifically address Southern participation in nanotechnology. Participant data was drawn from an online database of INC presenters, an online database of IDRRDN attendees and a list of NSDN attendees, emailed to me by the organiser.

In terms of the limitations of assessing conference data, as with Martens and Saretzki's (1993) study of biotechnology conferencing, my study was quite constrained in terms of the data it explored and could have considered aspects such as who was organising each conference and the orientation of content explored at each. Furthermore, my results show an obvious, yet expected, bias towards host country that needs to be taken into consideration in analysis of the data. Moreover, having used a non-representative sample size, my results can only be considered illustrative. However, given the lack of any baseline data, illustrative results can still be considered a useful measure in the initial mapping of the 'state of play'.

2.1.2.3 International patenting

Patent data provides a key indicator of both a country's R&D capacity (UNDP, 2001; Huang *et al.*, 2004) as well as who will wield control in the shaping of a technology's trajectory (Foladori and Invernizzi, 2007). Therefore, as acknowledged by previous studies (see, for example, Compañó and Hullman, 2002; Marinova and McAleer, 2003), patent data is important for assessing nanotechnology's foreseen global implications (Carroll, 2001; Crow and Sarewitz, 2001; Roco and Bainbridge, 2001; ETC Group, 2005c), with the Meridian Group (2007) stating:

> Intellectual property rights will be a key factor in determining which nanotechnologies are developed, who controls existing and emerging markets, and who can access nanotechnology products and processes at what price (p. 10).

Whilst a number of contemporary patent studies share similar methods and results and are of significance to my work (see Huang *et al.*, 2004; ETC Group, 2005b), when I commenced my patent research, I found only two studies that had assessed global nanotechnology patent data in any significant way (see Compañó and Hullman, 2002; Marinova and McAleer, 2003). However, these two studies were of a broad, rather than sector-specific scope, restricted to an exploration of data relating to national engagement and, therein, limited in the extent to which nanotechnology's foreseen implications could be considered. Furthermore, Marinova and McAleer (2003) restricted their review to the U.S. Patent and Trademark Office (USPTO) database, whilst both studies used assessment dates that drew on data up until only the year 2000.[16]

To collect my data I therefore examined patents[17] registered between 1975 and 2004[18] using the widely encompassing esp@cenet database[19] that, as of 2003, incorporated records from over 70 countries including: the European Patent Office (EPO), the USPTO, the World Intellectual Property Organisation's Patent Cooperation Treaty and the Japanese Patent Office (JPO).

Commencing in December 2003, I conducted a title and abstract search[20] of patents using the specifically chosen truncated term 'nano*'[21] combined with selected health-related[22] terms. I focussed on an assessment of health-related patents given healthcare is presented as both an important condition and outcome of development (World Bank, 1993) and that patenting in this sector is central to many of the contemporary disputes surrounding equitable development (as will be explored in Chapter 3). Furthermore, as I shall explore in Chapter 4, nanotechnology is said to present a range of applications for Southern healthcare (see, for example, Juma and Yee-Cheong, 2005; Mnyusiwalla *et al.*, 2003; Ratner and Ratner, 2002; Salvarezza, 2003), and healthcare is one of nanotechnology's fastest growing sectors (Kalam, 2004; El Naschie, 2006), with pharmaceuticals a leading area in nanotechnology patenting (Compañó and Hullman, 2002). Correspondingly, there were over 250 health-related nanotechnology applications in preclinical, clinical or commercial development phases as of 2006 (ETC Group, 2008). Furthermore, sustained growth is expected for this sector that would lead to big impacts within global trade (Gross, 2003; Cientifica, 2007), such as nanotechnology affecting half of the world's drug production by 2011 (LaVan and Langer, 2001). In developing the framework for my search, I used the European Classification system[23] to distinguish health-related areas. Discovering some limitations within this system,[24] I revisited and completed my patent research in April 2005. Here I took similar steps[25] to Huang, Chen, Chen and Roco (2004), expanding my search by identifying the ten, primary health-related terms to appear as keywords in my existing titles which I then combined with the term 'nano*'. This produced a further 197 secondary health-related terms that I subsequently used in a combined search with the term 'nano*' (for examples of searched classifications, primary terms and secondary terms, see Table 2.2. For further details on health-related patent classifications see *Appendix B*).

Table 2.2 Classifications and Terms Used for Health-Related Nanotechnology Patent Searches.

European Classification System Areas	Primary Health-Related Terms	Examples of Secondary Health-Related Terms
medical or veterinary science; hygiene; foodstuffs; water purification; antibacterial paints	health*; medic*; disease*; diagnos*; detect*; drug*; delivery; therap* cosmetic*; treat*	antibacterial; antiseptic; prescription; bone; prophylaxis; pharmaceutical; genetic; vaccine; targeted; vitamin; skin

* Signifies truncation.

As of April 24, 2005, there were approximately[26] 16,940 patents on the esp@cenet® database that included the term 'nano*' in combination with health-related areas and the primary and secondary health-related terms I had identified. As I was only interested in recording an entry for each unique patent, and many patents appeared in my search on multiple occasions due to cross-over in search terms or filings across various patent offices, I created a number of specific rules to avoid duplication (see *Appendix C*).

In order to get some concrete data on global ownership and control associated with nanotechnology R&D, I conducted a 'basic analysis'[27] of country and sectoral patent engagement. To present the national distribution of patents, I first divided the collated data based on 'patent holder nationality'. Given a significant concentration of patent ownership amongst the top seven countries, I then showed this split separately before disaggregating the entire data by continental groupings to enable a broader assessment of patent distribution. I continued by analysing the data by 'sectoral representation' using the following five categories: private (company); private (individual); academic; government; and independent/not-for-profit. I refined my analysis further by using a number of discriminatory rules (see *Appendix C*). Given a significant concentration of patent ownership amongst private companies, I showed this split separately by producing a list of the top 20 patenting institutions in health-related nanotechnology. As a number of Chinese applications were written in Mandarin, I utilised the assistance of a Chinese-born research colleague, Dr Xu Xioada, as well as the free services at http://freetranslation.com, for clarification of each entity's sectoral grouping.

Given content analysis of nanotechnology patents can be an important way to measure foreseen social impacts (Carroll, 2001), the second stage of my patent research involved a more in-depth analysis of patent orientation. I first analysed the 'general utility' of health-related patents by assessing the text in abstracts for indications of each patent's orientation towards one of three functional categories alluded to in research by White (2003): 'therapeutic', 'diagnostic' and 'consumer health'. Along with this disaggregation, I provided examples from each category. To continue this exploration, I analysed the health-related 'specific utility' of patents by assessing the text in abstracts for references to various health conditions, ranking these by citation and, again, adopting discriminatory rules for what was included in this analysis (see *Appendix C*).

Because of the previously mentioned overlap in patent data stemming from initial searches, the in-depth nature of the second half of my patent analysis, and, given that manual reading and interpretation of patents is an accepted analytical method (Huang *et al.*, 2004), I analysed all my data without special software[28] and, instead, visually reviewed the title, abstract and bibliographic information for each of the 16,940 patents.

Considering quantitative data needs to be generalisable, valid, reliable and replicable (Cresswell and Plano Clark, 2007), a number of limitations should be noted with respect to my data collection. The first relates to the identification of relevant patents. Definitional ambiguity for nanotechnology means patent identification is imprecise (ETC Group, 2005b). The new use of the term 'nano' means certain relevant patents would not yet be classified using this language, whilst, additionally, the esp@cenet® database cannot claim to cover data registered at every single patent office in the world. Conversely, broad search terms, such as 'nano', can result in exaggerated counts (*ibid.*). As Bai (2005) explains for certain Chinese cases: "...because of the sudden popularity that the term 'nano' enjoys, some firms in China have been finding that they can raise their profits simply by adding the label 'nano' to their products" (p. 63). Evidence of this practice emerged during my data collection, with one Chinese national holding over 500 health-related nanotechnology patents by "...simply turning traditional plants into fine powders with particles under 100 nanometres...and claiming a new invention" (Coalition Against Biopiracy, 2004). In another example, a patent held by the firm Stirling Winthrop refers to tumour targeted particles of around 1000 nanometres, which is on a micro-, not nanoscale. These, and other similarly questionable results, were excluded.

The second limitation involves the ownership and control of patents. Here, transnational rights to patents, a lack of updated information on corporate mergers and acquisitions within patent databases, as well as the absence of mandatory exclusive licensee disclosure in the U.S. (ETC Group, 2005b) made it difficult to attribute a specific entity or nationality to some patents.

The final limitation involves my methods. In terms of the data collection, Huang *et al.* (2004) note that a full-text patent search provides a more complete survey than just a title and abstract keyword search. However, they simultaneously accept the validity of a partial-text patent search, given studies have shown that "... the number of patents searched by 'title claims' are in the same range with the data published by other groups" (p. 327). Whilst any research using my selected keywords and timeframe would produce data identical to mine, my analytical categorisation of patents by 'general utility' was subjective and is therefore not replicable.

2.1.3 Qualitative

In order to explore a range of perspectives within both the South and North about nanotechnology's foreseen implications for global inequity, in the qualitative phase I interviewed 31 key informants from Thailand and Australia. Despite this phase of the research focussing on qualitative methods, I supplemented the interviews with surveys of 24 members of the Thai nanotechnology research community.

In terms of the interviews, gauging in-depth participant perspectives is said to be a particularly good way of conducting exploratory research (Yin, 1994; Seidman, 1998). As with biotechnology (Fransman, 1994), interviewee perspectives involving participants from the South are seen as useful for understanding nanotechnology's foreseen consequences more fully (Carroll, 2001; ETC Group, 2004c; Singer, Salamanca-Buentello and Daar, 2005), ensuring a richer set of information that can be shared by the global community (Carroll, 2001).

In terms of the surveys, Johnson and Onquegbuzie (2004) note the acceptability of supplementing a qualitative research study with "...a closed-ended instrument to systematically measure certain factors considered important in the relevant research literature" (p. 19). In this light, surveys have been noted as a useful approach for gaining insights into the nature of nanotechnology's emergence (Carroll, 2001) and a great way to augment nanotechnology-related interview data (Roco and Bainbridge, 2001).

2.1.3.1 Key informant interviews

From March to September, 2004, I interviewed 16 Thai and 15 Australian key informants, each in their respective country. Key informants are defined as "...those who can provide relevant input to the process, have the highest authority possible[29] and are committed and interested" (Gutierrez, 1989, p. 33). As influential people, their place in the study of technological innovation is said to be particularly insightful, given:

> Understanding the implicit assumptions, values and visions — or 'imaginaries' — of key actors has been recognized as a central part of the social science challenge, because of their significant role in shaping research and innovation trajectories (Brown and Michael; Hedgecoe and Martin; Kearnes *et al.*; Rose; van Lente in Kearnes *et al.*, 2005, p. 297).

Furthermore, when a number of 'expert' views are combined, they are said to provide "useful snapshots" (Carroll, 2001, p. 191). Subsequently, the key informant approach has been successfully used in the sociological study of biotechnology (see Rezaie, Frew, Sammut, Maliakkal, Daar and Singer, 2008) and nanotechnology (see Mee *et al.*, 2004).

Ensuring interviewee diversity is an important component in selecting a key informant sample to explore nanotechnology's implications (Mee *et al.*, 2004). I therefore focussed on diversity across three areas (for a full summary see *Appendices G* and *H*. For interviewee biographies see *Appendix I*). The first was sectoral diversity, whereby I ensured interviewees spanned academia, NGOs and private and government sectors — with these four sectors previously identified as important for studies of the foreseen implications arising from biotechnology and nanotechnology (see Meridian Institute, 2006; Singer, Berndtson, Tracy, Cohen, Masum, Lavery and Daar, 2007; Throne-Holst and Stø, 2008).

As a further extension of this categorisation, the second area included diversity of occupations[30] and the associated range of expertise. Reflecting on previous research, a number of authors (2005), say that studies of nanotechnology's foreseen implications must go beyond science- or business-based consultations, with Reid (1996) adding that maximum variation by occupation is a good way to ensure representation for a range of experiences relating to a phenomenon. I therefore deliberately included interviewees from both countries with expertise in fields as broad as science, innovation, engineering, foresight, business, finance, ethics, law, education and social science. Whilst such diversity fits with my broad research focus, I selected interviewees, where appropriate, whose work engages with healthcare (ahead of other fields) in order to enhance my focus via the expected use of similar scenarios and examples. As part of this diversity within occupations and expertise, interviewees also worked at differing levels relevant to technology and its implications: from frontline service delivery and grassroots activism, through to technological design, manufacturing, marketing, planning, policymaking, education and public outreach.

The final area in which I ensured diversity was that of experience with the subject matter. Thus, 19 interviewees (slightly more than half of my sample) were engaged in work that involves nanotechnology ('nano-engaged'), with the remainder having very little to no pre-existing engagement or understanding of the field ('nano-disengaged'). I believe this final aspect of diversity is critical to ensuring an assessment that is more reflective of the realities associated with technological development, whilst fulfilling my desire for a mix of people who can bring knowledge about social context to debates about nanotechnology as well as those who can bring knowledge about nanotechnology to debates about social context.

The process for indentifying key informants began with my initial review of the literature on nanotechnology, development and inequity, in which I identified 16 fields of interest.[31] I thus established both the kind of knowledge proficiencies I needed for my interviewee sample as well as an appropriate sample size to ensure informational saturation whilst maintaining a deep individual analysis. Identifying the specific key informants involved web and literature searches — a technique used in one prominent study of biotechnology's implications for Southern development (see Daar, Martin, Nast, Smith, Singer and Thorsteinsdóttir, 2002). I also used a 'snowballing' method, whereby I asked already identified key informants to recommend other interview candidates.[32] Such a method had already been used in relevant nanotechnology research[33] (Court *et al.*, 2004) and is said to be particularly useful for "...studies of difficult-to-find populations...[and] any small population for which it is impossible to construct a sampling frame" (Bernard, 2000, p. 179). However, snowballing is also frequently critiqued as favouring sample participants who have a large number of interrelationships, whilst excluding those who are socially isolated (Berg, 1998; Salganik and Heckathorn, 2004). Furthermore, Salganik and Heckathorn (2004) point to a general belief that any bias in the initial participant selections will be compounded as the sample grows. To manage this, I made a particular effort to source initial key informants — whom I then drew upon for contacts via the snowballing process — from the more diverse fields of knowledge I had identified as important. Moreover, as already noted in my opening chapter, my qualitative research should be viewed as illustrative, rather than representative, given the various limitations inherent in my sample.

In terms of engaging with the interviewees, my first step was to gain research ethics approval from the University of Technology, Sydney.[34] As part of this process, I developed documentation for the interviewees which covered issues of informed consent, confidentiality, and access to their data. In this respect, I obtained consent from all interviewees to be named in this research. As examples of the sensitivity to language and cultural issues required in a cross-cultural study of this nature, I had all the key documents for Thai interviewees formally translated from English into

Thai,[35] I offered translation services for each Thai interview and I accepted verbal consent when written consent proved customarily inappropriate.[36] I also drew on the services of Dr Nares Damrongchai, from the National Science and Technology Development Agency of Thailand (NSTDA), who assisted with introductions, scheduling and customary information in relation to each Thai interview. However, before proceeding with the Thai interviews, I also gained approval from the National Research Council of Thailand[37] through a process as rigorous as that presented by the University of Technology, Sydney.

Shown to be an appropriate length of time for an interview-based study of nanotechnology (see Throne-Holst and Stø, 2008), my average interview was 45 minutes in duration, with the shortest being 20 minutes and the longest being 80 minutes. Interviews were face-to-face, a method noted by Garrett (1999) as advantageous for future-oriented research.

When designing my interviews, I tried to use open-ended questions as much as possible, given their usefulness in assessing interviewee perceptions about nanotechnology (Roco and Bainbridge, 2001). Following a broad focus at the commencement of each interview, my questions gradually became more specific and relevant to the interviewees' expertise, demonstrating what has been described as a 'funnel' approach (Brenner, Brown and Canter, 1985). Additionally, my interviews were semistructured, allowing flexibility and adaptability in my questioning, both within and between interviews, whilst remaining guided by a broad framework (Britten, 1995). The semistructured approach to interviewing is seen as particularly important for nanotechnology, given the high variation in levels of associated knowledge amongst stakeholders (Throne-Holst and Stø, 2008).

A number of issues arose regarding limitations pertaining to my interview methods. Firstly, the diversity of my sample was restricted due to financial and temporal limitations as well as my decision to interview 'key informants'. As with previous work assessing nanotechnology's foreseen global implications (see Salamanca-Buentello, Persad, Court, Martin, Daar and Singer, 2005), I was only able to identify three females for my study. In a similarly limited fashion, all the Thai interviewees were Bangkok-based, whilst all but one of my Australian interviewees were based in large cities, meaning a lack of consideration for rural perspectives. In terms of age and educational diversity, the vast majority of my interviewees were over 40 years of age and held PhDs having, at some stage, received educational training in a Northern country. Subsequently, most Thai key informants spoke fluent English. However, this is not to deny the second limitation: the cross-cultural constraint that meanings can be lost in translation, no matter how fluently an individual speaks a language other than their native tongue.

To commence my analysis of the data, I transcribed recorded interviews in full. Interviewees were given the opportunity to read and edit the transcripts of their interviews in order to correct misinterpretations and refine sentiment, with some choosing to make minor adjustments and one of the Australian interviewees withdrawing from the research project in 2005 at the behest of their workplace, although the reasons for withdrawal were not disclosed.

Whilst not strictly following Glaser and Strauss' (1967) 'grounded theory' approach,[38] I drew on many of the concepts and methods central to their work. For example, I used an inductive approach for my initial analysis of the data by going through and 'open coding'[39] all the transcripts and setting up relevant 'free nodes'[40] in NVivo™ — a program noted for its ability to assist in investigating and developing an emergent analysis (Gibbs, 2002; Reid, Wood, Smith and Petocz, 2005). As Olssen (2003) notes, when there is a large volume of research, such a coding framework is needed in order to "facilitate the systemic analysis of the participant's accounts" (p. 108). I then continued with a review of the entire data in which I refined and sorted these nodes into categories, known in NVivo™ as 'tree nodes'.[41] Categorised information was then placed into a logical sequence within my overarching frames of analysis (understandings, innovative capacity, technological appropriateness and approaches to governance), and analysed in terms of similarities and differences, both within categories and across themes.

The final limitation to my interviewee research involves the extent to which my data collection and analysis were overly prejudiced by concepts and theoretical frameworks I introduced to the study. Here I acknowledge that my questions, coding and analysis were undoubtedly influenced by understandings I developed through my review of the literature. Furthermore, although I made efforts to ensure open-ended questioning, nanotechnology's nascent nature, combined with the lack of nanotechnology knowledge held by certain interviewees, demanded leading questions be part of some of the interviews. In response, some (ETC Group, 2005b; Meridian Institute, 2006) controversially[42] argue that studies of nanotechnology will be best served by sector- or application-specific analyses.

2.1.3.2 Thai nanotechnology practitioner surveys

My qualitative study was supplemented by a two-page survey that I emailed[43] to 55 members of the Thai nanotechnology research community between August and September, 2004. Members of this community were identified through my review of a report by researchers at Chulalongkorn University on the situation of nanotechnology researchers and R&D in Thailand (see Unisearch, 2004). As with my interviews, a consent form accompanied each survey, with translated documents also made available. Each respondent was required to print and complete the survey before mailing it back to me in Thailand or Australia, depending on the date of completion.

Twenty-four surveys (44%) were returned and deemed valid, constituting a high response rate[44] given I provided no incentives for survey completion. Eleven of the survey respondents were self-identified nanotechnology researchers, correlating with approximately 11% of the Thai nanotechnology research community at the time.[45] Thirteen of the survey respondents were Thai researchers who claimed to be working in nanotechnology-related areas. Given nanotechnology's wide scope and the fact that all of the respondents working in nanotechnology-related areas used terminology common to nanotechnology, I merged the survey data for my analysis.

My survey focussed on closed-questions of a personal and professional demographic nature (for the synthesised results, see *Appendix J*). To analyse the descriptive data, I used a number of basic statistical measures, similar to those used in my patent analysis. Whilst surveys were initially identified for verification, when combined, the results were de-identified.

In terms of the limitations of this method, my response rate is likely to have been constrained by the material means of survey completion, given the surveys were posted and postage costs not covered. However, the benefit was that I recruited participants with a genuine interest in the topic. With respect to the data I gained, and as is typical with surveyed responses, the rigidity of my questions offered little scope for clarification or qualification by those surveyed.

However, this rigidity ensured data continuity, thereby enabling a more consistent analysis.

2.2 Conclusion

In this chapter I have outlined my approach for addressing my research questions and aims. I have shown the appropriateness of a mixed methods methodology for an exploratory study assessing the extent to which nanotechnology offers hope for a more equitable world. I have identified the need for a comprehensive review of both historical and contemporary literature to precede my quantitative and qualitative research phases, commencing with the broad debates about technology, development and inequity before progressing to more specific debates about nanotechnology in relation to these issues. Such reviews will provide me with a context in which to assess nanotechnology's promises and early trends against those experienced with the emergence of previous technologies, such as biotechnology. They will also allow me to develop and test an interpretive framework through which to assess relevant claims.

My use of search engines in the quantitative section of my research has been shown to be a method by which greater clarity about the nature of global engagement with nanotechnology R&D may be established. Here I highlighted the value of generating descriptive data and analysis covering national R&D engagement, conference hosting and participation and, more comprehensively, matters of ownership and orientation associated with health-related global patenting. By using these methods I am also able to provide a preliminary assessment on the likelihood that nano-innovation and innovative capacity will be globally decentralised.

Finally, I have identified interviewing as an appropriate method for exploring a range of perspectives within both the South and North relating to nanotechnology's foreseen implications for global inequity. I have also shown the benefit of supplementing interviews with targeted surveys of Thai nanotechnology practitioners. These methods allow me to explore my research questions surrounding the commonality of nanotechnology understandings, the decentralisation and localisation of nanotechnology capacity and innovation, nanotechnology's appropriateness for the South and the level of participatory governance in nanotechnology's development.

Throughout this chapter I have acknowledged many of the limitations and constraints of my methodology and methods. Overall, it may be argued that my research is constrained by the non-representative samples and data with which I engaged, as well as the accuracy of the quantitative information I collected. Such limitations are symptomatic of investigating a new field of research and the desire to be methodologically innovative. Having outlined my methodology and methods, in the coming chapter I will assess key historical and contemporary literature surrounding technology, development and inequity, thereby laying the grounds for establishing a framework, including criteria, for evaluating nanotechnology's potential to provide hope for a more equitable world.

Endnotes

1 Although, as mentioned (Chapter 1), I seek to reference other technology in order to provide contextual benchmarks against which I can assess nanotechnology.

2 A process, whereby three or more sources of data are used to strengthen the reliability and validity of an analysis.

3 As proposed by Invernizzi *et al.* (2008).

4 http://google.com.

5 Here I use 'national nanotechnology activity' to refer to the level of a country's commitment to nanotechnology R&D.

6 http://google.com.

7 According to the 2004 Search Engine Yearbook (le Roux, 2004).

8 A search based on Boolean logic, whereby use of the term 'and' between primary search terms presents records in which the primary search terms are both present.

9 Truncating a term by using a wildcard symbol, such as '*', allowed me to search for different endings of the same root word. The use of the keyword 'nanotechnology' has been used previously to great effect in a study assessing national nanotechnology R&D activity across multiple countries (Court *et al.*, 2004).

10 For the purposes of this book, the terms 'economies' and 'countries' include 'territories' recognised by the World Bank.

11 In their study, Court *et al.* (2004) categorised countries as either: 'front runner', 'middle ground' or 'up and comer', according to their level of engagement with nanotechnology R&D.

12 Accessible at: http //nanotech-now.com/events-2004.htm. As Martens and Saretzki (1993) highlight, it is extremely difficult to identify established and acknowledged databases for interdisciplinary fields.

13 See: http://pharmabiz.com/article/detnews.asp?Arch=&articleid=24 040&sectionid=9 [conference site no longer available).

14 The meeting report may be found at: http://tinyurl.com/yfyu9mc.

15 See http://ics.trieste.it/Nanotechnology/ for more details.

16 Compañó and Hullman's (2002) review encompassed the period: 1990–1999, whilst Marinova and McAleer (2003) considered the period: 1975 – 2000.

17 Considering Rader's (1990) notes on comparability among U.S. and foreign success rates for biotechnology patent applications, I used both patent applications and assigned patents for my research of 'patent data'.

18 Since 1974 was the official time at which the word 'nanotechnology' was coined, and given it takes up to eight months before 90% of the European Classification System data is confirmed (European Patent Office, 2005), 1975 was taken as the point for the patent search to commence. The range of data was also chosen so as to facilitate as broad an assessment as possible and given that, in 2004, efforts were being made to harmonise the nanotechnology-related classifications between the USPTO, the EPO and the JPO (Huang *et al.*, 2004).

19 The esp@cenet® database was established by the European Patent Office in 1998. It can be accessed at: http://ep.espacenet.com.

20 According to Huang *et al.* (2004), title-searches form the basis of most of the data published in literature.

21 A term noted as one of the best ways to gauge enthusiasm for nanotechnology (Mooney, 1999), with Huang *et al.*, (2004) showing that 92.5% of the nanotechnology patents registered with the USPTO office between 1976-2003 included the term 'nano*' and Marinova and McAleer (2003) also using the main keyword 'Nano$' to extract data but similarly excluding the term 'nanoseconds' and the chemical compound 'NaNO' (here the symbols '$' and '*' are substitutable forms of search wildcards). Furthermore, nanotechnology patent classes were not established until 2004 for the USPTO and 2006 for the EPO (ETC Group, 2008).

22 Others, such as Srivastava and Chowdhury (2008), have used the terminology 'health-related sectors' in their research, including food and cosmetics in their review.

23 See: http://ep.espacenet.com/help?locale=en_EPandmethod=handle HelpTopicandtopic=ecla.

24 For example, many Chinese patents without an abstract could not be included in the areas defined by the European Classification System, but could be identified as health-related, via their title.

25 Combinations with basic keywords were also the basis for Huang *et al.*'s (2004) patent study.

26 As with Zhou and Leydesdorff's (2006) scientometric analysis, the esp@cenet® system of patent retrieval has certain limitations. The system does not provide an exact number when the recall is larger than 10,000 and the limit for each viewing is 500. This was overcome, in part, by my use of 207 search terms.

27 Such an approach is said to be a standard measure used in assessments of technology and development (Huang *et al.*, 2004).

28 Microsoft Excel excluded.

29 The vast majority (27) of the interviewees held PhDs. Most often, these qualifications were relevant to the specific areas of questioning around which I engaged each interviewee respectively.

30 Each key informant's response in this research represents their own views and does not necessarily represent those of the organisations with whom it is noted that they were affiliated. Where mentioned, each key informant's title and affiliation has been used to add credibility to their statements and allow for deeper analysis. Stated titles and affiliations were those held at the time of each interview.

31 The 16 identified fields included 10 broad areas: development; science, technology and society; appropriate technology; ethics; education; international collaboration; technology transfer and NGOs; IP and regulation; capabilities assessment; investment and innovation policy; and six sector-specific (health) areas: international health; biopolitics; bioethics; service delivery; diagnosis; and treatment.

32 In order to provide greater depth to my research, as part of this snowball process I specifically and successfully sought recommendations for Australian interviewees whose work had, in some way, engaged with Thailand or Thai issues.

33 Court *et al.*'s (2004) study involved: "...personal communication with approximately 30 government officials, academic researchers and industry representatives in developing countries [who] provided further information on contacts and nanotechnology activity in the country. Academic researchers from developed countries who had attended nanotechnology conferences in developing countries

were also contacted to provide information on developing world researchers".

34 See: http://gsu.uts.edu.au/policies/hrecguide.html.

35 Two interviewees used the translated consent form and overview materials.

36 One interviewee used the option of verbal rather than written consent.

37 See: http://nrct.go.th.

38 For example, my basic review of the literature was conducted in advance of the taped interviews and I did not use memos to document thoughts on the text throughout my analysis.

39 A process by which I marked each transcript margin with the name(s) of the phenomena found in each distinct passage.

40 Nodes without a formally recognised relationship to other nodes in the project.

41 Nodes organised into a formal hierarchical structure.

42 Such arguments would seem unnecessary given the Meridian Institute's (2006) research into nanotechnology, water and development shows that the issues raised by nanotechnology are likely to be cross-cutting and transferable.

43 Sent electronically, given I was able to access relevant email addresses and this made the survey-distribution process affordable and easier to manage.

44 Although such a rate is less important given I was seeking practitioner insight, rather than representation, from my surveys.

45 This figure is estimated, based on claims in previous reports that there were approximately 100 nanotechnology practitioners in Thailand around 2003/04 (Lin-Liu, 2003; Unisearch, 2004).

Chapter 3

Development, Technology and Inequity

In this chapter I will review the literature surrounding development, technology and inequity in order to establish a context and create an analytical framework for assessing the extent to which nanotechnology offers hope for a more equitable world. In doing so, I will commence my contextualisation phase, building on the need outlined in my introductory and methodology chapters: that any contemporary assessment of nanotechnology's implications for the global South must be grounded in an assessment of historical and theoretical underpinnings to development, technology and inequity.

3.1 The Creation of Inequity and Establishment of Development Debates

Historically, technology has been inextricably linked with changes in global dynamics (Crow and Sarewitz, 2001). Highly relevant to my research is the industrial revolution that stemmed from the accumulated wealth created by merchant capitalism and the origins of the world market. The resultant technologies, developed in Europe, such as mass manufacturing, communications and transportation, were foundational in establishing and maintaining hegemonic control over colonies in the global South (Alvares, 1997). The nature of industrial capitalism demanded foreign control and an international division of labour,[1] resulting in the

Nanotechnology and Global Equality
Donald Maclurcan
Copyright © 2012 by Pan Stanford Publishing Pte. Ltd.
www.panstanford.com

exploitation of cheap, unskilled labour and low levels of regulation and compliance within the various colonies (Goldthorpe, 1996). In such circumstances, capital was transferred by colonialists, out of the colonised South, via the extraction of raw materials that then had 'value' added in the North before being repurchased by the colonies at much higher prices (Sagasti, 1980). The Indian leader, Vinoba Bhave (in Lanza del Vasto cited in Prime, 2002), explains this phenomenon clearly through the following vignette:

> The field opposite grows cotton. The owner of the field sells it to a man who collects it. This man sells it to a dealer who sells it to another who transports it to Bombay, where it is sold to a shipper who ships it to an English port where it is sold to a factory which turns it into spun cotton and sells it to another factory which turns it into woven cloth and sells it to a dealer who ships it to Bombay where it is sold to a dealer who sells it to a pedlar who sells it in the village to the owner of the cotton field (p. 84).

The industrial revolution dramatically redefined international trade, markets and competition. The concentrated exploitation of technologies accelerated and exposed new inequities, 'catapulting' some nations and world regions ahead of others, whilst coinciding with redistribution in the global balance of power (Lucas Jr and Sylla, 2003; Juma and Yee-Cheong, 2005). Global inequities, it is therefore argued, are closely linked with technological inequities, and can be seen through the different levels of accumulation, creation, mastery and utilisation of modern technology (Salam, 1991; Lalor, 1999; Juma and Yee-Cheong, 2005; UNCTAD, 2006).

As the speed of technological innovation increased towards the latter part of the nineteenth century, so too did societal transformation (Crow and Sarewitz, 2001). The industrial revolution had shown the ability for emerging technology to displace labour on a massive scale (*ibid.*) and transfer power from marginalised rural populations to city-based elites (Moore Jr, 1966).

In the wake of the Second World War, an emerging focus on national self-determination stimulated greater attention to the growing gaps between country wealth, signalling the commencement of the 'development' era (Sachs, 1997a; Rist, 2002) and the emergence of 'development economics' as a distinct discipline (Vernengo, 2006). Considered foundational to this era is the 1949 speech of the then U.S. President, Harry Truman (in Esteva, 1997), in which he said:

We must embark...on a bold new program from making the benefits of our scientific advances and industrial progress available for the improvement and growth of underdeveloped areas. The old imperialism — exploitation for foreign profit — has no place in our plans. What we envision is a program of development based on the concepts of democratic fair dealing (p. 6).

Three points that are central to the emergence of development debates can be noted in this excerpt: Truman's demarcation between 'developed' and 'underdeveloped' countries (Sachs, 1997a); the conspicuous linking between technology and development; and that greater equity is the ultimate goal of development, stemming from the universal ideals of the Enlightenment (Simms *et al.*, 2009).

The constructs for development debates formed as these, and similar philosophies, were contested by individuals such as Gunder Frank and Mohandas Gandhi. As I shall soon explore, critical challenges were made to the underlying assumptions of developmentalism, the means and ends to development, and where the 'development problem' lies (see Baran, 1952; Prebisch, 1959; Frank, 1966; Schumacher, 1973; Wallerstein, 1974). Of particular centrality to this book are further contestations around the relationship between technology and development, given "visions of the role of science and technology in development have always been diverse, as diverse as visions of development itself" (Leach and Scoones, 2006, p. 15).

3.2 Foundational Approaches

Early development debates provide a foundational context in which contemporary contestations continually emerge. This is particularly true for discussions about the relationship between technology and development, as highlighted by Eckhaus' (1977) comments:

> In the early, optimistic period of development analysis and policy as well as in the more recent pessimistic days, technology has been attributed a critical role both in resolving and creating development problems (p. 7).

Thoroughly investigating the early debates therefore allows the establishment of a relevant, contextual framework in which to assess nanotechnology's emergence. As Feenberg (2004) questions, "...how can one study specific technologies without a theory of the larger society in which they develop?" (p. 73). Lucas and Sylla (2003) add that emerging technologies must also be placed within a framework of major innovations in modern economic history in order to adequately assess issues of equity.

3.2.1 Modernisation Theory

Emerging after the Second World War as an explanation for how societies change, modernisation theory was presented, by many, as a strategic blueprint for Southern economic development and the best means by which to reduce global inequities (Hoogvelt, 1997). In the wake of reconstruction projects in Europe, known collectively as the Marshall Plan, modernisation theory was central to the geopolitical rivalry for influence over the global South during the Cold War (Escobar, 1995). As Escobar (*ibid.*) notes: "in the late 1940s, the real struggle between East and West had already moved to the Third World, and development became the grand strategy for advancing such rivalry" (p. 34).

Modernisation theory is founded upon philosophies of 'modernity', bound up in the Enlightenment principles of idealism, progress and liberty (Hettne, 2009). Drawing on Durkheim's theories of social evolutionism, modernisation theory claimed that countries progress from 'undeveloped' to 'developed' through a universal, linear pattern of stages (Escobar, 1995; Alvares, 1997; Sachs, 1997a; Hettne, 2009). Most famous is Rostow's (1960) 'stages of growth' theory that identified five stages through which all societies are said to pass on their path to development: "...the traditional society, the preconditions for take-off, the take-off, the drive to maturity, and the age of high mass-consumption" (p. 4). In Rostow's model, the countries of the South are in the first three stages of development, extending the colonial attitude that Southern societies are inferior (Packard, 1997). As an example of this attitude, Packard (*ibid.*) cites a speech in 1950 by Pierre Dorolle, then Deputy Director General of the World Health Organisation (WHO), to the First Malaria Conference in Equatorial Africa, in which she said: "it is true that a great part of the peoples of Africa south of the Sahara are still in an underdeveloped state so far as degree of civilisation and culture and social development are concerned" (p. 109).

Implicit in Rostow's (1960) claims are beliefs that barriers to Southern development are endogenous, that poverty and inequality can be explained by late modernisation, and that Southern countries should, therefore, disregard existing means of cultural existence. Subsequently, the corresponding proposal is that the North, considered to be in the latter stages of development, provides the blueprint for progress, and that Southern countries "...will only 'advance' if they take on the characteristics of the rich industrialised countries" (Clark, 1985, p. 164). As Meier and Seers (1984) note, such theory assumed that the conditions found in Southern countries would be the same as those found in post-war Europe, with Fukuda-Parr, Lopes and Malik (2002) summarising that early modernisation theory promoted a view of "...development as displacement" (p. 8).

Various international regulatory frameworks were used to engender Southern development. Having proven successful in assisting Western European reconstruction through the Marshall Plan, the 1944 Bretton Woods Agreement refocussed its efforts upon the former European colonies, known as the 'least developed countries' (LDCs). In the 1950s and 1960s, the provision of financial and technical 'aid' to the South by the established instruments of the Bretton Woods Agreement — including the International Bank for Reconstruction and Development, the International Development Association[2] and the International Monetary Fund (IMF) — coincided with the emergence of bilateral aid programs (Finnemore, 1997). Here aid was seen as a means to achieve modernisation and drive Southern economic growth (Schumacher, 1973; McRobie, 1981), with the belief that the South could easily emulate the North by taking advantage of its experience and aid (Fukuda-Parr *et al.*, 2002).

Of critical importance to the development approach was the grounding of modernisation in theories of economic growth, with capital accumulation presented as a universal ideal (Rostow, 1960). More specifically, understandings of 'development' were based on raising 'gross domestic product' — GDP (Finnemore, 1997), and focussed on "...resources for investment...to increase production or incomes and, thereby, consumption levels" (UNCTAD, 1999, p. 150). This is perhaps best highlighted by the World Bank's Articles of Agreement that emphasised: "...productivity, investment, capital accumulation, growth, and balance of payments" (Finnemore, 1997, p. 206).

Adopting a growth-based model was presented as a moral imperative for development, given two assumptions. The first was that the economic growth of countries having access to the same technology, population growth rate and savings' propensity — and differing only in terms of their initial capital-labour ratios — will converge (Yao, 2005). Emerging from the work of Alexander Gerschenkron,[3] it was believed that the more 'backward' a country the greater its potential to catch up (Radoševic, 1999; UNCTAD, 2002b). The South was seen as able to accumulate capital faster than the industrially pioneering North (Fukuda-Parr *et al.*, 2002) because invested capital has a diminishing return (see Solow, 1956; Swan, 1956). Furthermore, it was believed that the South could 'leapfrog' technological capacity by implementing technologies developed in the North (Lucas Jr and Sylla, 2003) and harnessing 'comparative advantages', such as large amounts of labour and natural resources (Khan, 1979).

The second assumption was that national economic growth would automatically produce benefits for all through a 'trickle-down' effect, with social progress assumed a natural by-product of the inherent drive to expansion and incorporation promoted by capitalism (Hoogvelt, 1997).

As variants of modernisation theory, strategies for Southern economic growth ranged between neo-classical, laissez-faire approaches and Keynesian interventionism, as well as compromises between the two. The Harrod-Domar model was one such compromise, proposing that economic growth depended on policies to increase investment by increasing savings (Peet and Hartwick, 1999). Subsequently, others (see Solow, 1956; Swan, 1956) explored methods to increase savings-based capital accumulation and labour transition in an effort for countries to improve their GDP. On the other hand, structural change theories, as espoused by Lewis's (1954) 'dual-sector model' and Chenery's (1975) 'patterns of development theory', proposed that Southern countries shift away from a subsistence, agrarian focus and reorient economic structures to a "...more modern, more urbanized, and more industrially diverse manufacturing and service economy" (Todaro and Smith, 2002, p. 73). Here it was believed that new industries should replace traditional agriculture as the engine of economic growth (Meier and Seers, 1984). According to the dual-sector model, if the right amount of surplus labour shifts from the agricultural to manufacturing sector, the outcome should be greater productivity and improved social welfare (Lewis, 1954).

3.2.1.1 Western technology and development

Technology and innovation lie at the heart of modernisation theory. Technology was seen as the means by which to improve people's working capacities, allowing "...long-term growth in real wages and the standard of living" (Solow, cited in Peet and Hartwick, 1999, p. 40). Given the neo-classical assumption that capital is subject to diminishing returns, the uptake and diffusion of modern technology becomes critical in order to perpetuate growth (Solow, 1957). Sustained growth is achieved through technological advances that lower the capital-output ratio (see Domar, 1946; Harrod, 1948) and by technology producing a sequence of quality-improving innovations, each of which destroys the costs generated by previous innovations (Schumpeter in Correa, 1998).

This approach to innovation is heavily grounded in modernity and the Enlightenment view that technology is inherently beneficial and progressive (Bush, 1945; Shallis, 1984; Escobar, 1995; Sbert, 1997). In terms of development, these factors resulted in a situation in which industrial development was viewed as the means by which modern and rational societies emerge (Simms *et al.*, 2009), with Western science projected as: "...a universal, value-free system of knowledge which has displaced all other belief and knowledge systems by its universality and value-neutrality" (Shiva, 1989, p. 15). As Ullrich (1997) reflects, the starting assumption of European modernity was:

> ...that unremitting diligence, constant progress in the production of material goods, the unbroken conquest of nature, the restructuring of the world into predictable, technologically and organizationally manipulable processes will automatically and simultaneously produce the conditions of human happiness, emancipation and redemption from all evils (p. 278).

From a modernisation perspective, therefore, technology is seen as determining society's values, social structure and history:

> Social systems are functions of technologies; and philosophies express technological forces and reflect social systems. The technological factor is therefore *the* determinant of a cultural system as a whole. It determines the form of social systems,

and technology and society together determine the content and orientation of philosophy (White, 1949, p. 366).

In this sense, modernisation theory draws on Rostow's (1960) evolutionary model, with technological revolutions seen as both central to progress and fixed in their paths:

> ...moving along one and the same track in all societies. Although political, cultural and other factors may influence the pace of change, they cannot alter the general line of development that reflects the autonomous logic of discovery (p. 138).

Subsequently, technological progress is viewed as inevitable, unavoidable and irreversible (Shallis, 1984; Feenberg, 2002). According to Rostow (1960), lags in Southern technological capacity are seen as restricting the evolutionary process; the traditional society is equated with the 'pre-modern', with a productivity limit due to "...the inaccessibility of modern science, its applications, and its frame of mind" (p. 5). In the 'preconditions for take-off' phase there is an increase in manufacturing productivity, although modern manufacturing enterprise remains sporadic and the society is "... still mainly characterized by traditional low-productivity methods" (*ibid.*, p. 7). In the 'take-off' phase, a surge in a relatively narrow form of industrial development sees growth begin to become steady. Resulting in revolutionary changes in productivity, this growth is particularly dependent on the commercialisation of agriculture and the exploitation of "...hitherto unused natural resources and methods of production..." (*ibid.*, p. 8). In the 'drive to maturity' phase, growth is secured and modern technology becomes both broadly applied to production across the whole front of an economy's activity and, simultaneously, "...more refined and technologically often more complex" (*ibid.*, p. 9). In this phase it is believed that:

> ...an economy demonstrates the capacity to move beyond the original industries which powered its take-off and to absorb and to apply [technology] efficiently over a very wide range of its resources (*ibid.*, p. 10).

Concurrently, choices over the areas of technological development emerge as "...an economy demonstrates that it has the technological and entrepreneurial skills to produce not everything, but anything that it chooses to produce" (*ibid.*, p. 10). Dependence is said to become "...a matter of economic choice or political priority rather than a technological or institutional necessity" (*ibid.*, p. 10). Rostow's (*ibid.*) final claim is that, in the age of high mass-consumption, leading sectors shift towards manufacturing and providing consumer durables and services, accompanied by societal reorientation towards social welfare.

Hence, from a modernisation perspective, accumulation of technology was considered critical to economic growth (Schumacher, 1973), with the Marshall Plan having generated the view that, when combined with capital, short 'injections' of knowledge and adoption of proven technology from outsider countries would automatically lead to swift economic growth (Fukuda-Parr *et al.*, 2002). The receipt and deployment of Northern technologies was viewed as a relatively passive process involving absorption, primarily through State-run utilities (UNCTAD, 1999; Wilkins, 2002).

Early development efforts were thus focussed on grants and loans for acquiring and replicating the capital infrastructure of the North (Schumacher, 1973; McRobie, 1981; Fukuda-Parr *et al.*, 2002) and the transfer, from the North to the South, of the expertise, information, skills and knowledge needed to run a modern industrial society (Fukuda-Parr *et al.*, 2002). In such circumstances, Schumacher (1973) notes there was a belief that:

> The latest was obviously the best, and the idea that it might not serve the urgent needs of Southern countries because it failed to fit into the actual conditions and limitations of poverty, was treated with ridicule (p. 142).

Such an attitude built on the residual outcomes from the colonial era in which Western development officials manufactured a belief in the superiority of Western goods (Waddell, 1993), as highlighted by this quote from the economist Nicholas Kaldor (cited in Schumacher, 1973): "there is no question from every point of view of the superiority of the latest and more capitalistic technologies" (p. 152). This approach supported the universal displacement of traditional cultures, with:

> ...a convergence in liberal and Marxist scholarship that development means industrialization and that the process involves a more or less clean sweep of all previous patterns and relations of production (Galli, 1992, p. 1).

One of the driving motivations presented for technology transfer to the South was the claimed ability for technologies to solve societal challenges. Prominent declarations in the late 1940s, such as the imminent conquest of all infectious diseases (Marshall in Najera, 1989), were indicative of a common approach to technology, with the then Indian Prime Minister, Jawaharlal Nehru (in Ghose and Ghosh, 2003), stating, in 1961:

> It is science alone that can solve the problem of hunger and poverty, insanitation and illiteracy, of superstition and deadening custom and tradition, of vast resources running to waste, of a rich country inhabited by starving people... (p. 4).

Thus, in the 1970s, Mowshowitz (1976) reflected on the creation of an environment in which "...needs and conflicts are almost invariably formulated as technical problems requiring technical solutions" (p. 257).

Some of the prominent technologies of early development were those emerging from the green revolution. Building on Norman Borlaug's famous semidwarf wheat varieties,[4] the case of 'IR8 rice'[5] has been presented as a critical example of the green revolution's success (Bell, 2004). More commonly referred to as 'miracle rice', IR8 was suitable for heavy fertilisation and mechanical harvesting, with its introduction resulting in South East and Southern Asian rice yields increasing by 30% between 1968 and 1981 (*ibid.*).

The green revolution was indicative of other capital-intensive, large-scale projects of early development, such as the building of dams and highways (Finnemore, 1997). The State and centralised planning were key features of such technological development (Finnemore, 1997; Pieterse, 2001), with the green revolution ushering in State training of Southern scientists and control over scientific activities and priorities (Jalali, 1999).

Implicitly and explicitly, the most commonly applied criterion for the choice of these technologies was net output maximisation or cost minimisation (Eckaus, 1977). In this light, there was a strong focus on export-orientation, such as the development of 'market-oriented agriculture' (Engdahl, 2007). This philosophy was reinforced by the international division of labour (Vernengo, 2006) and based on assumptions manufactured throughout colonialism about the importance of shifting from subsistence economies to primarily export-oriented economies (Waddell, 1993), with rural commercialisation viewed as a pre-cursor to industrialisation (Moore Jr, 1966).

3.2.2 Dependency Theory

Whilst clear agreement on the need for some sort of development is said to characterise the first decades of the development era (Escobar, 2000),[6] in the 1960s and 1970s, a new set of theories emerged as a rebuttal to modernisation theory (Wallerstein, 1974; James, 1997). Reacting to the failure of post-war efforts to make real progress in alleviating poverty (Eckaus, 1977), dependency theory brought into focus the "...structures of power; systematic patterns of inequality; [and] practices and ontologies of dependence" (James, 1997, p. 207). Incorporating a wide range of perspectives on the problems faced by the South, the two main schools of dependency thought were the neo-Marxist critique of economic and social history and the Latin American structuralist critique of inequitable relations within global capital (Vernengo, 2006). Fundamental to both critiques was the belief that development and underdevelopment are relational states, with Southern development paths dependent on the actions of the North (Frank, 1966; James, 1997). The understanding of such dependency is best summed up by Dos Santos (1970):

> By dependence we mean a situation in which the economy of certain countries is conditioned by the development and expansion of another economy to which the former is subjected (p. 231).

Such a proposition particularly challenged the prevailing neo-classical economic view that capitalism and economic growth were inextricably linked (James, 1997).

Accordingly, dependency theorists across both schools were critics of "...the ethnocentric tendency to assume that the Third World would simply follow the fivefold path taken by the West" (*ibid.*, p. 209). As one of the early dependency theorists, Andre Gunder Frank (1969), elaborates:

> It is fruitless to expect the underdeveloped countries of today to repeat the stages of economic growth passed through by modern developed societies...this expectation is entirely contrary to fact and beyond all real and realistically theoretical possibility... (p. xvi).

Frank continues by suggesting it is:

> ...necessary instead scientifically to study the real processes of world capitalist development and underdevelopment

and to develop a realistic political economy of growth in the underdeveloped part of the world (*ibid.*, p. xvi).

From a neo-Marxist perspective, contemporary under development can be partly explained by the economic and social history of the past, with Frank (1966) claiming:

> We cannot hope to formulate adequate development theory and policy for the majority of the world's population who suffer from underdevelopment without first learning how their past economic and social history gave rise to their present underdevelopment (p. 17).

In this sense, underdevelopment is seen not as an original condition but, rather, a relationship beginning through colonisation in the 16th century (James, 1997) and a country's long-term, external relationship to global capital (Frank, 1972; James, 1997). Here, the 'trickle-down' philosophies of neo-classical economics are challenged (see, for example, Baran, 1952; Clâemenôcon, 1990), whilst exploitation of the poor is said to be compounded by Northern-aligned bourgeois within the South:

> ...in the poor countries, the educated people, a highly privileged minority, all too often follow the fashions set by the rich societies... and attend to any problem except those directly concerned with the poverty of their fellow-countrymen (Schumacher, 1973, p. 168).

Building on neo-Marxist thought, underdevelopment was explained from a structuralist perspective by exogenous obstacles and ongoing Southern engagement with the global economic system (Frank, 1966; Biel, 2000), with the world economy considered part of a single, hierarchical system (Wallerstein, 1974). According to structuralists, the world economy can be divided into the developed, industrialised 'core' that produces manufactured goods for itself and the remainder of the world, and the underdeveloped, non-industrialised 'periphery' that produces commodities, mainly for the core, whilst maintaining a largely subsistence system (Prebisch, 1959; Vernengo, 2006). This 'world-system' stems directly from "... very great disparities in technological densities" (Prebisch, 1959, p. 261) and capitalism's creation of the international division of labour, via market relations (Wallerstein in Galli, 1992; Vernengo, 2006). According to Wallerstein (1974), inequities are the ongoing predicament of global capitalism:

...it is not possible theoretically for all States to 'develop' simultaneously. The so-called 'widening gap' is not an anomaly but a continuing basic mechanism of the operation of the world economy...the some that rise are at the expense of others that decline (p. 7).

3.2.2.1 State-led, endogenous innovation

According to dependency theorists, the potential for Southern innovation is profoundly restricted within a world system. As Vernengo (2006) summarises: "at the core of the dependency relation between center and periphery lies the inability of the periphery to develop an autonomous and dynamic process of technological innovation" (p. 552). This inability is compounded by an inhibitive dependence on foreign knowledge and imported technology (Radoševic, 1999), given the development of endogenous scientific and technological capacity was neglected throughout colonisation and early post-colonial development (Sagasti, 1980; Vernengo, 2006).

In response, dependency theorists presented a number of policy prescriptions. Central to these was the belief that the South could only develop if there was either "...a crisis of capitalism or a delinking of those countries from the global economy" (James, 1997, p. 210). Given dependency theorists supported the fundamental principles of modernity and some form of development, the premise of delinking was therefore to avoid 'dependent development' (Pieterse, 1998). To do so, it was argued, countries must promote State-led development, pursuing internal growth by building on internal strengths (Wallerstein in Yergin and Stanislaw, 2002). Such development, it was believed, must focus on national innovation,[7] with technical advance in primary production, as part of a broader move to industrialisation, "...an inescapable part of the process of change accompanying a gradual improvement in per capita income" (Prebisch, 1959, p. 251). Here the relationship between technology and development is presented as mutually supportive, with technology considered to be either ethically neutral or beneficial. To support endogenous innovation, nations should adopt import substitution industrialisation — in which the proportion of goods supplied from domestic sources is increased in order to reduce dependence on foreign imports (*ibid.*). To assist, industrial development must occur behind protective, high import barriers

(UNCTAD, 1999; Wallerstein in Yergin and Stanislaw, 2002) whilst limiting the importation of luxury and manufactured items from the North, forbidding foreign investment and subsidising domestic nursery industries.

3.2.3   Alternative Development Theory

Alternative development theory encompasses a radical critique in reaction to dissatisfaction with mainstream development and its outcomes (Pieterse, 1998) and the failure of dependency theory to adequately detach itself from the modernist discourse (Manzo, 1991). It covers a broad body of literature, some of which I will address more fully, and travels under many aliases including: "...participatory development, people-centred development, human scale development, people's self-development, autonomous development [and] holistic development..." (Pieterse, 1998, pp. 351–352).

Alternative development squares its main criticisms at greater inequity, unemployment and mass urbanisation arising from the early mainstream approaches to development in the South (Schumacher, 1973). Focussing development efforts on the cities is said to bypass the needs of 85% of the population who are members of the 'non-modern' sector (*ibid.*). Reflecting on mainstream development, Schumacher (*ibid.*) believes Southern countries are forced "...into the adoption of production methods and consumption standards which destroy the possibilities of self-reliance and self-help" (p. 163). According to Schumacher (*ibid.*), this equates to unintentional neo-colonialism, leading to "...hopelessness for the poor" (p. 163). Moreover, modernisation theory is seen as utilitarian and ethnocentric (Waddell, 1993; Szirmai, 2005), promoting development that not only ignores local contexts and needs but proves destructive for indigenous cultures (Ullrich, 1997).

However, as with dependency theory, alternative development theory holds a "...retaining belief in development" (Pieterse, 1998, p. 364). In this light, Pieterse (*ibid.*) says it focusses on "...introducing alternative practices and redefining the goals of development..." (p. 344). Whilst alternative development supports variable paths to social change, it also emphasises the power of human agency ahead of goods or capital (Schumacher, 1973; Pieterse, 1998). Such agency is said to offer a vision of "...development *from below*" (Pieterse, 1998, p. 346), with a belief that local and grassroots' activity can become vehicles for redefining the agents, processes and goals of development (*ibid.*).

With the emergence of the modern environmental movement in the 1970s, ecological considerations also became essential to alternative development theory. The outcomes of the green revolution disproved Ehrlich's (1986) prediction of unavoidable mass starvation from population outstripping resources in countries such as India in the 1970s and 1980s. However, the Club of Rome's (Meadows, Meadows, Randers and Behrens, 1972) assessment of the world's finite raw materials supply in 1972 — predicting that the combined implications of population increase, environmental degradation, food shortages and the disappearance of non-renewable energy and metal resources would ultimately lead to collapse — has proven prescient.

More generally, alternative development has embraced an expansive understanding of equity as an "...ethical and usually people-oriented concept with primarily social, and some economic and environmental dimensions..." (Munasinghe, 1999, p. 17).

3.2.3.1 The critique of technology

Problematising modern technology was central to the responses of alternative development theorists to the technological advances accompanying modernisation theory, given "...the way it [technology] has developed, is developing, and promises to further develop, is showing an increasingly inhuman face" (Schumacher, 1973, p. 126). As Alvares (1997) notes: "generally speaking, development was merely modern science's latest associate in the exercising of its political hegemony" (p. 221). He continues:

> If one attempts to live close to the peasant or within the bosom of nature, modern science is perceived differently: as vicious, arrogant, politically powerful, wasteful, violent, unmindful of the other ways (*ibid.*, p. 232).

In terms of how technology is generally appraised, the alternative development approach maintains a belief that technology is far from neutral (Grimshaw, 2008). Rather, it is believed that technology "...carries an inbuilt social, political and ethical structure" (Hallen, 1991, p. 42). Whilst certain technologies, such as aspects of organic farming, may be intrinsically decentralising, democratising and helpful (Mooney, 1999; ETC Group, 2003a), other technologies, such as nuclear power, can be seen as highly undemocratic and centralising and, thus, inherently "...pollute, imperil or otherwise threaten our environment, health and security" (Mooney, 1999, p. 64).

As a result of societal conditioning, some tools "...insist on being used in particular ways" (Mowshowitz, 1976, p. 8). Everts (1998) adds that some technologies have an in-built gender bias, positing that the central influence of men over the development process, and particularly technological development, has, in many cases, led to "...the introduction of technologies that are beneficial and suited to men, but much less so to women" (p. xii); this despite indigenous and rural women tending to be "...the major repositories for local scientific knowledge as well as the major innovators in community-based research systems" (*ibid.*, p. 125). In this light, Everts (*ibid.*) claims that technological development has been promoted under the false premise of leading to gender equity, noting:

> The opposite has been the case: sometimes women's positions have worsened as a consequence of mechanization processes which made them lose control of profitable income-generating activities...entire sectors can lose their markets to new, better or cheaper goods" (p. xii; 12).

In addition to gender bias, technology can be culturally loaded. In this respect, Western technology, within the modernisation framework, is said to provide a culturally chauvinistic, ethnocentric view of progress (Brugger and Hannan, 1983). Schummer (2007) explains this argument further:

> ...any technological product made for the improvement of life is based on and confers an idea of what a good life is. Since countries differ to some degree in their ideas of a good life, the wide use of imported technological products can impact the cultural value system (p. 295).

Rather than the mechanistic views of modernisation theory — where physical technologies are seen as the ultimate answer for many of society's problems — from an alternative development perspective, technology is viewed as socially embedded, with the social context seen as playing a mutually shaping role with respect to technology's implications. In terms of equity, here it is argued that social and political issues can be more important than technical ones, requiring an integrated response that considers aspects of technology's emergence such as ownership, input and distribution (Mooney, 1999). Compounding the need for contextualisation, the effects of a technology can be vastly different across cultures and settings, even when the technology and the way it emerges are identical (Kolm, 1988). As Everts (1998) summarises: "...effects never arise from a technology itself, but always from the interplay between the technology and the complicated social, cultural and economic patterns that form its context" (p. 21).

Furthermore, many Western technologies considered for Southern development are said to have produced a great deal of hype that has proven overstated, in light of unanticipated or unacknowledged consequences. For Kearnes, Macnaghten and Wynne (2005), this phenomenon is part of a more general trend: "the emergence of new technologies is characterised by complex and heterogeneous cycles of hope, expectation, hype and disappointment, which are connected with material realities" (p. 286).

Amongst the technologies most heavily critiqued by the alternative development movement have been those associated with the green revolution, given scientific developments, such as miracle rice, produced a host of interrelated problems (Bell, 2004). Despite acknowledgement that the green revolution led to an initial increase in crop yields across certain countries, many authors, such as Chrispeels (2000), highlight the overall inequity of its global distribution, noting the limited gains in Africa and parts of Asia (Nuffield Council on Bioethics, 2004). Others (see Hallen, 1991; Goldstein, 1992; Rosset, Collins, Lappé and Luis, 1998) question the green revolution's long-term ability to address hunger and malnutrition, with the Nuffield Council on Bioethics (2004) claiming that "the initial rate of improvement of the Green Revolution was not sustained between 1985–90" (p. xiii). According to Engdahl (2007), the scientific claims made by many proved short-sighted, with long-term impacts on yields proving detrimental:

> The mono-culture cultivation of new hybrid seed varieties decreased soil fertility and yields over time. The first results were impressive: double or even triple yields for some crops such as wheat and later corn in Mexico. That soon faded.

In the case of miracle rice, for example, critics such as Bell (2004) claim that it has actually increased pest damage.

Underlying many of these problems was the fact that unsuitable technologies were foisted upon the South. According to Malloch Brown (in Fukuda-Parr et al., 2002), the modernisation view of technology transfer "...ignored — or at least underestimated — the importance of local knowledge, institutions, and social capital in the process of economic and social development" (p. vii). As Galtung (cited in Ullrich, 1997) claims:

> The total picture...is one of transfer of technology as a structural and cultural invasion, an invasion possibly more insidious than colonialism and neo-colonialism, because such an invasion is not always accompanied by a physical Western presence (p. 288).

In the case of the green revolution, Evenson and Gollin (2003) noted that varieties of seed, unsuitable for local conditions, continued to be transplanted into countries within Africa up until the 1980s. More broadly, the green revolution drove a rapid shift from agricultural subsistence to commercial agriculture, whereby "... farming ceased to be a way of life and became a commercial activity" (Goldstein, 1992, p. 279). Subsequently, traditional knowledge, varieties and practices were lost, given a failure to consider the value of indigenised technologies (Latham-Koenig, 1974; Shiva, 1992; Dano in ETC Group, 2004e).

Western technology has also been criticised in terms of its scale and costs. Engdahl (2007) believes the green revolution was particularly inappropriate in terms of its capital-intensive nature, saying that, accompanying expensive chemical inputs, were "... large irrigation projects which often included World Bank loans to construct huge new dams, and flood previously settled areas and fertile farmland in the process". These large infrastructural projects, based on sophisticated, capital-intensive, high-energy-input-dependent technologies, were a feature of Northern aid and technology transfer across the South during the 1960s given "the industrialists, and most official development agencies, were interested only in selling the latest and most sophisticated hardware to the poor countries" (Schumacher, 1981, p. xii). As an example, Schumacher (1973) recalls how nuclear reactors were established in countries including Indonesia, Iran, Venezuela and Vietnam by the U.S. and their allies and, in some cases, the former Soviet Union, at the expense of addressing overwhelming basic problems in these countries relating to "...agriculture and the rejuvenation of rural life" (p. 140).

Such technology, created with Northern contexts in mind, often required support services such as modern transport, communications, accountancy and marketing, thus only serving a minority of the total population residing in the metropolitan sections of some countries in the South (Schumacher, 1981). Understandably then, 'transferred' technology has rarely proven sustainable (Kumaraswamy and Shrestha, 2002; Singleton, 2003). As Schumacher (1973) explains:

> In every 'developing country' one can find industrial estates set up in rural areas, where high-grade modern equipment is standing idle most of the time because of a lack of organisation finance, raw material supplies, transport, marketing facilities, and the like (p. 149).

Such technologies, Schumacher adds, were commonly human labour-saving via greater mechanisation and a shift to mass production, eliminating skilful production of the hands and wiping out 'smaller' technologies. As Jackson (2009) puts it, a case of "doing more with fewer people" (p. 489) given:

> Continuous improvements in technology mean that more output can be produced for any given input of labor. But, crucially, this also means that fewer people are needed to produce the same goods from one year to the next (p. 488).

For the green revolution this involved a shift to fossil fuel-based, industrial farming machinery, such as tractors, characteristic of the shifts in the nature of agriculture across many parts of the world in the twentieth century (Scrinis and Lyons, 2007). However, Schumacher (1973) notes the inappropriateness and irony of expensive, labour-saving technologies given that, in direct contrast to the North, the South has a considerable surplus of labour and a shortage of capital. Thus, the process of modernisation is said to correlate with more sophisticated technologies that maintain the exclusive nature of high-technology, whilst extinguishing jobs and traditional workplaces faster than it can create jobs and modern workplaces (Schumacher, 1973; Latham-Koenig, 1974; McRobie, 1981).

Another aspect central to the alternative development critique of technology has been that many transplanted Western technologies have had dramatic impacts on social structures and power in the South. New technologies are said to often bring "...the loss of social norms, value systems and roles, and changing expectations" (Lopes, 2002, p.122), resulting in financial polarisation and social stratification (Schumacher, 1973; Eckaus, 1977; McRobie, 1981). As Grimshaw (2008) notes, technologies under the banner of 'development' often reinforce existing power hierarchies and serve to "...disturb traditional practices and cultures in the name of 'progress'" (p. 10). The green revolution followed this trend, creating heavy indebtedness among subsistence farmers who were pressured to borrow money for chemical and other modern inputs (Schumacher, 1973; Engdahl, 2007). Increases in local power inequities and the concentration of land ownership amongst lenders and speculators accompanied the disenfranchisement of small land owners forced to default on loans (Hallen, 1991; Goldstein, 1992). Such inequity was encouraged by the export-orientation of production for the global market that privileged large-scale production at the expense of small- and medium-scale production for local needs (Max-Neef in Simms *et al.*, 2009). Moreover, rural and urban imbalance spurred the emergence of dual economies (Latham-Koenig, 1974; Eckaus, 1977), wherein the patterns of living between rural and urban areas become highly distinct (Schumacher, 1973) to the point that "growth in an urban economy does not benefit the rural poor due to mobility restrictions across these economies" (Topalova cited in Kilby, 2007, p. 118). In many countries the social structures of agriculture changed (Goldstein, 1992; Strand, 2001), leading to increased social injustice and community disruption (Hallen, 1991). Inappropriate technologies drove mass urbanisation and rural unemployment and underemployment as peasants sought jobs in the cities (Simms *et al.*, 2009), leading to an increasing number of metropolitan slums, whilst the cost of rural subsistence living increased (Schumacher, 1973).

Others, such as Latham-Koenig (1974), have written critically about the implications of Western technology in terms of creating greater international dependency for the South. Many (Hallen, 1991; Shiva, 2002a; ETC Group, 2003a) believe the need for chemical inputs, animal feed, fossil fuels and machinery associated with the green revolution has perpetuated the cycle of Southern dependency on the North, with miracle rice demanding irrigation and "...a costly package of chemical fertilizers and pesticides..." (Dano in ETC Group, 2004e, p. 2). For Shiva (1989), the dramatic shift in power has occurred along geographical, class and gender lines, "...removing control of plant generic resources from third world peasant women and giving it over to Western male technocrats in multinational seed corporations" (p. 121).

An important aspect of the alternative development critique was questioning the ecological and human health consequences of Western technologies. Publications, such as Rachel Carson's (1965) 'Silent Spring', spurred critical thinking, with some (see, for example, Commoner, 1971; Coleman, 1976) arguing that capitalist technologies are chiefly responsible for environmental degradation. In terms of the green revolution, extremely high input requirements for chemical fertilisers, herbicides and pesticides are said to have 'exhausted' the land (Engdahl, 2007). The Western-driven focus on monoculture crops has also been criticised in terms of creating genetic uniformity and ecological damage (Hallen, 1991), highlighted by the loss of genetic diversity associated with miracle rice (Bell, 2004). Furthermore, many technologies are said to have had severe impacts on ecosystems, with paddy fish, for example, allegedly killed from the introduction of miracle rice (*ibid.*). Finally, questions have been asked in terms of the green revolution's impact on human health, specifically in terms of the carcinogenic effects of working with certain petrochemicals (Kimbrell, 2002).

3.2.3.2 Intermediate, appropriate technology

Many of the critiques of technology brought forward by alternative development theory are present in Schumacher's (1973) seminal publication 'Small is Beautiful: A Study of Economics as if People Mattered'. In this work, Schumacher describes the 'law of the disappearing middle', whereby technology is either too 'big', and therein "...inherently violent, ecologically damaging, self-defeating in terms of non-renewable resources, and stultifying for the human person" (p. 149), or too 'small', encapsulating low level technology that keeps people poor through inferior productivity. Schumacher's search, therefore, was for an alternative path that would empower people towards sustainable practices. He proposed a shift to 'technology with a human face' through "...a new orientation of science and technology towards the organic, the gentle, the non-violent, the elegant and the beautiful" (*ibid.*, p. 27). More formally, Schumacher's proposal was for 'appropriate technology'. Grounded in hope for greater human equity, what Schumacher proposes could also be considered "...democratic or people's technology" (*ibid.*, p. 128). Such technology, he claims, would address the 'law of the disappearing middle' by proposing an 'in-between' technology that would be: "...vastly superior to the primitive technology of bygone ages but at the same time much simpler, cheaper, and freer than the super-technology of the rich" (*ibid.*, p. 128). In doing so, Schumacher considered the potential to draw from both the best of the North's modern knowledge and experience as well and the South's traditional, local wisdom.

As an overview, appropriate technology can be said to have five broad characteristics. Although not negating the potential for high- or hybrid-technologies, the first characteristic is that technology must be simple and "...suitable for small-scale application" (*ibid.*, p. 27) within a relatively unsophisticated environment. Incorporated within this characteristic, a technology must be easily understandable, user-friendly and "...suitable for maintenance and repair on the spot" (*ibid.*, pp. 150–151).

Secondly, Schumacher envisages appropriate technology as considerably more productive than indigenous technology. High productivity would translate into cost-effectiveness, with appropriate technology being "...immensely cheaper than the sophisticated, highly capital-intensive technology of modern industry" (*ibid.*, p. 150). In terms of fulfilment, although recognising the usefulness in occasionally reducing labour requirements of traditional technologies in the South, appropriate technology would generally generate employment through its labour-intensive nature, with Schumacher drawing on Gandhian philosophies in reiterating that "the poor of the world cannot be helped by mass production, only by production by the masses" (*ibid.*, p. 128). Furthermore, Schumacher proposes that appropriate technology draws and builds upon a community's existing skills and knowledge, satisfying the human need for creativity and serving humans, instead of making them the servant to machines.

Thirdly, technology must be locally owned and widely distributed. To be truly appropriate, technology should enhance local or regional capacity rather than creating or amplifying dependencies on systems beyond local control. As an example, Schumacher says that "most agricultural populations would be helped immensely if they could themselves do the first stages of processing" (*ibid.*, pp. 155–156). Schumacher sees low-costs increasing the potential for widespread accessibility and argues that appropriate technologies are those conducive to decentralisation and common ownership, helping every individual rather than simply a concentrated few.

Fourthly, technology must be sensitive to local needs and cognisant of cultural norms. Schumacher says that appropriate technology will particularly target those with the greatest material needs, suggesting that what such people need most are things such as building materials, clothing, household goods, agricultural implements and a better return from agricultural products.

Finally, technology must be without harm, presenting benign or benevolent impacts for humans and the environment. Schumacher explains this in terms of technology being gentle in its use of scarce resources, compatible with the laws of ecology and recycling materials where possible.

In reviewing this section on the foundational approaches to development, modernisation, dependency and alternative development theories present different positions relating to the goals and means of development, with particular disagreement about issues of autonomy and agency. The respective approaches to technology are indicative of the underlying philosophies of each school of thought, with dramatically different beliefs having emerged regarding whether technology has embedded values, what technology offers and how technology should be approached, with respect to development. Yet, for all these differences, appropriate technology appears to offer a bridge between theories by simultaneously supporting emerging technology, endogenous innovation and a focus on local, human needs. As Schumacher (1973) notes "the applicability of intermediate technology is extremely wide, even if not universal" (p. 157).

3.3 Contemporary Approaches

3.3.1 Neo-Liberalism

In the 1970s, economic neo-liberalism, underpinned by neo-classical economics, emerged as a global trend affecting development (Davies, 2004). Characteristic of the approach was the reduced role of the State in economic management via deregulation, privatisation, trade liberalisation and an openness to investment flows (Radoševic, 1999;

UNCTAD, 1999), as well as support for the emerging intellectual property rights regime (North, 1981). This approach, it was claimed, would advance economies towards full, global economic integration via the most efficient allocation of resources (UNCTAD, 1999). In turn, this was seen as increasing competitiveness, raising GDP and resulting in a trickle-down of wealth (Foladori *et al.*, 2008), promoting the global convergence of living standards and reducing the incidence of poverty worldwide (United Nations Department of Economic and Social Affairs, 2001).

The causes behind the rise of neo-liberalism as the dominant model for the global economy and Southern development were varied. In part, they were a response to the claimed failings of the Keynesian policies of strong State intervention, including those of the structuralist dependency theorists (Telfer, 2002), spearheaded by the failure of centrally planned, State-led development in Soviet countries and elsewhere (World Bank, 1997). According to Davies (2004), a slowing global economy, the war in Vietnam, the oil shocks of the 1970s and the success of newly industrialising countries also paved the way for neo-liberalism. The market-friendly approach was further aided by deteriorating terms of trade[8] for commodity-dependent countries (Ocampo and Parra in Vernengo, 2006; Meridian Institute, 2007; Foladori *et al.*, 2008), with a number of Southern countries emerging within a new international division of labour as manufacturers — initially of labour-intensive activities for the clothing industry (UNCTAD, 1999). The composition of Southern exports shifted rapidly away from primary commodities. Labour-intensive manufactures rose from 20% of Southern exports at the beginning of the 1980s to 70% at the end of the 1990s (UNCTAD, 2002c), with most Southern countries shifting to market-oriented and private sector-led economies (UNCTAD, 1999). Globally, the South was seen as offering low-cost, semiskilled labour and attractive fiscal incentives, resulting in the proliferation, from the 1970s onwards, of custom-free areas for manufacturing such as 'export processing zones' (UNCTAD, 1999; Oman, 2000). These facilities served the predominantly Northern multinational corporations (MNCs) that were now central to the global economy (Castells, 1999).

Of particular note during the 1960s and 1970s were the 'East Asian tigers', a group of countries that included Taiwan, Singapore, South Korea and Hong Kong. During these decades, 'the tigers' raised GDP dramatically, shifting from import substitution industrialisation to export-oriented growth (Stallings, 1995; Fukuda-Parr *et al.*, 2002). Further reasons for the success of the tigers are said to include supportive policies for business, linking industry with academia, and strategic investments in education, infrastructure (particularly absorptive capacity) and human resource development, with a focus on skills development and industrial specialisation as well as 'opening up' to international human resource flows (UNCTAD, 2006). Such liberalisation also included openness to foreign capital and technology transfer (*ibid.*).

Supported by the 1995 establishment of the World Trade Organisation (WTO), another outcome of the neo-liberal model was the rise of IPRs, within an emerging context of the 'knowledge economy' (Radoševic, 1999). Of particular importance to development was the introduction of the WTO's Trade-Related Aspects of Intellectual Property Rights (TRIPs) Agreement that came into effect at the commencement of 1996 with a phase in for 'developing countries in transition from a planned to market economy' by 2000 and the LDCs by 2006. The TRIPs Agreement is a method for standardising intellectual property law protection across all WTO member states (Sterckx, 2004), with the pharmaceutical industries in India and Thailand as particular targets (Lee and McInness, 2003). The introduction of TRIPs was based upon three lines of argument: 'natural rights' to ideas; 'distributive justice' that rewards the service provided to society by inventors; and 'utilitarian', economic arguments that claim the necessity to incentivise innovation (Sterckx, 2004). It is the latter — and fear of copycat companies overriding patents and manufacturing drugs at bargain prices — that has proven the main argument industry has made for pharmaceutical innovation (see Klotzko, 2003; Sterckx, 2004).

Conversion of knowledge into intellectual property (IP) played a critical role in the emergence of tiger economies, with Enriquez (2005) noting that, in 1985, the US patent office granted Argentina, Venezuela, Brazil, Mexico and South Korea between 12 and 50 patents each. In 2003 the same office granted between 20 and 180 patents for each, except South Korea, which received 4132.

According to neo-liberal theory, the tiger economies highlighted the ability for the 'periphery' countries to industrialise which, along with the demise of the Soviet Union and subsequently buoyed enthusiasm for market-friendly development strategies,[9] contributed to dependency theories losing favour in the second half of the twentieth century (James, 1997; Vernengo, 2006). With development practices having been the subject of many East-West ideological disputes (UNCTAD, 1999), the neo-liberal cause was further strengthened by the collapse of Communism in the former Soviet Union, leaving liberal democracy and the capitalist-free market economy, according to authors such as Fukuyama (1992), as the only remaining ideological alternative for nations in a post-Cold War world.

To accommodate, many dependency theorists reoriented their views. Said to be led by Cardoso,[10] the neo-Marxist position was abandoned through a new emphasis on internal rather than external barriers to development (So, 1990). With the experience of the tigers seen as exemplifying the ability of free trade and export-orientation to reduce technological dependence (Vernengo, 2006), neo-liberalism was embraced by "...redefining market ties between the North and South, increasing South-South relations and promoting regional ties" (Petras and Morley, 1990, p. 42). The focus of the Latin American structuralists had changed dramatically, particularly with respect to technology and development:

> The importance of technology, the role of multinationals in the process of technology transfer and the role of the State in promoting technological innovation through industrial policy then became the foci of the Latin American structuralists (p. 558).

The logic underpinning these shifts was that of 'associated-dependent development', in which Southern development was seen to co-exist alongside dependence on Northern capital, technology and governments (So, 1990; Vernengo, 2006).

Utilising this historical opportunity, neo-liberals devised the Washington Consensus: a set of ten policy prescriptions for 'crisis-wracked' countries in the South. The Consensus promoted a belief that global economic convergence was conditional, requiring "...open trade, export-led growth, greater deregulation, and more liberalized financial markets" (Palley cited in Pieterse, 2001, pp. 164–165). Through the mechanism of 'structural adjustment programs', the provision of Northern aid became conditional on Southern countries adopting the Consensus' neo-liberal policy prescriptions and was accompanied by an increasing emphasis on 'tied aid', whereby donor money had to be used to make purchases from the donor country (Perkins, 2004).

3.3.1.1 Corporate technologies

Complementing the ascendency of neo-liberalism has been the emergence of 'corporate technologies' — technologies largely owned by corporations that, in turn, provide further control over markets (Scrinis and Lyons, 2007). Since the 1990s, one corporate technology — biotechnology — has stood out as the 'magic bullet' dominating the public imagination (Mooney, 1999). Both pharmaceutical and agricultural biotechnology have rapidly affected the South in terms of environmental, regulatory and social impacts (ETC Group, 2003a).[11]

Whilst many promises have been made for pharmaceutical biotechnology (see, for example, Daar, Thorsteinsdóttir, Martin, Smith, Nast and Singer, 2002; UNCTAD, 2002a; Ogundiran, 2005), it is the claim that GM crops are a 'universal fix' that has received most attention. Believing conventional agriculture and non-mechanised farming systems to be inherently limited (see Ogundiran, 2005), those promoting GM for agriculture speak of it as the "...new Green Revolution" (TajaNews, 2004). Here it is seen as the only means by which to feed the three billion people to be born over the next 30 years (World Bank, 1998), whilst enhancing farm income, crop sustainability and food security for the South (WHO, 2005b). These claims stem from a belief that GM can address:

> ...specific health, ecological and agricultural problems which have proved less responsive to the standard tools of plant breeding and organic or conventional agricultural practices (Nuffield Council on Bioethics, 2004, p. xiv).

More specifically, GM crops are said to be able to increase crop yields in the South by up to 25% (World Bank, 1998), partly due to reduced crop loss to insects (WHO, 2005b). Furthermore, GM crops are seen as enhancing crop nutritional values (UNCTAD, 2002a; WHO, 2005b) and reducing agricultural chemical usage (WHO, 2005b). As with the case of miracle rice that emerged from the green revolution, GM crops have produced examples said to hold great promise, such as 'golden rice' (UNCTAD, 2002a). Here, rice that is genetically modified to produce pro-vitamin A has been touted as "...a miracle cure for malnutrition and hunger" (Shiva, 2002a, p. 58).

When it comes to the issue of risk, according to many such as Paarlberg (2003), the evidence suggests agricultural biotechnologies present little harm, with 15 years of trials having failed to produce

"...any evidence of added risk to human health or the environment from any GM crop to date" (p. 86). This is qualified in a review of studies by the Nuffield Council Trust (2004) that found no evidence that GM crops, or food produced from GM crops pose a significant risk for human consumption. Similarly, the Royal Society (2002) found that: there is no current evidence that GM foods cause allergic reactions; risks to human health associated with the use of specific viral DNA sequences in GM plants are negligible; and consuming DNA poses no significant risk to human health.

As with the green revolution, the mainstream approach to crops utilising genetically modified organisms (GMOs) is grounded in neo-classical economic theory and utilitarian rhetoric. According to DaSilva (2002), for example, "controversial or not, GMOs could be the breakthrough technology for economic progress in developing countries" (p. 15). Similarly, Fukuda-Parr (2007), previous Director of the United Nations Development Program (UNDP), speaks of genetically modified (GM) crops as "...a source of increasing productivity that opens opportunities for people and for developing countries to become internationally competitive" (p. preface).

That such claims emerged within an environment in which there was growing recognition that the liberalisation of the 1980s had not always delivered anticipated economic growth, strengthened calls by institutions such as the World Bank for endogenous Southern innovation, within a neo-liberal framework (Foladori *et al.*, 2008). Intricately linked to the innovation process, technology was increasingly presented as "...the engine of economic growth" (Kumaraswamy and Shrestha, 2002, p. 7), with some, such as World Bank representatives (see Watson, Crawford and Farley, 2003), claiming there is a direct correlation between innovation and growth.

The composition of Southern exports was rapidly shifting towards labour-intensive manufactures. However, some (see, for example, UNCTAD, 1999; Enriquez, 2005) were already arguing for Southern countries to reorient their economies towards endogenous knowledge generation, shifting from low- to high-technology activities for productive structures across all sectors, including the primary sector. Such calls were buoyed by examples such as Chile, Indonesia, Malaysia and Thailand successfully increasing their commodity earnings as a result of:

...expanding their export markets to include non-traditional, high-growth commodities, including value-added and processed goods, and for increasing their productivity in traditional commodities production (Nabuki and Akiyama cited in Meridian Institute, 2007, p. 5).

Moreover, the knowledge economy is said to make cutting-edge innovation "...not a luxury but an absolute necessity" (InterAcademy Council, 2004, p. 2) if Southern countries wish to increase their competitiveness (Robertson, 1991; Fukuda-Parr *et al.*, 2002; InterAcademy Council, 2004; Enriquez, 2005). This mantra spread, with the UNDP (2001) stating: "all countries, even the poorest, need to implement policies that encourage innovation, access and the development of advanced skills" (p. 5).

Endogenous innovation was also presented as the most important mechanism for countries to ultimately reduce international inequalities (McArthur and Sachs, 2002). From the perspective of one World Bank official (Farfan, cited in Meridian Institute, 2007), this is particularly true for the 'technology divide', given the dynamics of global trade:

> Unless countries develop robust science and technology capacity, the technology gap will widen as developing countries are reduced to the production of manufactured goods whose prices increasingly behave as those of primary commodities, with declining terms of trade relative to knowledge-based goods of developed countries (p. 7).

Similarly, the participation of Southern-based scientists in cutting-edge research is seen as a way to help avoid the U.S. monopoly on R&D that sidelines global needs (Hassan, 2005), with the Global Forum for Health Research (2002) claiming that appropriate endogenous Southern research capacity will be required to bridge the 10/90 research orientation gap[12] in an area such as healthcare. In this sense, Yonas and Picraux (2001) believe that, "on the national level, redistributions of global technological strength could result in realignments of global prosperity and influence" (p. 43).

According to Leach and Scoones (2006), such approaches return to the linear view of 'development as modernisation', with technology seen as an accelerator through developmental stages, possibly facilitating 'leapfrogging', whilst the Asian tigers, India and China provide the models for such development.

However, the prominent 1960s concept of leapfrogging took on a different meaning in the latter stages of the 20th century. Whilst countries such as Brazil, Mexico, South Korea and Taiwan demonstrated an ability to leapfrog their development by absorbing Northern-generated innovations (Juma and Yee-Cheong, 2005), the new understanding suggests technology transfer is becoming less meaningful as emerging technology presents new opportunities for Southern innovative capacity, both in terms of contributing to global markets and addressing local needs (UNCTAD, 2002a; Meridian Institute, 2005; Ogundiran, 2005; Osama, 2006). According to some (Radoševic, 1999; Brown, 2001; Watson *et al.*, 2003; UNESCO, 2006), the impact of ICT, in terms of the globalisation of knowledge and networks, combined with declining prices in technological hardware, provides new platforms for Southern countries to access scientific and technological knowledge. This has stimulated a belief that "the technology divide does not have to follow the income divide" (Brown, 2001, p. iv), whilst building on the knowledge that a country's GDP is not necessarily indicative of its global R&D capabilities (Third World Academy of Sciences, 2004).[13] In a knowledge economy, founded on opportunities provided by ICT, Enriquez (2005) believes "...you need ever fewer people, time, or capital...to build a nation" (p. 10). This can be seen for the case of biotechnology, where knowledge-based infrastructure is opening up new opportunities to develop R&D capabilities in emerging markets (Kenney and Buttel cited in Fransman, 1994; Correa, 1998). In pharmaceuticals, India and Argentina provide evidence of this new phenomenon (Correa, 1998), whilst Cuba and Mexico, for example, are seen as major players in the genomics arena (Ogundiran, 2005). Some, such as Hapgood (cited in Anton, Silberglitt and Schneider, 2001), see opportunities opening up for all countries: "low cost and wide availability of basic genomic equipment and know-how will likely allow practically any country, small business or even individual to participate in genetic engineering" (p. 9).

In this light, the WHO (2005b) reports 'widespread diffusion' of biotechnology R&D:

> Sixty-three countries have been involved in some phase of biotech plant research and development, from laboratory/greenhouse experiments, to field trials, to regulatory approval and commercial production (p. v).

More than half of the countries engaging with biotechnology R&D are from the South (see Table 3.1), including one LDC — Bangladesh.

Table 3.1 Countries with Biotechnology Production or Research Activity (drawn from WHO, 2005b).

Africa/Middle East	Asia/Pacific	Latin America	Western Europe	Eastern Europe
Egypt	Australia	Argentina	Austria	Armenia
Kenya	Bangladesh	Belize	Belgium	Bosnia and Herzegovina
Morocco	China	Bolivia	Denmark	Bulgaria
South Africa	India	Brazil	Finland	Croatia
Tunisia	Indonesia	Chile	France	Czech Republic
Zimbabwe	Japan	Colombia	Germany	Georgia
North America	Malaysia	Costa Rica	Greece	Hungary
Canada	New Zealand	Cuba	Ireland	Moldova
United States	Pakistan	Guatemala	Italy	Romania
	Philippines	Honduras	Netherlands	Russia
	South Korea	Mexico	Portugal	Serbia and Montenegro
	Thailand	Paraguay	Spain	Slovenia
		Peru	Sweden	Ukraine
		Uruguay	Switzerland	
		Venezuela	United Kingdom	

Similarly, Salamanca-Buentello *et al.* (2005) believe evidence shows that Southern countries have been able to play a significant role in the global advancement of information and communications technology.

Hassan (2005) claims that examples of Southern engagement with emerging innovation are indicative of a shift in the global R&D dynamic since the mid 1980s, with a number of Southern countries now broadly embracing innovation as a critical element in overall economic growth strategies. As Watson *et al.* (2003) confirm, countries including Brazil, China, India, Malaysia, Mexico, Thailand and the Philippines are "...using technological capabilities to capture growing percentages of expanding global high tech export markets" (p. vi). Although the investments of most Southern governments are comparatively very small (WHO, 2002a), Hassan (2005) cites some examples where funding has been quite considerable:

> China, devotes approximately 1.1% of its GDP to science and technology, having recently become the world's third-largest investor in research and development (in absolute terms); India, that invests approximately 1.2% of its GDP to science and technology and has emerged as one of the world's leading countries in the application and, increasingly, the development of information technology; and Brazil that spends an estimated 1.1% of GDP on science and technology (p. 65).

Hassan (*ibid.*) adds that a desire to compete in global science through cutting-edge R&D explains China's extensive investment in biotechnology and information and communications technology, the Brazilian parliament's acceptance of stem cell research and the Nigerian launch of a remote-sensing satellite in 2003.

However, whilst neo-liberal theorists claim that there are increased opportunities for endogenous Southern innovation, they also see the greatest challenges for Southern innovation as endogenous, with bodies such as the IMF focussing on "...issues such as corruption, barriers to private enterprise, budget deficits, and State ownership of production" (Sachs, 2005, p. 79). Wilkins (2002) speaks of additional challenges relating to business investment and currency stability, technological absorptive capacity and domestic investment in science and technology, with Badran (1999) noting that the South, combined, contributes a mere 10% to gross world expenditure on R&D. Osama (2006) adds to these arguments, saying that, for endogenous innovation to be possible, Southern countries "...must first solve tough, often unglamorous, problems like illiteracy, social mobility, government inefficiency and corruption, and a lack of economic opportunities".

Critical to the claimed ability for the South to engage in global R&D is the role of the State in technological governance. In the mid 1990s, the State returned, not as a "...direct provider of growth but as its partner, catalyst, and facilitator" (World Bank, 1997, p. 1). New roles were said to include regulating trade and promoting R&D-enabling policies that incentivise competition and local and foreign investment (Wilkins, 2002; Ogundiran, 2005). Some of the more familiar tools favoured by the State have included the provision of tax advantages, reduced custom tariffs and special permits to foreigners (Sahai, 1999; InterAcademy Council, 2004). The State was also seen as having a direct role in the coordination of national strategic planning for science and technology (UNCTAD, 2002a; InterAcademy Council, 2004). Central to this task was ensuring adequate institutional infrastructure by focussing on human resource development (Bhumiratana, 1991; Watson et al., 2003; Mani, 2004). Of equal importance is the creation of an appropriate environment for innovation by stimulating entrepreneurship and venture capital (UNDP, 2001) and strengthening 'national systems of innovation'[14] (Gabriele, 2001; InterAcademy Council, 2004; UNCTAD, 2006). Within this model, high priority is given to private sector engagement and the linking of science with technology-generation in industry (UNCTAD, 1999; UNDP, 2001; Watson et al., 2003). Common mechanisms for such interactions to take place have included 'technology parks'[15] (Brown, 2001; InterAcademy Council, 2004), 'virtual networks of excellence' (InterAcademy Council, 2004) and geographically located 'centres of excellence' (UNCTAD, 2002c; InterAcademy Council, 2004; Juma and Yee-Cheong, 2005).[16] The latter have received considerable attention in recent years and are seen, by some (InterAcademy Council, 2004; Hassan, 2005), as mandatory for every country in the South, with a 2005 report from the Commission for Africa recommending that three billion dollars be invested in developing such centres (Leach and Scoones, 2006).

3.3.2 Alter-Globalisation

In response to increasing inequity accompanying the ascendency of neo-liberalism and corporate globalisation, 'anti-globalisation' or, more aptly, 'alter-globalisation' critiques emerged, bound by a common focus on global injustices (Pieterse, 2001; Lee, Fustukian and Buse, 2002; Sachs, 2005; Leach and Scoones, 2006; Jamison, 2009). Whilst the alter-globalisation movement is bound by a common desire for 'another world' (see, for example, Raskin, Banuri, Gallopin, Gutman, Hammond, Kates and Swart, 2002; Cavanagh and Mander, 2004; George, 2004; Grimshaw, 2007; Pleyers, 2009), its proposals for how that world should look and the means by which it can be reached are far from homogenous. Thus, I will focus here on the commonly emerging critiques that relate to global inequities.

Whilst the alter-globalisation movement focusses broadly on injustices, I argue that many inequities stem, in part, from inequalities relating to capital wealth and modern technology, resulting in inequities of power. Furthermore, despite important critiques (see, for example, McRobie, 1981; Salleh, 2009),[17] indicators relating to income and aspects of engagement with modern technology are worth exploring, given they have long been the ground on which neo-liberal claims for greater equity have been laid, and they are some of the indicators for which longitudinal data exists.

The unprecedented increase in material wealth and prosperity, experienced from the 1970s onwards, has been highly uneven (UNDP, 2006; Jackson, 2009), with evidence suggesting global economic divergence (UNCTAD, 1997; UNDP, 2006; Simms *et al.*, 2009). As Jackson (2009) notes, "far from raising the living standards for those who most needed it, growth let much of the world's population down. Wealth trickled up to the lucky few" (p. 487). According to the UNDP (1996), the extent of this divergence is dramatic: "if present trends continue economic disparities between the industrial and developing nations will move from inequitable to inhumane" (p. iii).

Globally, the ratio of income between the world's richest and poorest citizens increased dramatically between the 1960s and 1990s (UNCTAD, 1997; UNDP, 1998).[18] Similarly, the global 'Gini coefficient' — an indicator of income inequity, in which higher values refer to greater inequity — "...stood at 0.66 in 1965, rose slightly to 0.68 in 1980 and reached 0.74 in 1990" (UNCTAD, 1997, p. 79). Enriquez (2005) shows that this inequality affected the poorest countries most, stating: "...from 1970 to 1995, the world's richest countries grew 1.9 percent per year on average...the middle-income countries grew 0.7 percent...the poorest third did not grow at all" (p. 143). Moreover, global income inequality continues to rise (UNCTAD, 1997; Mooney, 1999; UNDP, 2005),[19] with the UNDP (2003) adding that 54 countries were 'poorer' in 2002 than in 1990.

Regionally, certain areas have felt the brunt of inequality more than others. According to Fukuda-Parr *et al.* (2002), "since 1990, the number of income-poor people has increased every year in sub Saharan Africa, South Asia, and Latin America and the Caribbean" (p. 2). As noted by the UNCTAD (1997), between 1965 and 1995, Africa's average per capita income[20] halved from 14%to 7% in relation to that of the North, whilst Latin America's dropped from 36% in 1979 to around 25% in 1995.

With increasing acknowledgement that international indicators of inequality can mask a country's internal situation (Vermeulen, Garside and Weber de Morais, 2009), it is important to note that inequality is said to be rising *within* a large number of countries across both the North and South (UNDP, 2001; United Nations Department of Economic and Social Affairs, 2001). According to James (1997), development in the 'economic miracles' of the South has been associated with an increasing domestic division between rich and poor.[21] In China, for example, domestic inequality has risen simultaneously to the country becoming more 'competitive' (Foladori *et al.*, 2008). Likewise, as Mexican competitiveness increased from the mid-1980s to the mid-1990s, so too did the country's internal Gini coefficient, which rose from 0.49 to 0.55 (Delgado Wise and Invernizzi in Foladori *et al.*, 2008). Similarly, income inequality grew during the 'Brazilian economic miracle', in which the country's GDP increased annually between 1968 and 1973 by more than 10% (Vernengo, 2006). Finally, alongside national economic growth in India, Shiva (2008) claims "...the majority of Indians have become poorer" (p. 10).

Fundamental to alter-globalisation critiques is the claim that, through neo-liberal globalisation, the South is 'locked-in' to structural injustices that challenge Southern development (Blowers, 1997). As James (1997) notes, global capitalism "...now frames the various forms of dependency and exploitation" (p. 214), with the current form of global capitalism substantially undermining de-linking attempts by locales, regions and states as part of independent, or even semiautarchic development (James, 1997; Pieterse, 1998).

Authors most commonly refer to three inequitable systems propounded by neo-liberal globalisation. The first is global finance, where the liberalisation of financial markets and currencies has been said to lead to new forms of dependency (Vernengo 2006). Northern neo-imperialism has left Southern governments hamstrung by debts, with Ellwood (2001) noting:

> In six of the eight years from 1990 to 1997 developing countries paid out more in debt service (interest plus repayments) than they received in new loans: a total transfer from South to North of $77 billion (p. 49).

Associated dangers for debtor countries were particularly exposed near the end of the 20th century by the contagion of the East Asian financial crisis that decimated the Asian miracle economies (Corbett, Irwin and Vines, 1999; Vernengo, 2006).

The second inequitable system is that of global trade. Southern countries have been forced to liberalise whilst their share of global exports has decreased from 4% in the 1960s to barely 1% in the 1990s (Mooney, 1999) and their share of imports, both globally and in relation to exports, has increased dramatically (Mooney, 1999; UNCTAD, 2002c). The U.S. and E.U. maintain selective subsidies and other hypocritical protectionist policies, presenting severe structural barriers for Southern development (Petras and Morley, 1990; Fishburn and Green, 2002). For example, La Prairie (2005) astutely notes that pigs and cattle currently have greater purchasing power than most of those living in the global South. In this light, the alter-globalisation movement largely protests bilateral and multilateral moves towards 'free trade', with a belief that "lowering import duties [by Southern governments] encourages the destruction of domestic markets and domestic production" (p. 12). According to many, such as Shiva (2008), the effects of free trade policies have been disastrous, with Shiva claiming that: "most Indians are, in fact, eating less today than a decade ago, before the era of globalisation and trade liberalisation...economic growth has gone hand in hand with growth in hunger" (p. 10).

The third inequitable system is that of global labour. As Southern countries liberalise their economies, they are forced to open up to international human resource flows. Workers become tradeable knowledge commodities, fought for in the global marketplace. Given great inequalities exist between the North and South in terms of the incentives on offer to potential workers, there has been a constant migration, or 'brain drain', of researchers and professionals from the South to the North, undermining attempts at building critical endogenous capacity (Robertson, 1991; Pang, Lansang and Haines, 2002).

3.3.2.1 Emerging technologies and inequity

Emerging corporate technologies are seen as a means by which Southern countries are locked into inequitable global networks of capital, trade and labour (Thompson cited in Grimshaw, 2008). According to Stilgoe (cited in Grimshaw, 2008):

> We can judge new technologies according to the extent to which they lock people into certain systems (as, for example, GM crops and centralised nuclear power do) or provide an open platform for new sorts of use (for example, Linux or micro-renewable energy) (p. 6).

Despite rapid development of new technologies since the 1970s, such as microelectronics, information and communication technologies and biotechnologies, these technological advances have had "...questionable impacts on the technology-inequality gap" (Invernizzi *et al.*, 2008, p. 137). In this light, many (see, for example, Crow and Sarewitz, 2001; ETC Group, 2003a; Gould, 2005; Meridian Institute, 2005; Grimshaw, 2008) believe that the gap between rich and poor will, at least initially, exacerbate with any major new technology introduced into an unjust society,[22] with the tendency for emerging technology to produce "...larger disparities between cultures and subcultures..." (Hook, 2002, p. 11).

I will now, therefore, progress to some of the more detailed alter-globalisation critiques of emerging technology and inequity. Here I will explore four problems for technology and development that arise within the current global context: the privatisation of knowledge and concentration of innovative capacity; the restriction of technological access; the over-hyped, inappropriate nature of universal techno-fixes; and the externalisation, inequitable distribution and non-participatory governance of modern risk.

With respect to the privatisation of knowledge and the concentration of innovative capacity, some of the most pressing structural challenges for equitable Southern development are said to relate to the nature of technological ownership and control (Mooney, 1999; ETC Group, 2003a). Knowledge creation is said to be increasingly inequitable (Cetto, Schneegans and Moore, 2000; Fukuda-Parr, 2007), with nearly all technological innovations coming from less than 15% of the world's population (Giddens, 2006). Furthermore, there is a growing divergence emerging in national innovative capabilities between the 'information rich' and 'information poor' (Jalali, 1999; UNCTAD, 1999; Enriquez, 2005). Inequalities relating to 'high-tech' are even starker. In 1998, low-income countries did not export any high-technology, whilst for middle-income countries the figure was 18% of exports and for high-income countries, 33% of exports (World Bank, 2001b). Runge and

Ryan (2004) note that, despite Southern countries making up more than half of the 63 countries engaged in agricultural biotechnology R&D and/or production, five nations[23] account for 98% of the total biotechnology crop land, with a significant gap to the 'second tier'. In this light, Hassan (2005) speaks of the emergence of a 'South-South gap' in capabilities between scientifically proficient countries, such as Brazil, China, India and Mexico, and scientifically lagging countries, many of which are located in sub-Saharan Africa and the Islamic world.

Technological inequalities span all components of knowledge creation (as highlighted in Fig. 3.1), with 28% of the world's scientists[24] residing in the South, accessing only 12% of the research funds, producing 8% of the peer-reviewed papers and being granted less than 2% of all patents (Mooney, 1999). As Mooney (*ibid.*) boldly states, the people making the key decisions about future technology invariably come from "...a handful of corporations, a still smaller number of countries, and tend to be male, white, middle-aged and middle-class" (p. 122).

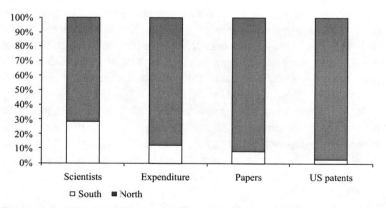

Figure 3.1 Who Decides Future Science? Comparison between Northern and Southern Countries (Adapted from UNESCO, cited in Mooney, 1999, p. 122).

When it comes to the structures supporting such inequalities, a central target of the alter-globalisation critique is the WTO and its activities (Sachs, 2005). Under particular scrutiny has been the TRIPs Agreement, a mechanism criticised by many (see, for example, Intermediate Technology Development Group, 2002;

UNDP, 2003; Birdsall, Rodrik and Subramanian, 2005; Schummer, 2007) in terms of creating greater inequities. Accession to the TRIPs Agreement requires that all Southern countries honour the Northern interpretation of patent rights and the Western industrialised model of innovation, denying or ignoring the more informal, community-based systems of innovation through which Southern populations have long innovated without any property rights or patent protection (Shiva and Holla-Bhar, 1996; Third World Network, 1997). According to Shiva (1993), the TRIPs Agreement has led to the privatisation of knowledge flows via a restricted framework in which patents "...are recognised only when knowledge and innovation generate profits, not when they meet social needs" (p. 115).

Intellectual property has thus become "...a non-tariff barrier to market entry for smaller innovators" (Mooney, 1999, p. 81). With only about 2% of global patents ending up being applied to a product (Foladori and Invernizzi, 2007), Northern R&D, particularly patent monopolies, block Southern innovation (Sterckx, 2004; ETC Group, 2005c; Schummer, 2007). Furthermore, emerging industries have limited informational and financial resources (Schummer, 2007), with such circumstances proving more difficult in the face of increasing costs for patent filing and litigation (*ibid.*). Moreover, farming communities and even public research institutes have struggled to claim patents under the rules set out by the TRIPs Agreement, given the inventor must be named, the patent must involve an inventive step, the cost of advice from patent lawyers is prohibitive and patent holders must defend their patents under civil law (UNDP, 2001).

Moreover, the TRIPs Agreement has been used as a mechanism to support the now broad interpretation of what knowledge is open to privatisation. Crucial to widening the realms of proprietary knowledge was the 1980 U.S. Supreme Court ruling[25] that living organisms are patentable (ETC Group, 2008). Thus, there are now "...patents on genes, gene sequences, entire species, on human cell lines and on indigenous knowledge" (ETC Group, 2002, p. 2). In many cases this has equated to 'biopiracy',[26] whereby patents have been falsely claimed on characteristics relating to natural resources, such as the neem tree, turmeric and the Mexican Enola bean (UNDP, 2001). Shiva and Holla-Bhar (1996) outline, for example, how the establishment of 'plant breeder rights'[27] led to U.S. and Japanese companies accessing and patenting synthetic forms of naturally occurring compounds from India. In the case of the neem tree this meant exploiting local knowledge relating to "...age-old village techniques for extracting the seed oil and pesticide emulsions [that] do not require expensive equipment" (*ibid.*, p. 149). The methods on which the Northern-held patents are based are merely "...an extension of the traditional processes used for millennia for making neem-based products" (*ibid.*, p. 151).

Accompanying such developments has been a widespread enclosure of 'the commons',[28] whereby natural resources and local knowledge have been appropriated in order to generate profit, often for Northern interests (see, for example, Shiva, Jafri, Bedi and Holla-Bhar, 1997; UNDP, 1998; ETC Group, 2005c). Monsanto's move to privatise water in India in the mid 1990s provides a prime example of this phenomenon (see Shiva, 2002b). For Shiva (1997), these developments are nothing less than a 'colonisation of the commons': "through patents and genetic engineering, new colonies are being carved out. The land, the forests, the rivers, the oceans, and the atmosphere have all been colonised, eroded, and polluted" (p. 5). Hallen (1991) agrees, saying that the modern approach to biotechnology represents a mercenary, reductionist assault whereby "nature is reduced to a pool of genetic 'raw materials' which can be spliced and recombined..." (p. 38).

Offering some of the strongest examples of moves to enclose nature are 'genetic use restriction technologies' (UNCTAD, 2002a), such as the 1998 patented 'terminator technology' that would prevent farmers from reusing patented commercial seed via a genetically in-built mechanism that causes the seed to commit 'suicide' after one harvest.

In all these examples, what is highly evident is that the control of knowledge has been shifting dramatically from the public to the private domain (ETC Group, 2005c; Schummer, 2007). As Chrispeels (2000) notes, research contributing to the green revolution was conducted in the public domain and included free distribution of resultant technologies "...without concerns for the intellectual property rights of those who produced them" (p. 4). However, Rader (1990) shows that for the early stages of health-related biotechnology[29] (1986–1989), the private sector held 74% of patents, universities held 13.3%, independent and not-for-profit entities held 9.4% whilst the government held just 3.3%. In more recent years, this trend has continued, with 90% of the patents related to high-technologies now held by global enterprises (UNDP, 2000). As Enriquez (2005) notes, "in 1998, a single company, IBM, obtained more U.S. patents...than the total granted to 139 countries" (p. 147). Referring to the case of *Bacillus Thuringiensis* — a bacterium commonly used as pesticide for genetically modified crops — it can be seen how the privatisation of knowledge accelerated rapidly in the 1990s (see Fig. 3.2). However, it is interesting to note that this has not been a universal trend, with over 80% of Chinese biotechnology patents, for example, still held by public research institutes and universities (Zhenzhen, Jiuchun, Ke, Thorsteinsdóttir, Quach, Singer and Daar, 2004).

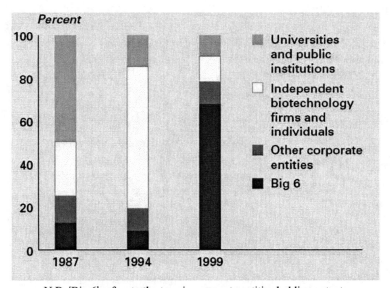

N.B. 'Big 6' refers to the top six corporate entities holding patents

Figure 3.2 Share of *Bacillus Thuringiensis* Patents by Type of Holder (de Janvry and Others cited in World Bank, 2001b, p. 185).

One key aspect indirectly highlighted by the case of *Bacillus Thuringiensis* is the increasing levels of mergers in the life sciences that have created ever more powerful multisector corporations (Mooney, 1999; Nuffield Council on Bioethics, 2004). In the 1980s, the top 20 drug houses held barely 5% of the world patented drug market. Today the top 10 companies control 35% of the market (Mooney, 1999). For agricultural biotechnology, these trends are even stronger:

> In a space of twenty years, the seed market shifted from many
> thousands of breeding enterprises (public and private) to a
> market where the leading ten companies currently control half of
> global commercial seed sales (ETC Group, 2005c, p. 6).

Restricted access to the fruits of technological development has
accompanied the privatisation of knowledge and concentration
of innovative capacity (Mooney, 1999; Cetto *et al.*, 2000). Whilst
one third of the world is said to lack access to the technologies
developed by other nations (Juma and Yee-Cheong, 2005; Giddens,
2006), the uneven distribution encompasses "...countries, regions
and social groups, and between the sexes" (Cetto *et al.*, 2000, p.
8). With limited purchasing power, market forces, as enshrined by
the TRIPs Agreement, mean orientation away from the priorities
of many in the South, particularly rural populations (UNDP, 2001;
Sterckx, 2004; Subramanian, 2004). Inequities of access have been
particularly striking in the area of healthcare (Global Forum for
Health Research, 2002), with only 13 of the 1233 drugs in the market
between 1975 and 1999 aimed at tropical diseases (WHO, 2002a).
On the contrary:

> The pharmaceutical industry develops profitable drugs for the
> wealthy regions in the world, and makes its biggest profits from
> hair tonics, anti-impotency drugs, drugs for cholesterol, ulcers,
> depressions, allergies and high blood pressure (Sterckx, 2004, p.
> 69).

Similarly, GM crops continue to be developed largely by private
enterprise to suit the needs of large-scale farmers in the North
(Nuffield Council on Bioethics, 2004; Meridian Institute, 2005).

This approach extends to technologies emerging from the South,
with the then Indian Minister of Science, Kapil Sibal (cited in Padma,
2006), saying that the professional conditioning of scientists has led
them to focus on:

> ...whatever is urban, industrial, high-technology, capital-intensive,
> appropriate for temperate climates and marketed and exported...
> to the neglect of what is rural, agricultural, labour-intensive,
> appropriate for tropical climates, retained by households and
> locally consumed.

Nonetheless, many technologies that would appear suitable
for addressing Southern needs relating to energy, shelter, water,
sanitation and nutrition already exist but are not accessible to the

people who need them most (Intermediate Technology Development Group, 2002; ETC Group, 2005c). In part, this is due to restrictive costs, legitimised and enshrined by IPRs (see, for example, UNDP, 1999; WHO, 2002a; Subramanian, 2004; Aubert, 2005). As the UNCTAD (2002a) summarises:

> The patenting of gene sequences and biotechnology techniques with broad applications means that developing countries in particular may be excluded from affordable access to technologies that they urgently need (p. 12).

Access to essential medicines has proven a bellwether issue with respect to the problems resulting from technological inequity (Brower, 2002; Subramanian, 2004; ETC Group, 2005c). Of particular note is the case of HIV/AIDS that caused international outrage when it was exposed that antiretroviral treatment cost up to $15,000 a year in the South (Fishburn and Green, 2002). The situation has been restrictive in another way for cases such as the neem tree where appropriation has:

> ...turned an often free resource into an exorbitantly priced one... As the local farmer cannot afford the price that industry can, the diversion of the seed as raw material from the community to industry will ultimately establish a regime in which a handful of companies holding patents will control all access and all production processes relating to neem as a raw material (p. 153).

But this issue goes beyond financial costs, as clearly demonstrated by the case of oral rehydration therapy. A simple salt-and-sugar solution used effectively since the 1980s to treat diarrhoea, each sachet costs approximately 10 cents yet is still unavailable for 38% of diarrhoea cases in the South (Healy, 2001).

Linked to the concentration of innovative capacity, the privatisation of knowledge and the restriction of access to technological outputs, the hype for emerging technology's ability to solve the 'development problem' has often failed to be matched by real progress. Rather, the inappropriate nature of universal techno-fixes has commonly created greater marginalisation. Even the Internet, anticipated by some (see, for example, Cecchini, 2003; Singer *et al.*, 2005; Osama, 2006) as an exciting platform for decentralised access to knowledge and distributed prosperity, is said to be creating greater inequity; a 'digital divide' (UNDP,

1999; Grimshaw, 2004). Furthermore, significant inequalities are emerging in diffusion rates between urban and rural, high- and low-income and male and female users (Lucas Jr and Sylla, 2003) in a diverse range of countries including China and Senegal (UNDP, 2001). For GM crops, claims of sustained increases in yields have been challenged by the Food and Agriculture Organisation (2007) who claim that there is now "...uncompromising evidence of diminishing returns on grains despite the rapid increases of chemical pesticide and fertilizer applications..." (p. 1). Shiva (2008) makes her objections even more clear, stating: "it is a myth that industrial, chemical agriculture produces more food" (p. 11).

Part of the explanation presented by the alter-globalisation movement for the emergence of these inequities continues the alternative development critique of technology. Here there is a rejection of the tendency towards linear models of innovation and diffusion and universalised views of poverty entrenched in technical fixes (see Leach and Scoones, 2006; Grimshaw, 2007; Kilby, 2007). As Brown (2001) notes: "within development circles there is a suspicion of technology boosters as too often people promoting expensive, inappropriate fixes that take no account of development realities" (p. foreword).

Under particular scrutiny is the failure of scientists to consider existing cultural and socioeconomic factors, such as: "...social acceptance, customs, and specific needs; moral, legal, economic, and political barriers; and social and environmental costs..." (Schummer, 2007, p. 294). With respect to information and communications technology, Lucas Jr and Sylla (2003) show the problem of such an approach by explaining: "the data suggest that what appear to be key determinants of the Internet's penetration in more developed country settings have almost no explanatory power for developing countries (p. 4).

In response, authors (Kimbrell, 2002; ETC Group, 2003a; Gould, 2005) reiterate that, given hunger, inequality, poverty and environmental degradation are the consequences of sociopolitical inequities, 'silver bullet' technologies cannot offer 'solutions'. On the contrary, technologies such as GM crops are said to be part of the inequity-generating problem (Kimbrell, 2002; Shiva, 2008), presenting "...a continuation and exacerbation of today's food production crisis" (p. 36). Like the green revolution, agricultural biotechnology is seen as increasing both commercial and technological dependency through centralised, large-scale, capital-intensive agriculture (Hallen, 1991; Invernizzi and Foladori, 2005). According to the ETC Group (2003a), there is little evidence that such developments have saved labour but, rather, have threatened the job security of the most marginalised. As Shiva (2006) notes: "the policies of corporate-driven globalised and industrialised agriculture deliberately destroy small farms, dispossess small farmers, and render them disposable" (p. 120). James (1997) gives us an insight into the resulting consequences, saying: "sixty percent of the people of Brazil now live below a harsh poverty line, but without the old means of agricultural subsistence" (p. 222). Shiva (2006) goes further, directly blaming the imposed, universal techno-fixes of global agribusiness for the 16,000 farmer suicides recorded in India during 2004.

With respect to risk, a number of new inequities have emerged. To provide background to this topic, I will first explore Beck's (1992; 1994; 1999) 'risk society' thesis. Beck (1992) argues that society has shifted from 'first modernity' — grounded in industrial society and scarcity — to 'second modernity' — grounded in a society in which contemporary risks are human-made and the unforeseen consequences of the victory of the first modernity and subsequent attempts to maintain control over risks. As Shiva (2001) highlights for the case of vitamin A deficiency:

> The reason there is vitamin A deficiency in India, in spite of its rich biodiversity and indigenous knowledge base, is that the green revolution technologies wiped out biodiversity by converting mixed cropping systems to monocultures of wheat and rice, and by spreading the use of herbicides that destroy field greens (p. 17).

Golden rice, the magic response to vitamin A deficiency, has led to "...major water scarcity since it is a water intensive crop and displaces water prudent sources of vitamin A" (Shiva, 2002a, p. 61).

With second modernity, the unprecedented pace of scientific discovery and technological change (Adams, 1998; UNCTAD, 1999; Lovgren, 2003) is accompanied by an increase in the magnitude of potential risks (Strand, 2001). This leaves contemporary risks more catastrophic than the previous risks associated with natural hazards (1992). Devastating impacts can be 'hidden', with the UNDP (2001) citing: nuclear accidents at Three Mile Island in the U.S. and Chernobyl in the Ukraine; thalidomide; and the long-hidden harms of chlorofluorocarbons. Such a scenario, "...for the first time, threaten[s] the capacity of natural systems at a global level to cope with the burden placed upon them" (Blowers, 1997, p. 849).

Characteristic of the risk society is a loss of faith in public institutions, a weakening of collective sources of certainty and meanings and greater scrutiny, scepticism and 'risk' associated with science and technology and its key drivers (Yearley cited in Blowers, 1997; UNDP, 2001; Mee *et al.*, 2004). Here Beck (1999) notes:

> Radicalized modernity undermines the foundations of the first
> modernity...the very idea of controllability, certainty or security
> — which is so fundamental in the first modernity — collapses"
> (p. 2).

Scientific uncertainty has plagued debates about the human and environmental impacts of agricultural biotechnology (UNCTAD, 2002a). For example, there is a recognised inability to extrapolate data from one field study to another, given different conditions of individual ecosystems (UNDP, 2001; UNCTAD, 2002a; Nuffield Council on Bioethics, 2004). Concerns have also been raised around the introduction of unknown allergens into the food chain (UNCTAD, 2002a). Whilst GM crops raise many similar kinds of environmental concerns to those of the green revolution, GMOs are also inherently different, carrying special risks and hazards (FAO cited in Mooney, 1999; Li Lin, 2000; Kimbrell, 2002).

Given the erosion of controllability, certainty and security, new technologies are understandably emerging in temporary regulatory vacuums (ETC Group, 2005c), with "...the accelerated introduction of new technologies...outrunning government's capacity to understand them" (Glenn and Gordon cited in ETC Group, 2005c, p. 23). In a risk society, unpredictable or unknown effects cannot be reconciled to institutionalised standards or interpreted by expert systems of the existing society's modernity (Beck, 1992; Blowers, 1997; Throne-Holst and Stø, 2008). Rather, experts provide false promises of technological safety and certainty and are portrayed as "...exercising control over technologies that cannot ultimately be controlled" (Blowers, 1997, p. 856). For genetic engineering, this translates into the assumption that humans can "...manage, improve on and even dispense with nature" (Hallen, 1991, p. 37). In this light, despite some examples to the contrary,[30] governments and industry in the North have largely rejected a strong adoption of the precautionary principle.[31] Here the main arguments are that such an approach potentially hinders the development of important technologies (Nuffield Council on Bioethics, 2004).[32] One particular claim is that the case of the E.U. moratorium on the development of GM food[33] has restricted Southern R&D (Singer, 2003; Salamanca-Buentello *et al.*, 2005) whilst leading some Southern countries to reject U.S. aid that might have included GM food, for fear of jeopardising export markets with GM-free countries (Paarlberg, 2003; Nuffield Council on Bioethics, 2004). Compounding these arguments is a belief that:

> There is not enough evidence of actual or potential harm to justify a blanket moratorium on either research, field trials, or the controlled release of GM crops into the environment at this stage (Nuffield Council on Bioethics, 2004, p. xvii).

The outcomes of such an approach from Northern industries and governments are now heightened, given risks are no longer geographically or temporally contained (Pieterse, 1998). The globalised, deferred risks resulting from anthropogenic climate change, created largely by industries in the North, are said to provide clear examples of this phenomenon (Blowers, 1997). Moreover, there is an uneven impact of risk (Blowers, 1997; Sachs, 1997c), with the globality of risk not necessarily equating to the global equality of risk (Beck, 1999). Rather:

> ...a minority of extraterritorial elites are enjoying a disproportionate share of the benefits of globalization, while the bulk of the world's population, a 'localised majority', bears the brunt of its risks and problems (Lee *et al.*, 2002, p. 7).

This situation is increasingly acknowledged in the case of climate change, where many in the South will suffer both first and hardest from its negative consequences (Blowers, 1997). The global inequity of risk is also an issue for GM crops, with Li Lin (2000) saying that the South faces even greater environmental concerns than the North "... because most of the global centres of crop origin and diversification are located in the South". Such concerns have been heightened following cases of unintentional cross-pollination of crops, such as with corn in Mexico (Invernizzi and Foladori, 2005). Compounding this situation, risk has a long history of being 'exported' to countries often lacking adequate infrastructure to support and maintain new technologies safely (Li Lin, 2000). This is a situation exploited by MNCs, establishing the context for disasters such as the 1984 Bhopal gas tragedy (Murphy-Medley, 2001).[34]

In addition to the externalisation and inequitable distribution of technological risk, certain critical voices are excluded from debates about technology and development (UNDP, 2001; Fukuda-Parr *et al.*, 2002). Such a phenomenon can be seen with the historical and contemporary debates relating to biotechnology (Singer, 2003; Court *et al.*, 2004). For example, OECD meetings in the 1980s and 1990s brought together scientific, technical, and national government experts solely from the North to establish definitions, exchange basic data and develop "...practical approaches for evaluating, harmonizing, and establishing national and international safety guidelines" (Roco and Bainbridge, 2005a, p. 78). Similarly, the E.U. and U.S. have, more recently, established "...a consultative forum on biotechnology that touches on issues of interest to developing countries. Yet the forum does not include any members representing the developing world" (UNDP, 2001, p. 77).

3.3.3   Mainstream Alternative Development?

Whilst the alter-globalisation movement seeks distance from mainstream practices, orthodox approaches to development have increasingly co-opted human development concepts, remorphing the alternative development agenda under the banner of 'greater equity' (Pieterse, 1998; Craig and Porter cited in Kilby, 2007). Such has been the influence in recent years, argues Pieterse (1998), that "...mainstream alternative development (or MAD), might not be an odd notion" (p. 350).

The basis for the co-option of alternative development seemingly emerged from mainstream acknowledgement, in the 1960s and 1970s, that rapid national growth fails to automatically reduce poverty or inequality (International Labour Organisation, 1976; Finnemore, 1997; McNamara in Rahnema, 1997; Woodward and Simms, 2006). On the contrary, Sachs (1997b) notes: "...when it became obvious, around 1970, that the pursuit of development actually intensified poverty, the notion of 'equitable development' was invented..." (p. 29). What emerged was a growing consensus about the insufficiency of development focussing only on economic growth (UNCTAD, 1999). Mainstream commentators and practitioners (see Meier and Seers, 1984; Finnemore, 1997) began to question the usefulness and accuracy of the capitalist emphasis on GDP. A 'purely quantitative' approach, it was argued, fails to consider issues such as unemployment or social tensions (McRobie, 1981).

Thus, during the 1970s and 1980s, development thinking shifted from a focus on the financial condition of States, to a focus on the condition of people, with the implied belief that "...being developed also required the guarantee of a certain level of welfare to one's population" (Finnemore, 1997, p. 205). Headlining this 'new' approach to development was a focus on 'poverty alleviation', adopted by both the World Bank and the IMF (Finnemore, 1997; World Bank, 2003; Kilby, 2007). Accompanying this method was particular consideration for programs that addressed inequalities in the fulfilment of 'basic needs' (Streeten *et al.*, 1981; Pieterse, 1998; World Bank, 2008b).[35]

In turn, such practices produced a deeper questioning of development's 'ends', ultimately leading in the 1990s to the more nuanced concept of 'human development'. Whilst adopting aspects of the 'growth and equity' approach of the 1970s,[36] the human development approach was critical of defining well-being as 'utility-maximisation' and challenged neo-liberalism's "...inherent neglect of rights, freedoms, and agency..." (Kilby, 2007, p. 121). Grounded in a belief that development is "...thoroughly dependent on the free agency[37] of people" (Sen, 1999, p. 4), Sen (*ibid.*) and Nussabaum (2003) proposed the realisation of substantive freedoms and the enablement of human capabilities as central to development's ultimate objectives. The UNDP rapidly adopted this approach, shaping new goals for development around issues of life expectancy, education, standard of living, political freedom, human rights and self-respect, all within "...a process of enlarging people's choices" (UNDP, 1990, p. 9). Accordingly, new indices emerged,[38] with the most prominent 'human development index' (HDI) a measure combining life expectancy, educational attainment, literacy levels and adjusted incomes (see *ibid.*). Importantly, the HDI highlighted that a country's level of social development does not necessarily correlate with its level of economic development (as shown in Fig. 3.3), supporting the conclusion that "the link between economic prosperity and human development is thus neither automatic nor obvious" (UNDP, 1998, p. 20). As further evidence of this disjuncture, despite limited poverty reduction and stagnant economic growth in most of the South during the 1980s, infant mortality was significantly reduced due to immunisation and oral rehydration therapy (UNDP, 2003).

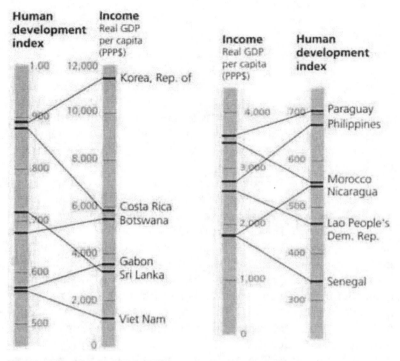

Figure 3.3 Similar HDI, Different Income and Similar Income, Different HDI (Adapted from UNDP, 1998, p. 20).

In addition to considering human freedoms and capabilities, understandings of development have expanded to more fully consider environmental, gender and indigenous needs (Finnemore, 1997; UNCTAD, 1999). In response to debates around environmental degradation and sustainability, the concept of 'sustainable development' emerged in the late 1980s as: "...development that meets the needs of the present without compromising the ability of future generations to meet their own needs" (Brundtland, 1987, p. 43). In terms of gender, from the 1980s onwards, the 'women in development' approach stressed the gender bias of development (Women's Environment and Development Organization, 1998), recognising that:

> While [women] represent 50 percent of the world population, and one-third of the official labor force, they [account] for nearly two-thirds of all working hours, receive only one-tenth of the world income and own less than one percent of the world property (United Nations in Chow, 2002, p. 2).

Critical, therefore, was recognition for the central role of women in the development process (Finnemore, 1997; Everts, 1998; Altwaijiri, 1999). From the 1970s there has also been greater attention to the plights and rights of indigenous populations, both nationally and internationally (Hannum, 1988).

In the late 1990s, the widening of development's 'ends' was followed by mainstream commentators (see, for example, UNCTAD, 1999; Fukuda-Parr *et al.*, 2002; Kilby, 2007) more fully questioning development's 'means', particularly the uniform development strategies and policy prescriptions of the previous half-century, such as structural conditionality. According to the UNCTAD (1999), the mainstream theories of the twentieth century had each ignored key factors in development:

> Import substitution, by removing the competitive spur to learning, led to technological inefficiency and lags. Liberalization helped technology development in the countries that had built up a strong base of absorptive capabilities, but by ignoring the needs of costly learning and by — incorrectly — assuming efficient markets, delayed or hindered it in others (p. 196).

Emerging evidence suggested few successes and generally poor results for those Southern countries conforming to Northern policy prescriptions (Birdsall *et al.*, 2005). Furthermore, the 'successful' cases of Southern development — South Korea, Taiwan, China and India — were each based on heterodox policy innovations and void of deep structural reform (UNCTAD, 1999; Birdsall *et al.*, 2005). Rejecting the claim that replicable mainstream policies contributed to growth in South Korea and Taiwan, Max-Neef (in Simms *et al.*, 2009) says:

> Countries like South Korea and Taiwan, frequently given as examples to be emulated, achieved their development through trade barriers, State ownership of the big banks, export subsidies, violation of patents and intellectual property and restrictions to capital flows including foreign direct investment. It would be absolutely impossible for any country to replicate these strategies

today, without severely violating the regulations of the World Trade Organization (WTO) and the International Monetary Fund… (p. 23).

Moreover, the case of India shows that poverty-reduction can occur irrespective of a country's level of national growth or trade liberalisation:

> …poverty levels in India have been falling at around 1 per cent per annum since the early 1970s, in times of both low and high growth rates, and closed and open economies (Ravallion cited in Kilby, 2007, p. 116).

As further evidence of the merits of autonomously driven development, Fukuda-Parr (2002) presents the successful cases of Botswana, Cape Verde, Costa Rica, Malaysia, Mauritius and Singapore.

What emerged from reflection upon these and other cases was recognition that "…sweeping beliefs are often incomplete" (World Bank, 2001a, p. 2). Mainstream development thinking thus shifted, in principle, "…away from debates over the role of States and markets, and the search for a single, overarching policy prescription" (*ibid.*), towards an increasing belief that 'development' is an open-ended process, with no single, fixed path (UNCTAD, 1999). The fundamental acknowledgement underlying the rethinking of development in the late 1990s was that each country has a unique social, economic and political context for development (Singleton, 2003), demanding 'differential diagnosis' (Sachs, 2005).

With universalist approaches indicative of the Cold War phenomenon of "…aid driven by politics rather than results" (Malloch Brown in Fukuda-Parr *et al.*, 2002, p. vii), it was only natural that questions were now increasingly raised about the effectiveness of aid (see, for example, Hancock, 1989; Easterley, 2006; Moyo, 2009). As Gibson, Andersson, Ostrom and Shivakumar (2005) summarise:

> Almost every part or process of the aid system has been criticized, from the geopolitical agenda of donors to the distributive politics of recipient countries; from the ties that bind aid to procurement from private firms in the donor's country to the constraints on aid bureaucrat's decision-making power; from the type of aid given to the type of accountability demanded (p. 3).

In claiming that "…aid has not been associated with the sustained increases in productivity and wages that ultimately matter…"

(Birdsall *et al.*, 2005, p. 142), Birdsall, Rodrik and Subramanian point to Sub-Saharan Africa which experienced negative growth during the 1990s, despite receiving annual amounts of aid equivalent to 12% of the region's GDP (p. 142). Amidst such happenings, it is claimed that many multilateral institutions shifted away from structural conditionality towards outcome-driven aid, tied to social indicators (*ibid.*). The effectiveness of aid was now seen, in part, to be based upon "...the ability of a recipient's economy and government to use it prudently and productively" (*ibid.*, p. 143), therein raising an issue that has come to figure prominently in development debates over recent years: 'capacity building'.

On the back of questioning universal policy prescriptions and exploring endogenous innovation, capacity building emerged in the 1990s as a prominent development philosophy and practice (Bhumiratana, 1991; UNDP, 2001; Fukuda-Parr *et al.*, 2002; Intermediate Technology Development Group, 2002; UNCTAD, 2002c; WHO, 2002a). Promoting a more holistic approach to development and innovation, capacity building focusses on strengthening a country's "...human, scientific, technological, organizational, institutional and resource capabilities" (United Nations Conference on Environment and Development, 1993, p. 450).

Given capacity building seeks to alter asymmetric power relationships, of critical importance to the emergence of its practice were the simultaneous shifts in development's actors and their associated roles (Fukuda-Parr *et al.*, 2002). Here, the structural conditionality imposed upon Southern governments gave way to support for the State as a 'steward' of capacity building (Brundtland, 2002). The new development focus was 'governance', and the creation of robust, effective and democratic institutions (UNDP, 1997; Watson *et al.*, 2003), with particular consideration for 'transparency' in all aspects of development (Wolfensohn, 2003). In greater detail, this new orthodoxy meant considering:

> Processes, convention, and institutions that determine how power is exercised to manage resources and societal interest, how important decisions are made and conflicts resolved, how interactions among and between the key actors in society are organized and structured and how resources, skills and capabilities are developed and mobilized for reaching such desired outcomes...[including] risk governance (Meridian Institute, 2006, p. 7).

Engagement with the private sector found new avenues, such as the materialisation of public-private partnerships, both locally and internationally (Leach and Scoones, 2006). This created shared responsibility for the equitable delivery of public services and infrastructure (see, for example, Global Forum for Health Research, 2002; UNCTAD, 2002a; WHO, 2002b; InterAcademy Council, 2004). However, with local participation increasingly understood by the mainstream as critical to development effectiveness, NGOs and civil society were recognised in the late 1990s as the key drivers of development (Pieterse, 1998; Fukuda-Parr *et al.*, 2002). As Mooney (1999) notes, NGOs were seen as offering a connection to the grassroots and a critical understanding of contextual political issues relating to trade, the environment and gender.

In reviewing the changes to mainstream development since the 1970s, Pieterse (1998) believes that co-option has extended to traditional alternative development philosophies around local participation, national democratisation, basic needs, women's concerns and capacity building. As Simms, Johnson and Edwards (2009) note, most definitions of development now have common characteristics that consider well-being, the environment, fair and just governance, inclusive economic and political freedoms and the ability to lead dignified and fulfilled lives (Simms *et al.*, 2009). For example, the 1996 UNDP Human Development Report states that, "to be valuable and legitimate, development progress, both nationally and internationally, must be people-centred, equitably distributed and environmentally and socially sustainable..." (p. iii).

Yet, according to Pieterse (1998), human development largely remains "...State-centred, top-down social engineering..." (p. 370), with Grimshaw (2007) adding that it continues to be underpinned by classical growth-centred development strategies. Reduced income inequity[39] is maintained as the ultimate goal of poverty alleviation (Chenery *et al.* cited in Pieterse, 1998), with people's living standards in the South still largely measured by their levels of income and consumption (Todaro cited in WHO Commission on Health and Environment, 1992; UNDP, 1998). New approaches by the Bretton Woods institutions are seen as merely "...structural adjustment with a human face" (Jolly cited in Pieterse, 1998, p. 359). According to Foladori, Rushton and Zayago Lau (2008), economic performance remains "...the core objective of World Bank and IMF policy prescriptions" (p. 6), justified by an ongoing belief in trickle-down economic theory (Lee *et al.*, 2002). The World Bank (2005), itself, states:

> The Bank has sharpened its support for the development agenda through a two-pillar strategy for reducing poverty that is based on building the climate for investment, jobs, and sustainable growth and on investing in poor people and empowering them to participate in development (p. 12).

In this light, new explorations that have been made into building 'social capital' have had the ultimate intention of building economic capital (Pieterse, 1998). One of the most lauded of human-centred approaches to development: microcredit, for example, arguably supports the neo-liberal growth agenda by creating a whole new cohort of credit-consumers (see Barker and Feiner, 2004). In such circumstances, public participation would seem to remain narrowly defined: "to participate is thus reduced to the act of partaking in the objectives of the economy, and the societal arrangements relating to it" (Rahnema, 1997, p. 120). In this sense, Hulme and Shepherd (2003) believe the majority of the South are viewed merely as "... those who are not effectively integrated into the market economy" (p. 404). A number of other new approaches are being explored to help facilitate the integration of social and market development. Non-governmental organisations, for example, are now seen by many businesses as 'partners', helping companies negotiate commerce in the South (World Business Council for Sustainable Development cited in Meridian Institute, 2005). On a larger scale, the re-embracing of the 'facilitating' State has seen the emergence of public-private partnerships, which are said to reinforce market-driven production and distribution of services (Foladori, 2006).

3.3.3.1 Techno-fixes for human development

Given the ongoing debates surrounding globalisation are "... essentially arguments about different facets of technology" (Juma and Yee-Cheong, 2005, p. 649), it is unsurprising that technology has been suggested as one of the tools by which the alternative development agenda has been co-opted (see Pieterse, 1998; Foladori *et al.*, 2008). Early signs of this phenomenon appeared in the 1970s with the apparent shift from:

> ...the construction of large infrastructure projects like dams and power plants to concerns about small farmers, renewable resources, and the provision of social services in urban areas (Finnemore, 1997, p. 203).

Drawing on aspects of appropriate technology philosophy, mainstream approaches over the following decades proposed applying to development a combination of simple, mature and emerging technologies, spanning both traditional and emerging knowledge (see, for example, Daar *et al.*, 2002b; Kumaraswamy and Shrestha, 2002; WHO, 2002a; UNCTAD, 2004). Moreover, there was increasing acknowledgement that "traditional and local knowledge systems...can make, and historically have made, a valuable contribution to science and technology" (Cetto, Schneegans and Moore, 2000, p. 10).

However, development realities have differed from the rhetoric, with the focus since the 1990s having been on "...big-hitting technologies with the potential for global scope and applicability" (Leach and Scoones, 2006, p. 13). According to the OECD (cited in UNDP, 2003):

> The number of projects drawing on low-cost technologies offering the best prospects of increased coverage for poor people — hand pumps, gravity-fed systems, rainwater collection, latrines — is very small (p. 106).

Rather, global leaders are intent on finding a "...magic bullet to 'make poverty history' and to neutralize global warming" (ETC Group, 2005a, p. 4). Reflecting the earlier claims of modernisation, revolutions in industry and production are said to make it possible to bring the benefits of progress to everybody and help solve our major social challenges, with Lopes[40] (cited in Fukuda-Parr *et al.*, 2002) claiming:

> We have finally reached a stage where we can potentially live without hunger, control major diseases and harmonize our relationship with nature. Technical and scientific knowledge allow all major material problems to be solved (p. 121).

More specifically, technology has been promoted as the key to poverty reduction (World Bank, 2008a), individual fulfilment and community and national well-being (Third World Academy of Sciences, 2004). Such claims continue, with contemporary mainstream development literature suggesting technology is central to achieving the bulk of the millennium development goals — MDGs (see, for example, Daar *et al.*, 2002b; UNDP, 2003; Watson *et al.*, 2003; Third World Academy of Sciences, 2004).

Most often cited in defence of mainstream development's almost exclusive focus on emerging high-technology are claimed successes, such as the mass vaccination campaigns that eradicated smallpox in the 1960s (Leach and Scoones, 2006) and the green revolution, that raised the health status of over 30 million preschool children (Evenson and Gollin, 2003). The World Bank (cited in UNDP, 2001) similarly claims the success of technologies in reducing mortality in the South by 40% to 50% between 1960 and 1990. In this light, some authors (Mnyusiwalla, Daar and Singer, 2003) say that high-tech is often misjudged and can actually be most appropriate to Southern development:

> ...what at first appears to be very 'high-tech' and costly and therefore perhaps irrelevant for developing countries, in the end might come to be of most value for those same developing countries (p. R11).

Others (Potoaçnik and Ezin, 2008) add that the example of using satellite imagery to measure desertification, plan sustainable farming and forestry and ensure food and water security in Africa demonstrates that high-tech can give Southern countries "...the tools and skills to attack poverty, drought, famine, water shortages and disease".

There have been few greater claims about the ability of high-tech to address global challenges than in the area of sustainability. However, here too is a realm in which certain language has been used as a guise for the expansion of neo-liberal interests. Morphing with growth theories, Korten (1990) says 'green' thinking is now institutionalised as 'sustainable development',[41] with the process of modernisation extended via 'ecological modernisation', in which it is believed that economic growth can be de-coupled from environmental degradation via changes in production processes and institutional adaptation (Rodrick cited in Blowers, 1997; Mol, 2001; Vermeulen *et al.*, 2009). Said to present solutions for de-coupling, efficient 'next generation' technologies include engine fuel from ethanol, electricity from biomass and zero-emission cars (Goldenberg cited in UNDP, 1998). Such technologies are thus believed to offer

Southern countries the ability to "...increase their consumption, industrialization and development without contributing to environmental damage" (UNDP, 1998, pp. 84–85). Here, as with modernisation theory, technology continues to be viewed as neutral, with levels of energy efficiency the defining characteristic of the sustainability discourse (Scrinis and Lyons, 2007).

From an ecological modernisation perspective, success requires embracing the free market as part of an overall expansion of the neo-liberal trade agenda, with challenges such as climate change only able to be solved by the further development of technological and market solutions, such as the deregulation of Southern energy sectors and promotion of energy trading systems (World Bank, 2006).

In such a system, hard fought gains: legal recognition for the value of indigenous knowledge (Mauro and Hardison, 2000); Northern liberalisation of trade in textiles, clothing, and agriculture (Subramanian, 2004); and the ability for countries to interpret the TRIPs Agreement in ways that allows protection for public health (Fishburn and Green, 2002; UNDP, 2003; Sterckx, 2004), have been circumscribed by the contingency of traditional knowledge being commoditised. In this light, Ford, Wilson, Costa Chaves, Lotrowska and Kijtiwatchakul (2007) claim: "...developing country governments have been pushed through trade pressure to implement much stricter intellectual property protection than required under international agreements" (p. S28).

The issue of IP demands from the North returns my review of the literature to the matter of domestic innovative capacity in the South. In this sense, greater acceptance seems to have emerged within mainstream development thinking for countries to pursue alternative paths (Fukuda-Parr *et al.*, 2002), with recognition that national innovation systems differ, both within and between the North and the South (Intarakamnerd, Chairatana and Tangchitpiboon, 2002). Hence, effective policies for building Southern science and technology capacity and the requisite form of government intervention are now seen as varying, based on a country's goals and level of development, size, population, previous traditions and existing capacity — including its skill base and the existence and maturity of local institutions and entrepreneurship (Kumar and Siddharthan, 1997; Watson *et al.*, 2003).

However, the literature and practice around Southern innovation has mostly focussed on customising Northern models, such as national systems of innovation (see Masinda, 1998; Intarakamnerd *et al.*, 2002; Aubert, 2005), in order to develop internationally competitive Southern industries (Aubert, 2005). Structures, such as centres of excellence, are suggested as perpetuating a bias towards high-tech (Rip and Laredo, 2008). Moreover, multilateral agencies have imposed, as a standard measure, the inclusion of Northern technical advisors within R&D capacity building teams in the South (Barretto and Rogov, 2000; Wilkins, 2002). In Africa, for example, 90% of the $12 billion a year spent on technical assistance goes to foreign expertise (Annan, cited in Jihui, Tisue and Volkoff, 2000).

Northern influence over Southern technological development has extended to regulatory regimes relating to risk. Northern approaches to regulation continue to guide capacity building in the South in order to ensure the development of appropriate regulatory and policy frameworks (UNDP, 2001; WHO, 2002a), with the added belief that developing a uniform approach to regulatory programs worldwide is advantageous (see WHO, 2002a). Here the UNDP (2001) suggests drawing on frameworks "...based on those established by early adopters" (p. 74). Such frameworks came to prominence with the emergence of modern bioethics in the 1960s (Benetar, Daar and Singer, 2005), 'science and technology studies' in the 1970s (Strand, 2001) and, from the 1990s onwards, the study of technology's ethical, legal and social implications — ELSI (Choi, 2002) — within a 'risk management' paradigm that seeks to minimise risks (see, for example, Blowers, 1997; UNDP, 2001; UNCTAD, 2002a). Here, Blowers (1997) claims that "...efforts are bent to minimise risks, to persuade society that the possibility of a major hazard is infinitesimally small" (p. 856). As part of a broad approach to technology's ELSI in the South, for example, the public is engaged via 'risk communication' (UNDP, 2001). This can occur through schools, public forums, the media and NGOs (Ogundiran, 2005) and is considered a necessary process in order to rectify unhelpful public perceptions (DaSilva, 2002; Fukuda-Parr, 2007) and raise public confidence in policy decisions (UNDP, 2001).

For some (Escobar, 1997; Kearnes *et al.*, 2005), these approaches form part of mainstream 'social engineering', whereby governments and industry implicitly seek to manufacture consent and acceptance about new technologies. Grimshaw (2007) adds that such an approach fits well within the traditional modernisation belief in technological determinism. In this light, Shiva (2002a) speaks of golden rice as "...a hoax...[whereby] public relations exercises seem to have replaced science in the promotion of untested, unproven and unnecessary technology" (p. 58). Whilst these processes aim to fill knowledge gaps, thereby laying the groundwork for debate in which the public can play a constructive role in technological development (UNDP, 2001; UNCTAD, 2002a), risk management is said to shut out reflexivity by framing debates in a way that ensures the inevitability of selected technologies (Kearnes *et al.*, 2005).[42] Kearnes *et al.* (*ibid.*) argue that, in the case of biotechnologies, there has been little institutionally sanctioned space for the social sciences to engage with wider issues. This, they say, has reduced projects to "...conceptualizing and evaluating and managing 'the impacts'..." (p. 271), resulting in a debate focussed on scientific risk, at the exclusion of bigger debates about alternatives and key questions around purpose, ownership, control, and responsibility (ETC Group, 2004b; Kearnes *et al.*, 2005).

3.3.4 Post-Development

Post-development is a radical reaction to modernity and the rhetoric and practice of mainstream development (Pieterse, 1998). However, whilst Pieterse (*ibid.*) claims post-development theoretically crosses over with areas such as critical theory and post-structuralism, he says it also expresses "...disillusionment with alternative development" (p. 360).

Post-development's overall criticism of development is that it has comprehensively failed:

> The idea of development stands like a ruin in the intellectual landscape. Delusion and disappointment, failures and crimes have been the steady companion of development and they tell a common story: it did not work (Sachs, 1997c, p. 1).

Three main reasons are provided to support this claim. Firstly, development is said to have created greater inequity and dependency, leading "...only to corruption, confusion and structural adjustment plans that turned poverty into destitution" (Latouche, 2004). As an example, Latouche (*ibid.*) reflects on the history of food security in Africa:

> Africa was self-sufficient in food until the 1960s when the great wave of development began. Imperialism, growth economics and

globalisation destroyed that self-sufficiency and make African societies more dependent by the day.

The second reason is that development comes at great costs, both financially and environmentally. As Woodward and Simms (2006) note: "...in the 1990s it took $166 of global economic growth, with all the associated environmental costs, to achieve just $1 of progress towards the MDG on poverty reduction" (p. 17). Moreover, development's attempts to dominate, fragment and dispossess, have demanded the destruction of the environment (The Ecologist, 1992), leading to the dangerous situation in which everybody's future is compromised (Rist, 2002).

The final reason is that mainstream development is said to have been a "...mechanism for the production and management of the Third World" (Escobar, 1991, p. 676); an ethnocentric form of cultural Westernisation and homogenisation (Pieterse, 2000; Latouche, 2004), depriving people of control over their own lives and shifting control to bureaucrats (Esteva, 1985). In this sense, a new imperialism is believed to have emerged since colonisation, with "... the nature and purposes of post-colonial 'development' [having] remained remarkably similar to those of colonial interventions" (p. 93). Constantino (in Pieterse, 1998) cites poverty alleviation as one such form of "subversive neo-imperialism" (p. 360), with Shiva (1989) problematising mainstream approaches to poverty along the following lines:

> Culturally perceived poverty need not be real material poverty: subsistence economies which serve basic needs through self-provisioning are not poor in the sense of being deprived. Yet the ideology of development declares them so because they don't participate overwhelmingly in the market economy, and do not consume commodities provided for and distributed through the market (p. 10).

Development's discourse, as practice, is thus believed to legitimise certain perspectives whilst de-legitimising others (see Foucault, 1972; Escobar, 1995). Moreover, according to post-development theory, mainstream development falsely places the development 'problem' with the South when, rather, Southern ways of life are believed to have much to teach the North's over-consuming societies in which the development problem actually lies (Lummis, 1997; Salleh, 2009).

In this light, there has been an increasing questioning of development's trajectory, with Rees (1990) claiming: "the possibility of sustainable development based on the growth-oriented assumptions of neo-classical economics is illusory" (p. 18). More broadly, there has been a revival of thinking around the limits to growth, with acknowledgement that exponential economic expansion in world of finite, biophysical limits is ultimately unsustainable (United Nations Environmental Program, 1999; Jackson, 2009) and can only end in disaster (Lowe, 2009). More alarmingly, Turner (2008), in revisiting the initial projections made by the Club of Rome, finds:

> ...the observed historical data for 1970–2000 most closely match the simulated results of the LtG [limits to growth] 'standard run' scenario for almost all the outputs reported; this scenario results in global collapse before the middle of this century (p. 410).

Whilst the legitimacy of Southern material aspirations, particularly around basic needs, is relatively unchallenged, there is acknowledgement that convergence at Northern standards of living is beyond the earth's carrying capabilities (Third World Academy of Sciences, 2004; Lowe, 2009). Furthermore, the vision of the South replicating the extractive production and trading processes of the North, in order to reach converge, is also said to be untenable (Rist, 2002; Andreasson, 2005; McKibben, 2007).

Thus, argue post-development theorists such as Sachs (1997c), development is no longer a useful concept:

> ...the historical conditions that catapulted the idea into prominence have vanished: Development has become outdated. But above all, the hopes and desires which made the idea fly, are now exhausted: development has grown obsolete (p. 1).

Development alternatives must therefore be replaced by alternatives to development (Escobar, 1995). At the heart of such alternatives would be greater autonomy via disengagement of the local from external dependency (Escobar cited in Pieterse, 1998; Esteva, 2006) and the removal of obstacles restricting different forms of development (Latouche, 2004). In this sense, post-development argues for disavowing universalism and embracing radical pluralism and cultural diversity (see Latouche, 1993; Escobar, 1995). Such processes should be grounded in reclaiming the commons (The Ecologist, 1992), support for locally rooted lifestyles (Pieterse, 1998;

Esteva, 2006) and associated values, such as conviviality and frugality (Pieterse, 2001). To support such shifts, claims Escobar (cited in Pieterse, 1998), will require "...the expansion and articulation of anti-imperialist, anti-capitalist, anti-productivist, anti-market struggles'" (p. 362).

For the North, says Jackson (2009), "...prosperity without growth is no longer a utopian dream. It is a financial and ecological necessity" (p. 489). Thus, building on strong foundations,[43] there are increasing calls (see, for example, Jackson, 2009; Stern, 2009) for the North to shift to a steady-state economy in which there would exist a "...constant population and constant stock of capital, maintained by a low rate of throughput that is within the regenerative and assimilative capacities of the ecosystem" (Daly, 2008, p. 3). To facilitate socially equitable convergence within limits, many (Allier, 2009; Latouche, 2009; Kerschner, 2010) argue that de-growth or contraction in the economies of the North must precede a steady-state, with Lowe (2009) noting: "...basic services in the poorest countries can only be provided if we curb wasteful consumption in the richest nations" (p. 92).

3.3.4.1 Technology as power, within biophysical limits

As with other theories critical of development, Western technology is a primary target of post-development's critiques. Foremostly, modern science is said to have failed to deliver on its claims:

> Its promise to transform the world into a materialist paradise and thereby put an end to poverty and oppression has lost all credibility. There is evidence indeed to show that it has accomplished just the contrary (Alvares, 1997, p. 220).

Furthermore, post-development presents technology as "...the imposition of science as power" (Nandy cited in Pieterse, 1998, p. 360), leading to 'laboratory states' in which technologies are trialed on vulnerable populations across the South (Vishvanathan cited in Pieterse, 1998).

The failure of Western technology is also seen, more broadly, in its inherent support for commodification and Capitalist rationality (Escobar in Pieterse, 1998). More importantly, ecological modernisation, it is claimed, does nothing to alter the impulses within the capitalist economic mode of production that inevitably lead to environmental degradation (Foster, 2002; Raskin *et al.*, 2002).

Moreover, given the contemporary extrapolation of the 'rebound effect'[44] (see Alcott, 2005; Polimeni, Mayumi, Giampietro and Alcott, 2008; Alcott, 2010), improvements in technological efficiency alone provide little hope for global prosperity. As Jackson (2009), in his comprehensive review of the possibilities for continued economic growth in the North, notes:

> ...evidence for overall reductions in resource throughput (absolute de-coupling) is virtually absent. Dramatic improvements in energy efficiency over the last 3 decades have been offset by massive increases in the scale of economic activity (p. 488).

In response, post-development proposes the need for exploring new power-knowledge regimes (Escobar cited in Pieterse, 1998), with particular consideration for ecologically sensitive indigenous knowledge (Pieterse, 2001).

3.3.5   Reflexive Development

Upon reviewing mainstream development and its critiques, Pieterse (1998, 2000, 2001) believes that development has reached an impasse. He firstly claims that mainstream development is stuck between the objectives of human development and neo-liberalism (Pieterse, 1998). Furthermore, alternative development is seen as too diverse to be considered a uniting counterpoint to mainstream development, having been:

> ...reinforced by and associated with virtually any form of criticism of mainstream developmentalism, such as anti-capitalism, green thinking, feminism, eco-feminism, democratization, new social movements, Buddhist economics, cultural critiques, and poststructuralist analysis of development discourse (*ibid.*, p. 346).

In this sense, alternative development is unable to generate "...a coherent body of theory" (*ibid.*, pp. 351–352). Pleyers (2009) highlights this reality for the alter-globalisation movement which, he says, is currently divided over its political and organisational direction, including those pushing local autonomy through participatory governance, those focussing on advocacy through single-issue networks and others focussing on the development of progressive policies through State leaders and institutions.

Correspondingly, the visible differentiations between alternative and mainstream development from the 1970s and 1980 are said to have disappeared (Pieterse, 1998). Post-development, too, is seen as failing to present clear, coherent alternatives to development and its associated social and political institutions (Beck, 1997; Pieterse, 1998). Moreover, Pieterse (2001) believes that none of its counterpoints are "...specific to post-development nor do they necessarily add up to the conclusion of rejecting development" (p. 107).

Importantly, Pieterse (1998) says that post-development fails to acknowledge that key aspects of its discourse, such as democratisation, difference and anti-development, actually arise, themselves, out of modernisation:

> Democratization continues the democratic impetus of the Enlightenment; difference is a function of the world `becoming smaller' and societies multicultural; and anti-Development echoes and elaborates the dialectics of Enlightenment of critical theory. More generally, the rise of social movements and civil society activism, North and South, are also expressions of the richness of overall development... (p. 365).

In this light, Pieterse (*ibid.*) concludes his critique by claiming that the diversity, complexity and potential reflexivity of mainstream development have been underestimated, with the different options for problematising and reworking modernity or exploring modernities in the plural having been ignored. Both alternative and post-development theorists, Pieterse says, have simplified Westernisation and mainstream development by viewing them as a "...single, homogeneous thrust toward modernization" (*ibid.*, p. 347). This has resulted in the construction of dichotomic thinking; theorists and practitioners are either: 'for' or 'against' modernity; pro- or anti-development. Pieterse (*ibid.*) reminds us that this is too simplistic, ignoring, for example, influences and business models from the East.

As a final criticism, post-development, Pieterse (*ibid.*) claims, fails to fully consider a shared global future and, therefore, ultimately falls short of addressing how a more equitable world might exist in connected circumstances.

However, in moving beyond yet another critique, one of Pieterse's (*ibid.*) main contributions is in presenting a concept of 'reflexive development' that provides a corollary to Beck's concept of 'reflexive modernisation'. According to Beck (1992), the North has entered a period of 'reflexive modernity', in which the preoccupation with mastering nature is replaced by an increasing focus on managing the side-effects stemming from modernity. However, rather than be restrictive, the uncertainty and crises common to a risk society are said to "...liberate[s] individuals opening up new possibilities for thought and action" (Blowers, 1997, p. 858). In this age, there is said to be a greater ability for self-reflection and self-criticism that can lead to the ability to act reflexively: in anticipation of the impacts of one's own actions (Beck, 1996). With the ability to 'rework' modernity, subsequent transformations are expected at the "...level of production, institutions and, consequently, in values and life-styles" (Blowers, 1997, p. 864). In this respect, Pieterse (1998) observes: "a new political culture is taking shape in which the separation between politics and nonpolitics becomes fragile and nonpolitics gives rise to subpolitics" (p. 368). Such subpolitics present "...new ways of conducting politics at social 'sites' that we previously considered unpolitical" (Beck, 1999, p. 93).

Although not 'new', one of the sites rapidly producing novel forms of political action is the increasingly connected international stage (Blowers, 1997; Goodman, 2003; Fisher, 2004). With the perplexities of progress now said to be shared across North and South (Beck, 1999), Pieterse (1998) suggests that "...[the] critique of science and of corporate policies and public relations...are increasingly being transnationalized...[through] North-South transfer of risk awareness..." (p. 367). With a precautionary approach having received uptake in parts of the South, albeit for contested reasons,[45] the global discourse surrounding GM food is said to provide a useful example of this new transnational reflexivity (see Throne-Holst and Stø, 2008).

For Pieterse (1998), reflexive modernisation, combined with transnational reflexivity, offers insights into a path beyond the dualism of developmentalism and post-development, and North versus South, towards a shared path of reflexive development that incorporates key insights from both sides of the debate. Here Pieterse (2001) highlights the existing nature of this approach: "... almost invariably, development theory stems from a reaction to and thus also a reflection on the limitations of preceding development policy or theory" (Pieterse, 2001, p. 161). Moreover, development's path is "...increasingly reflexive in relation to the failures and crises of development" (Pieterse, 1998, p. 367). New policies are "increasingly concerned with managing the hazards, risks, unintended consequences and side-effects brought about by development itself" (*ibid.*, pp. 367–368), with a heightened self-criticism (Gibson *et al.*, 2005) leading to "...a constant search for alternatives, a tendency towards self-correction and a persistent pattern of co-optation of whatever attractive alternatives present themselves" (Pieterse, 1998, p. 349).

Whereas the grounds for development's reflexivity have historically emerged through an uncontrolled process of conflict, for reflexive development to work in the future, a self-conscious development strategy must be in-built. In this way, Pieterse (*ibid.*) proposes moving beyond *any* alternative to the mainstream by speaking of a process grounded in participatory, popular reflexivity where all participants can shape and alter the strategies of development at the appropriate scale of intervention and as risks arise. In a social and political sense, such development could appear as "...broad social debates and fora on development goals and methods" (p. 345).

Internationally, Pieterse (*ibid.*) sees the need to shift to a process of 'world development', whereby reflexivity to the legitimate claims of equitable development emerges, across North-South divides, via pragmatic self-reflection. Here he suggests, as the challenge, finding the balance between polar constituencies:

> [bringing] separate and opposing interests and constituencies together as part of a world-wide bargaining and process approach...a global reform platform...The point is then to find a narrow path in which participatory approaches retain their meaning, the role of the State is reinvented through public sector reform, and the Washington agenda itself is reconsidered (*ibid.*, p. 168 and 66).

As exemplified by the work of people such as Schumacher (1973), holistic models for people-centred development are not new. Yet, in recent times, a number of important proposals have emerged for how to create a more equitable world.[46] Whilst is it beyond the realms of this book to explore these proposals in any depth, as Salleh (2009) notes, a common synthesis might be a three-fold focus on reflexivity towards ecological sustainability, socioeconomic justice and cultural autonomy.

3.3.5.1 Technology and reflexivity

The discourse around technology is also said to be reflexive, with a dialectical process of conflict viewed as "...part of development politics" (p. 343). Critiques of the function of science and techno-scientific development, for example, have "...led to appeals to indigenous technical knowledge and local knowledge..." (*ibid.*, p. 368). Yet, as with the development discourse more generally, some (UNDP, 2001; Fukuda-Parr, 2007) see the potential for such processes to assist in any shift to a more equitable world as constrained by highly polarised debates about technology and development. Here it is claimed that proponents of new technologies often fail to consider alternatives and opponents of new technologies often ignore the harms of the status quo (UNDP, 2001). According to Fukuda-Parr (2007), the case of agricultural biotechnology provides a prime example, with significant polarisation having developed between the 'naysayers and cheerleaders'. In such circumstances, she believes that:

> ...the commercial lobby overstates the near-term gains to poor people from the genetically modified organisms it develops. Meanwhile, the opposing lobby overstates the risk of introducing them and downplays the risk of worsening nutrition in their absence (p. 68).

However, there remains hope and theoretical willingness from some of emerging technology's toughest critics that the process of discursive engagement might be more reflexive. As Mooney (1999) says, "...we must still warn against techno-fatalism. All is far from lost. Much can still be gained from some new technologies" (p. 67). In this sense, the differing approaches to technology outlined in my review are not seen as mutually exclusive (Leach and Scoones, 2006). Moreover, Grimshaw (2008) points to the hope arising with new technologies before they are embedded: "older technologies are inevitably entrenched in existing systems of patents, production, and markets. There is an opportunity with new technologies to do things differently" (p. 3). According to Crow and Sarewitz (2001), our engagement with technology must therefore move towards "...a process of technology-supported societal progress where different sectors and activities can continually co-evolve in response to knowledge about one another's needs and constraints" (p. 58).

My review therefore arrives at what others, such as the UNDP (2001), have agreed is an outstanding question: what criteria should be used for technological assessment with respect to equitable development? In Beck's (1999) view, a new framework is needed that allows exploration of new social and political forces as well as emerging lines of conflict. However, just as the notion of reflexive modernisation "...is vague and offers no practical prescription or notion of how transformation will proceed" (Blowers, 1997, p. 867), so too is the concept of reflexive development. Pieterse's (1998) work proffers little in terms of how to assess whether emerging technology offers hope for a more equitable world, other than to suggest the need for feedback or 'participatory, popular reflexivity' around contemporary technological issues.

In recent times, however, a number of proposals have emerged regarding people-centred technology (see, for example, Sclove, 1995; Everts, 1998; Leach and Scoones, 2006; Grimshaw, 2008). Here I will explore the two of these that, like the proposal I will put forward in this chapter's conclusion, identify key themes for technological reflexivity, as well as criteria for assessing technology's ability to provide hope for a more equitable world. The first of these proposals is Sclove's (1995) outline for 'a democratic politics of technology' that hinges on four themes and nine criteria for technological assessment (see Table 3.2). Here Sclove is interested in exploring "the reconstitution of technology along more democratic lines" (p. ix). His approach is to synthesise literature on the social dimensions of technology and democratic theory and then outline themes and criteria for distinguishing technology that is compatible with democracy from that which is not.

The second proposal is Grimshaw's (2008) model of 'pro-poor innovation', in which he argues that, arising from various tensions,[47] there are four 'fault-lines' for emerging technology that each require reflexivity towards certain outcomes (see Fig. 3.5). Here Grimshaw is interested in unshackling from old ways of thinking in order to take new technologies in different directions; ones "...enabling choices to be made that fulfil the needs of people" (p. 28). His approach is to identify the attributes of new technology that need changing in order to "...[gain] insight into the kinds of actions that need to be embedded in international development efforts that aim to challenge poverty by the use of new technology" (*ibid.*, p. 3).

Table 3.2 Nine Criteria for Democratic Technologies (Adapted from Sclove, 1995, p. 98).

Theme	Criteria
Community	Seek a balance among communitarian/cooperative, individualized, and trans-community technologies. Avoid technologies that establish authoritarian social relations.
Work	Seek a diverse array of flexibly schedulable, self-actualising technological practices. Avoid meaningless debilitating, or otherwise autonomy-impairing technological practices.
Politics	Avoid technologies that promote ideologically distorted or impoverished beliefs. Seek technologies that can enable disadvantaged individuals and groups to participate fully in social, economic, and political life. Avoid technologies that support illegitimately hierarchical power relations between groups, organizations, or polities.
Self-Governance	Keep potentially adverse consequences (for example: environmental or social harms) within the boundaries of local political jurisdictions. Seek relative local economic self-reliance. Avoid technologies that promote dependency and loss of local autonomy. Seek technologies (including an architecture of human space) compatible with globally aware, egalitarian political decentralization and federation.
Social Structures	Seek ecological sustainability. Seek 'local' technological flexibility and 'global' technological pluralism.

Table 3.3 An Assessment Framework for New Technologies (Adapted from Grimshaw, 2008, p. 12).

Theme	Outcomes
Promise	Cheap energy, safe water
Poverty	A focus on poverty reduction
Price	Technology priced as a public good
Power	Open source technology: distributed community power

Common to both Sclove (1995) and Grimshaw's (2008) approaches is a concerted focus on local development via local ownership, orientation and governance, with underlying consideration for the limits to growth. There is mutual acknowledgement of the need to explore both generic and specific implications relating to the technology under assessment. Far from rejecting high-technology outright, both authors support technological pluralism. However, they promote critical engagement with emerging technology, noting the potential for technologies to have embedded ideologies as well as potential for such biases to be consciously shaped.

In this section I have built on my review of the foundational approaches to development by showing how, in the latter quarter of the 20th Century, development has been heavily influenced by neo-liberal ideology. The associated emergence of corporate technologies and endogenous Southern innovation has furthered neo-liberal aims, such as the creation of a more integrated, global market economy. However, the rise of alter-globalisation movements has highlighted the gross inequities produced by and evident through engagement with corporate technology within a neo-liberal framework. Compounding these challenges, the goals and means promoted by alternative development appear to have been co-opted, with a contested approach to technology and development having emerged under the guise of 'human development'. Such changes helped spur post-development theorists who reject development and expose the destructive use of technology as a tool of power within the broader development narrative. To overcome the subsequent 'development impasse' — given critiques are said to offer few realistic or thorough alternatives to development — Pieterse (1998) has suggested that development, and within it our approach to technology, must enshrine a conscious reflexivity across both North and South as part of a new approach to world development. To conclude this section, I have therefore raised some of the themes and criteria by which to assess an emerging technology's reflexivity to the legitimate requirements of equitable development.

3.4 Conclusion and My Framework for Technological Assessment

Debates about how to create a more equitable world have continued since the commencement of the development era, and contestations about the role and nature of technology have been central to these debates. Despite development practices presenting increased consideration for basic needs, participation and sustainability since the 1970s, an inequity-creating trajectory continues and is the overarching context in which new technology emerges. The ideals of alternative development have been largely co-opted into an ultimately unsustainable, neo-liberal framework, in which technology predominantly remains an instrument for economic growth that is falsely presented as 'the solution' to greater equity. The fundamental contradictions within the trajectories accompanying emerging technology have been ignored by mainstream alternative development, presenting critical barriers to overcome in order for a more equitable world to be possible.

Thus, there is a need for a clear break from the orthodoxy. As a reaction to modernity and development, post-development offers valuable critiques, but is said to fall short on providing realistic alternatives. However, given our approaches to modernity are increasingly reflexive, there is said to be hope in reframing development in a similarly reflexive manner, one in which popular development is viewed as a consciously adaptive process. In this way, development can be seen as an inclusive process, requiring that the impacts of technological change be considered across North-South divides.

The work of Sclove (1995) and Grimshaw (2008) suggests that there is value in looking at the themes and criteria by which to assess an emerging technology's reflexivity to the legitimate requirements of equitable development. As Grimshaw (*ibid.*) states "identifying the attributes of new technology that need changing is a first step towards the kinds of actions and policies that will ultimately lead to human need being enabled by new technology" (p. 28).

Given I see nanotechnology as a useful example through which to explore the latest approaches to the technology-development nexus, around what themes, issues and criteria would the transition to a more equitable world — a world of ecological sustainability, socioeconomic justice and cultural autonomy — require reflexivity from nanotechnology? Whilst my framework takes Sclove's (1995) and Grimshaw's (2008) work into consideration, my review of the literature suggests three overarching themes around which reflexivity needs to occur with respect to nanotechnology's development. For each theme I provide a criterion — a 'legitimate requirement' for equitable development against which I propose nanotechnology's trajectory should be assessed (for a summary of my themes and criteria see Table 3.4).

Table 3.4 Themes and Criteria for Assessing Technological Reflexivity towards a More Equitable World.

Theme	Indicator
Innovative capacity	Innovation and innovative capacity that is: Decentralised Autonomous
Technological appropriateness	Technologies that are sensitive to: Human needs Cultural norms Environmental Limits
Approaches to technological governance	Approaches to technological governance that (are): Democratic Empower people Influence innovation trajectories

In the foundational and contemporary literature that I have assessed, the nature of Southern engagement with global innovation is a constant theme for debate. Within debates, common explorations surround issues of ownership, influence and partnership, as well as the barriers to endogenous and local innovation. Throughout, centralised control over innovation has continually created greater inequity. I also believe that there is a strong argument to be made for redistributions in global technological strength resulting in realignments of global influence. I therefore see a more equitable world demanding greater reflexivity around *innovative capacity*. Here I see the indicator for progress as innovation and innovative capacity that is more decentralised and autonomous.

Schumacher's (1973) five criteria for appropriate technology: simple and user-friendly; productive and fulfilling; locally owned and widely distributed;[48] sensitive to local needs and cognisant of cultural norms; and benign or benevolent impacts for humans and the environment, have been both directly and indirectly central to critiques of the green revolution, biotechnologies and other technologies emerging within the development context. With the corporatisation of knowledge driving greater inaccessibility to technology's claimed benefits, and the world facing stark limits to growth — largely as a result of Northern levels of consumption and resource extraction — such philosophies would appear as relevant as ever (Wicklein and Kachmar, 2001; Aubert, 2005; Leach and Scoones, 2006). I therefore see a more equitable world demanding greater reflexivity around *technological appropriateness*. Here I see the indicator for progress as technologies that are sensitive to human needs, cultural norms and environmental limits.

Since the emergence of development debates, the roles of the State, business and, more recently, NGOs, have all been extensively explored with respect to innovation. Yet, the democratisation of science, and therein the ability of citizens to determine the goals of R&D, remain a critical struggle. Despite history showing the value of collective community wisdom ahead of exclusive expert views when it comes to addressing local needs, centralised control over guided and restricted technological trajectories remains the dominant form of engagement with emerging technology in both the South and the North. I therefore see a more equitable world demanding greater reflexivity around *approaches to technological governance*. Here I see the indicator for progress as more democratic approaches to technological governance that empower people and influence innovation trajectories.

It could be argued that all three of my themes and criteria fit within Schumacher's (1973) original framework for appropriate technology. However, my literature review on development, technology and inequity shows that there is enough contention around the themes of 'innovative capacity' and 'approaches to governance' to demand their unique consideration. Furthermore, as discussed in my opening chapter, to look holistically at technology requires consideration of its relevant artefacts as well as the whole package it presents, including its relationship to the context in which it emerges. I will therefore use my assessments of innovative capacity and approaches to governance to focus more on nanotechnology in its entirety, and consider the specific characteristics of nanotechnologies more fully in my assessment of technological appropriateness.

Part of assessing nanotechnology in its historical and contemporary context will be the need to judge its foreseen and present implications against the claims made and actualised for previous technology. As identified by many authors (see, for example, Einsiedel and Goldenberg, 2004; Mehta, 2004; Wolfson, 2004), biotechnology offers a useful reference in reflecting upon nanotechnology's emergence — irrespective of differences in implications — given its commonality as a 'strategic technology' (Einsiedel and Goldenberg, 2004).[49]

In reviewing the literature, I have established an historical and theoretical context in which to assess nanotechnology and development. I have also developed a framework for assessing nanotechnology that will prove foundational in answering my central questions. In the coming chapter, I will therefore test these three themes and criteria through an exploration of the literature surrounding nanotechnology, development and equity, whilst building a more in-depth context for my contemporary analysis of nanotechnology.

Endnotes

1 In which different sectors perform different functions within the same, overall production process (Wallerstein, 1974).

2 Along with the International Bank for Reconstruction and Development, the International Development Association now constitutes 'the World Bank'.

3 See Gerschenkron (1962).

4 Borlaug developed high-yielding, short-strawed, disease-resistant wheat whilst working, predominantly in Mexico, between 1944 and 1960.

5 Created by the International Rice Research Institute in the Philippines.

6 Whilst aspects of modernisation theory are heavily contested by socialist theorists (as shown in Sec. 3.2.2), its primary characteristic: a drive for national economic growth, cuts across both capitalist and socialist thinking, with Cleaver (1997) noting "...rather than abandoning the development project, socialists have consistently proposed the adoption of an alternative 'socialist development'" (p. 233).

7 For example, dependency theorists pushed for the national development of biotechnologies in Latin America.

8 'Terms of trade' refers to: "...the quotient between the export price index and the import price index" (Foladori *et al.*, 2008, p. 5).

9　See Klein (2007) for a detailed investigation of the promotion of economic 'shock therapy' in Eastern Europe following the fall of the Berlin Wall.

10　President of Brazil from 1995–2003.

11　For example, the Nuffield Trust (2004) reports a doubling of the total acreage of genetically modified crops in the South since 1999.

12　Whereby "…only 10 per cent of spending on health research and development is directed at the health problems of 90 per cent of the world's people" (UNDP, 2003, p. 158).

13　For example, the United Arab Emirates, with its GDP per capita close to $18,000, is considered by the World Bank as 'scientifically lagging', compared to Russia, with its GDP per capita less than $2500, that is considered 'scientifically advanced' (Third World Academy of Sciences, 2004).

14　In 'national systems of innovation' the elements and relationships which interact in the production, diffusion and use of new knowledge are embedded within the nation state (Lundvall, 1992). Put differently, government, industry and academia is linked through facilitated interaction (Masinda, 1998; Juma and Yee-Cheong, 2005).

15　A technology park is a physical space in which various individuals, organisations and companies congregate and work towards the commercialisation of scientific research.

16　'Centres of excellence' are "…physical locations in which research and advanced training are carried out, often in collaboration with other centres, institutions and individuals" (InterAcademy Council, 2004, p. 6).

17　Indeed, in a 1968 speech, the former U.S. Attorney General, Robert F. Kennedy (in Steer and Lutz, 1994), said that GDP"…measures everything in short, except that which makes life worthwhile" (p. 17).

18　Even when the data is updated to factor in purchasing power parity (UNDP, 2001).

19　Figures can be misleading, given different starting points. Ten per cent GDP growth in China versus three per cent in the U.S. can mean economic divergence for a considerable time before convergence occurs, given the different starting points in GDP.

20　Taking into account purchasing power parity.

21　Although unreliable data collection has often made any accurate assessment difficult (Forbes, 2000).

22 As evidence, authors (Mooney, 1999; ETC Group, 2003a) cite the disastrous impacts of the Renaissance on peasant farmers and the industrial revolution's outcomes for miners and textile workers in Great Britain and cotton growers and cloth weavers in India.

23 These are the U.S. (42.8 million hectares), Argentina (13.9 million hectares), Canada (4.4 million hectares), Brazil (3.0 hectares) and China (2.8 million hectares) (Runge and Ryan, 2004).

24 This is, perhaps, a Western understanding of 'scientist'.

25 See: *Diamond v. Chakrabarty*, 447 U. S. 303, 1980. Available at: http://caselaw.lp.findlaw.com/scripts/getcase.pl?court=us&vol=447 &invol=303.

26 The commercialisation of natural resources and associated traditional knowledge without equitable compensation to the people or nations in whose territory the materials or techniques were originally discovered.

27 A form of IP providing exclusive commercial rights to a registered variety of plant.

28 Resources that either are, or it is believed should be, collectively owned.

29 Using data from the USPTO.

30 In 1975, at Asilomar State Beach in the U.S., a group of professionals developed voluntary guidelines for conducting experiments using recombinant DNA technology. These guidelines were adopted by scientists, worldwide. In recent times, the precautionary principle has been used to approach the well-known cases of asbestos, radiation, chlorofluorocarbons and Mad Cow Disease (Throne-Holst and Stø, 2008).

31 Interpretations of the precautionary principle range between 'soft' and 'strong' readings (UNDP, 2001; Throne-Holst and Stø, 2008). As explained by the UNDP (2001): "soft formulations place the burden of proof on those who claim that harm will occur if a new technology is introduced. Strong formulations may shift the burden of proof to the producers and importers of a technology, requiring that they demonstrate its safety" (p. 70).

32 Miller and Conko (2000) contend: if the precautionary principle had been applied to polio vaccines and antibiotics, the prevention of occasionally serious, and sometimes fatal side- effects "…would have come at the expense of millions of lives lost to infectious diseases" (p. 49).

33   One of the most notable precautionary responses to scientific developments was the moratorium on the development of GM foods, first applied by European nations in the late 1990s.

34   Here 41 tonnes of deadly gas and toxins were released at a pesticide plant operated by the Union Carbide Corporation. The disaster led to the immediate death of at least 2000 people.

35   Governed by the principle that: "the basic needs of all should be satisfied before the less essential needs of a few are met" (Streeten, Burki, Ul Haq, Hicks and Stewart, 1981, p. 8). Such a process involved "...adding physical estimates of the particular goods and services required to achieve certain results, such as adequate standards of nutrition, health, shelter, water and sanitation, education, and other essentials" (*ibid.*, p. 3).

36   As Kilby (2007) argues, "a capabilities approach recognises economic growth is important, but as a means rather than an end, with equity becoming the major political objective, and measures of deprivation and distribution being central" (pp. 121–122).

37   Sen (1999) argues that certain factors influence the potential to achieve free agency, including: "...economic opportunities, political liberties, social powers, and the enabling conditions of good health, basic education, and the encouragement and cultivation of initiatives... [and] the liberty to participate in social choice and in the making of public decisions" (p. 5).

38   With the HDI acknowledged as limited in its "...attempt to capture a complex reality in a summary form with imperfect data" (Anand and Sen, 1995, p. 1), and questionable in its ability to measure inequality (UNDP, 2006), more wholesome measures emerged, such as the inequality-adjusted HDI (see Hicks, 1997), the human poverty index (see UNDP, 1997), the index of sustainable economic welfare (see Daly and Cobb, 1994) and the genuine progress indicator (see Cobb, Halstead and Rowe, 1995).

39   As judged by the increasingly used Gini co-efficient (Sen, 1976; UNCTAD, 1997).

40   At the time a member of the UNDP executive.

41   Although some authors now "...use 'growth' to mean quantitative change, and 'development' to refer to qualitative change" (Daly, 1977, p. 16).

42   As an example, Brown (cited in UNDP, 2000) suggests that we progress from the debate over technology and its role in to development "...to

identify the global and national policies and institutions that can best accelerate the benefits of technological advances" (pp. iii–iv).

43 See the foundational work of Stuart-Mill ([1848] 2001), Georgescu-Roegen (1971, 1977) and Daly (1973, 1977).

44 Also known as 'Jevon's paradox', whereby technological developments that increase the efficiency with which a resource is used are outpaced, with respect to environmental impacts, by general increases in the rate of consumption of that resource.

45 Mittal (2006) claims that Southern disengagement was a result of genuine health and safety concerns, whilst Juma and Yee-Cheong (2005) claim that anti-GM policies in the South stemmed from different reasons, such as Kenya's lack of regulatory capacity, legal and constitutional barriers posed by an NGO backlash in Brazil, and India's fear of MNC domination.

46 See, for example: Korten (2006) on 'earth community'; Max-Neef, (1991) on 'the human-scale economy'; Raskin (2002) on 'the great transition'; Shiva (2006) on 'earth democracy'; George (2004) on 'another world'; Bennholdt-Thomsen (2001) on 'subsistence in practice'; Henderson (1991) on 'win-win development'; Daly (1994) on 'stewardship'; Norberg-Hodge (2000) on 'local interdependence'; Hines (2000) on 'localization'; Mander (2007) on 'powering down'; Cavanagh (2004) on 'alternatives to economic globalization'; and Hopkins (2008) on 'transition towns'.

47 Grimshaw (2008) says that the issue of 'promise' arises from tension over the hype and expectations associated with new technologies, the issue of 'poverty' arises from tension over exclusion from the emerging knowledge-based global economy, the issue of 'price' arises from tension over the costs at which new technologies enter the marketplace and subsequent challenges to affordability, and the issue of 'power' arises from tension over control of technological agendas.

48 In developing my assessment criteria, local ownership and widespread distribution are assumed an automatic result, where desired, of decentralised and autonomous innovation that is governed in a participatory manner.

49 Described as a technology that has "...a firm footing in advanced research and an extremely wide applicability across industries and sectors" (Einsiedel and Goldenberg, 2004, p. 28)

Chapter 4

Nanotechnology, Development and Inequity

In this chapter I will review the literature surrounding nano-technology, development and inequity in order to test the framework I have developed in the previous chapter and further establish a context for my assessment of nanotechnology.

4.1 Understanding Nanotechnology

From the outset, a great deal of literature highlights the importance of looking at how nanotechnology is understood. Understandings are said to establish the framework for engagement and debate (UNESCO, 2006; van Amerom and Ruivenkamp, 2006), determining "...what people will pay attention to, worry about, ignore or investigate" (UNESCO, 2006, p. 4). According to many (Lau in Mason, 2003; ETC Group, 2005c; Meridian Institute, 2006; van Amerom and Ruivenkamp, 2006; Bawa, 2007; Bowman and Hodge, 2007b), universal definitions, standards and a common language will be needed to ensure appropriate assessment of nanotechnology's implications, suitable regulation and the successful facilitation of international relations and trade. Conversely, a lack of common understandings, internationally, is seen as potentially disastrous (Grossman, 2008). As summarised by the ETC Group (2005c):

Nanotechnology and Global Equality
Donald Maclurcan
Copyright © 2012 by Pan Stanford Publishing Pte. Ltd.
www.panstanford.com

> A common description, terminology and measurement for nano-
> scale materials will have a major impact on trade in commodities
> (e.g. carbon nanotubes), international norms for nano-patent
> regimes, technology transfer, liability and labeling as well as
> international agreements and national regulations relating to
> control or safety-testing of nanomaterials (p. 48).

In this respect, authors often assume universal understandings about what nanotechnology is and is not, despite evidence of definitional differences. Some (see, for example, Roco, 2001; Palmberg, 2007) present, as commonly accepted, the U.S. National Nanotechnology Initiative's (NNI) definition (presented in Chapter 1).[1] In this light, nanotechnology is said to have three universally acceptable characteristics: the size range; the ability to measure and restructure matter; and the ability to exploit properties and functions specific to the nanoscale (Roco, 2007).

However, Roco (*ibid.*) also acknowledges that nanotechnology has "no globally recognised definition" (p. 3.2), and research suggests that, in the North, mixed definitions are already resulting in hype, ambiguity and disagreement (Nature Nanotechnology, 2006; UNESCO, 2006). Internationally, the use of different nanotechnology definitions has been obvious:

> Definitions vary around the world, depending on national
> strengths. China and Korea emphasise the focus on materials
> and especially electronics, while researchers in Africa and Latin
> America often emphasize the materials in the context of medicine
> and environmental science (UNESCO, 2006, p. 5).

Flexibility in defining nanotechnology is seen as a deliberate move by advocates and opponents to serve respective causes (Drexler, 2004; 2006; Selin, 2007). According to Selin (2007), this is a critical means by which groups create legitimacy: "the story of the rhetorical development of nanotechnology reveals how speculative claims are powerful constructions that create legitimacy in this emerging technological domain" (p. 196).

Presenting fertile ground for ambiguity is the fundamental clash of nanotechnology paradigms between the speculative: 'advanced nanotechnology', focussed on Feynman's original vision of broad control utilising nanomachines at the level of individual atoms; and 'near-term nanotechnology', focussed on present-day applications and an expanded version of the NNI's vision that is said to include anything smaller than microtechnology (Peterson, 2003).

Despite the meaning of the word 'nanotechnology' shifting away from Feynman's vision within the Northern discourse (Drexler, 2004), the literature about nanotechnology and the South suffers from confusion in two main ways. Firstly, whilst most writing presents near-term nanotechnology as mainstream, there are instances where advanced nanotechnology is presented as 'the reality' for the South. Bruns (2004), for example, sees nanotechnology presenting answers for global poverty through a future of 'accessible abundance'. Similarly, an early international report by the Meridian Institute (2004) refers to nanotechnology's potential to "build anything out of anything" (p. 5). Furthermore, a briefing document for the 2003 United Nations Industrial Development Organisation Expert Group Meeting on Nanotechnology, refers to the ability for nanotechnology to address medical, energy and environmental challenges via "… factories operating at the nanometer level, including nanoscale conveyor belts and robotic arms bringing molecular parts together precisely…" (United Nations Industrial Development Organization, 2003, p. 2). Similarly, in terms of nanotechnology's impacts, Al'Afghani (2006) focusses on the need for future environmental laws in the South to incorporate "…mechanisms for licensing, supervision and control of emissions and disposal methods for both MNT [molecular nanotechnology] products and nanofactories".

The second source of confusion arises from those in the North who describe the benefits of near-term nanotechnology by drawing on the terminology and imagery of advanced nanotechnology without distinguishing between the two paradigms. One common way is to refer to recent nanotechnological innovation in areas of social development whilst presenting nanotechnology as part of a manufacturing revolution that will result in material abundance (see, for example, Choi, 2002; Samson and Symington, 2004; Barker *et al.*, 2005). The Association for Women's Rights in Development (in Samson and Symington, 2004), for example, highlight nanotechnology's near-term benefits for water purification, energy and medical treatments, whilst interweaving the language of 'nanobots' and visions of a world where "…many of the material dreams of humanity can be fulfilled" (p. 2). This rhetoric has also appeared in articles by Southern researchers, with one African author focussing on near-term nanotechnology despite opening by explaining nanotechnology as the process of "manipulating atoms and molecules into minuscule engines of creation,[2] maintenance and repair" (Etkind, 2006).

Whilst I remain open to accepting the reality of definitional ambiguity, my explorations with Thai and Australian key informants (Chapter 6) will attempt to shed light on how nanotechnology is understood across cultures. In the meantime, for the sake of clarity, I will avoid the literature in which nanotechnology's definition is compromised by reference or inference to advanced nanotechnology.[3]

4.2 The State of Play

Indicators of the global 'state of play' with respect to nanotechnology R&D have included studies on levels of national engagement and funding, as well as data outlining national publication and patent output. The first notable review of global engagement with nanoinnovation was a study by Siegel *et al.* (1999a), in which national nanotechnology activities were outlined in 22 countries, including China, India, South Korea, Singapore, Russia, Belarus, the Ukraine and Georgia. Five years later, another study reported that the number of countries having adopted national projects or programs had grown to 40 (Huang *et al.*, 2004). Simultaneously, Court *et al.*'s (2004) study, whilst highlighting the "surprising amount of nanotechnology activity" within the South, reviewed the level of engagement demonstrated by 10 Southern countries. Categorising nations in one of three ways (see Table 4.1), they reported that: as 'front-runners', China, India and South Korea had established government-funded, national nanotechnology activities; as 'middle-ground', Brazil, Chile, Philippines, South Africa and Thailand had some form of government support, with national funding programs under development; and as 'up-and-comers', Mexico and Argentina had some form of organised nanotechnology activity but no specific government funding.

Table 4.1   Selected Developing Countries and Their Nanotechnology Activity (Adapted from Singer *et al.*, 2005, p. 60).

Front Runner	China India South Korea	- National government funding program - Nanotechnology patents - Commercial products on the market or in development
Middle Ground	Brazil Chile Philippines South Africa Thailand	- Development of national government funding program - Some form of existing government support (e.g., research grants) - Limited industry involvement - Numerous research institutions
Up-and-Comer	Argentina Mexico	- Organized government funding not yet established - Industry not yet involved - Research groups funded through various science and technology institutions

Whilst funding levels are difficult to analyse (Tealdi cited in UNESCO, 2003),[4] a number of studies have attempted to attribute dollar figures to national nanotechnology initiatives. In this light, Lux Research (cited in ETC Group, 2008) reports that, as of 2006, global government spending on nanotechnology[5] was almost evenly split between Europe ($2.1 billion), North America ($1.8 billion) and Asia ($1.7 billion), leaving the 'rest of the world' well behind (see Table 4.2).

Table 4.2   Global Government Funding* for Nanotechnology, 1997–2006 (Adapted from Lux Research cited in ETC Group, 2008, p. 10).

	1997	1998	1999	2000	2001	2002	2003	2004	2005	2006
North America	116	191	256	339	512	753	1044	1579	1750	1836
Europe	128	128	152	203	532	657	766	925	1897	2073
Asia	133	162	204	359	656	1028	1265	1551	1510	1671
Rest of World	—	—	—	30	61	68	77	101	170	204
Total	377	482	613	932	1761	2507	3152	4156	5327	5785

* Figures are in $Million.

Whilst there are big differences in funding levels between the North and most of the South, there are also big differences within the South. Here, funding can be loosely grouped into three tiers. The top tier incorporates: Russia, with its plans to invest $1.1 billion in equipment for nanotechnology research from 2007–2010 (The Associated Press, 2007); South Korea, where projected funding from 2003–2007 was $1 billion (Choi, 2002); and China, where 2008 funding is about $180 million per year (Bai, 2008). The second tier includes: India, where a five-year national strategy for advancing nanoscience and nanotechnology will cost $220 million (*ibid.*); South Africa, where the national nanotechnology strategy budget over three years is approximately $60 million (Department of Science and Technology, 2006); and Brazil, that had planned to invest $30 million by 2006 (Lemie, 2005). On the bottom tier are countries including Malaysia and Thailand, each investing between $2–5 million per year (see Changsorn, 2004; Hamdan, 2005).

As with funding figures, research into publication output shows a strong concentration amongst a limited number of countries, with the top seven nations reportedly producing around 70% of the global scientific papers on nanotechnology (Compañó and Hullman, 2002). According to Compañó and Hullman's (*ibid.*) study of nanotechnology publication data (see Table 4.3), as of 2003, the U.S. led Japan and Germany in publication output. Russia is also prominent, with its high output in nanotechnology publications said to demonstrate the "...relatively strong significance of nanoscience in their research systems" (p. 245). Perhaps of most interest, however, is fourth-placed China, with over 6% of global nanotechnology publications (*ibid.*). Whilst not substantiating this figure, Hassan (2005) makes the startling claim that, "in 2004, scientists in China published more articles on nanoscience and nanotechnology in international peer-reviewed science journals than scientists in the United States" (p. 65). In support, one 2006 report (see Zhou and Leydesdorff, 2006) does comprehensively show that China has emerged as a major player in nanotechnology.

Highlighting a recurring trend, early analyses of patent distribution (see, for example, Compañó and Hullman, 2002; Marinova and McAleer, 2002, 2003; Huang *et al.*, 2004) have shown a concentration, yet internally uneven spread, of ownership amongst a select group of countries. In Huang *et al.*'s (2004) analysis of patenting with the USPTO (see Table 4.4), the U.S. holds a commanding lead (66.7%), followed by Germany (8.7%), Japan (7.5%), France (2.7%)

and South Korea (2.4%), with a small number of patents held by Venezuela and fast growth said to be occurring in China. However, as with publication data, the case of China presents wildly varying evidence, with one 2003 source (see Xinhua News Agency, 2003) claiming China was ranked third in overall nanotechnology patenting, behind the U.S. and Japan, whilst others (Marinova and McAleer, 2002) claim there is no evidence between 1975 and 2000 of Chinese nanotechnology patenting on the USPTO.[6] India, on the other hand, is seen as performing below expectations, considered to be trailing world patenting trends by approximately six years (Sastry, cited in Patil, 2005).

Table 4.3   Global Nanotechnology Publications by Nationality, 1997–1999 (Adapted from Compañó and Hullman, 2002, p. 245).

Rank	Publications	(%)	Rank	Publications	(%)
1	U.S.	23.7	9	Switzerland	2.3
2	Japan	12.5	10	Spain	2.1
3	Germany	10.7	11	Canada	1.8
4	China	6.3	12	South Korea	1.8
5	France	6.3	13	Netherlands	1.6
6	United Kingdom	5.4	14	India	1.4
7	Russia	4.6	15	Sweden	1.4
8	Italy	2.6			

Table 4.4   Top 20 Patent Assignees, by Country, for Nanotechnology Patents Registered with USPTO, 2003 (Adapted from Huang *et al.*, 2004, p. 331).

Rank	Country	No# of Patents	Rank	Country	No# of Patents
1	U.S.	1011	11	United Kingdom	11
2	Germany	132	12	Belgium	10
3	Japan	115	13	Sweden	9
4	France	41	14	Australia	8
5	South Korea	37	15	Italy	8
6	Canada	36	16	China	7
7	Taiwan[7]	26	17	Ireland	5
8	Netherlands	19	18	Singapore	5
9	Switzerland	15	19	Venezuela	4
10	Israel	13	20	Denmark	3

Of particular interest to authors, such as the ETC Group (2003a), is that, as compared with biotechnology patenting, many of the MNCs are engaging right from 'the beginning' in nanotechnology. In 2005, using a more restrictive USPTO classification for nanotechnology patents,[8] the ETC Group (2005b) published research about the disaggregation of nanotechnology patenting[9] with the USPTO (see Table 4.5). Reflecting results in earlier studies (see, for example, Compañó and Hullman, 2002; 2003), they found that almost 90% of patents are held by the private sector, with the majority of corporations being U.S., then Japanese and then German. Of the 30% of patents assigned to companies outside the U.S., representation from the South extended only to Venezuela (ETC Group, 2005b).

Table 4.5 Top Nanotechnology Patent Assignees with the USPTO, by Nationality and Sector, 2005 (Adapted from ETC Group, 2005b, p. 9).

Company/Institution	Nationality of Primary Ownership	Sector	Patents Issued
Canon Kabushiki Kaisha	Japan	Private	49
International Business Machines Corporation	U.S.	Private	47
Silverbrook Research	Australia	Private	28
U.S. Government	U.S.	Government	16
Hitachi, Ltd.	Japan	Private	16
Seagate Technology	U.S.	Private	16
Micron Technology, Inc.	U.S.	Private	14
Eastman Kodak Company	U.S.	Private	13
Olympus Optical Co., Ltd.	Japan	Private	10
University of California	U.S.	University	10
Rohm and Haas Company	Germany	Private	9
Polaroid Corporation	U.S.	Private	9
Sony Corporation	Japan	Private	8
Molecular Imaging Corporation	U.S.	Private	8

Consequently, companies predominantly from the U.S., Japan and South Korea, as well as universities from the U.S., Taiwan, Japan and the U.K., hold patents for key nanotechnology materials, including carbon nanotubes, quantum dots, dendrimers and scanning probe microscopes, such as the STM and AFM (ETC Group, 2005b). Given the latter is "a fundamental tool indispensable to the development of nanotechnology" (p. 30), the ETC Group express particular concern that this technology remains largely controlled by the companies IBM and Veeco.

4.3 Interpreting the State of Play: Instrumentalist versus Contextualist Perspectives

Although both sides of the debate claim, as their vision, a more equitable world based on justice and fairness (UNESCO, 2006), the debate about whether nanotechnology's emergence offers hope for a more equitable world is highly polemical (Wood *et al.*, 2003; Munshi *et al.*, 2007; Invernizzi *et al.*, 2008). According to Invernizzi, Foladori and Maclurcan (2008):

> The different positions on the role that nanotechnology can play in alleviating poverty, or in promoting development, reflect particular interpretations on the relationship between science, technology and society (p. 124).

In this light, they say that it is useful to organise positions "...under a theoretical framework" (*ibid.* p. 124). Drawing on Feenberg (1991), Invernizzi *et al.* (2008) therefore classify perspectives and interpretations into 'instrumentalist' and 'contextualist' positions.[10]

The instrumentalist view presents a reductionist, 'mechanical' vision of the relationship between science and society. From this viewpoint, poverty and social problems are largely blamed on a lack of technical capabilities. With emerging technologies seen as inevitable and necessary to overcome inequality, strong Southern engagement is seen as an imperative. Whilst globalisation opens up new opportunities for prosperity, endogenous barriers are seen as holding the South back. If the South can become more technologically competitive, this will lead to growth that will mechanically trickle-down and deliver social outcomes. From this perspective, emerging technology is critical to generating competitiveness as well as solving poverty and other social problems; if a problem can be identified correctly, then the application of a suitable technology is all that is required for the problem to be solved. Paradigm-shifting revolutions present the assurance of superior, new technologies over existing alternatives. Furthermore, technologies are neutral and can be transferred, unproblematically, through a one-size-fits-all approach. Social concerns can be addressed as a by-product of market development, whilst risks can be managed, through further research. Finally, decision-making is largely a matter for expert consultation with public engagement merely a necessary means to ensure smooth adoption.

On the other hand, the contextualist view presents a holistic vision of the relationship between science and society. From this viewpoint, poverty and social problems are part of a complex web of socioeconomic trends involving systemic inequities at the global, national and local levels. Engagement with emerging technology is not seen as inevitable or as an imperative, nor is it seen as necessarily desirable. Whichever way, in its current form, globalisation is viewed as reinforcing the technology-inequality gap. From this perspective, it is largely exogenous obstacles that restrict Southern R&D and access to the fruits of technological development. Market entry and increasing competitiveness do not guarantee reduced internal inequalities and can mask a lack of social development. Technology alone does not solve sociopolitical problems, but rather reflects them. The technology-push, especially one that privileges a single, technological trajectory at the expense of alternative trajectories that could be more context-friendly, is largely considered detrimental. This contextualist assessment is founded upon the historical experience of increasing inequity resulting from hyped emerging technologies, given an orientation away from Southern needs, driven by the corporate bottom-line. In this sense, appropriate applications will likely be inaccessible. Furthermore, technologies are socially conditioned, not neutral, embodying and reinforcing the social structures, relations, political power and values in which they are created. To have a positive effect, the global system must accommodate autonomous Southern technological trajectories that reflect local needs. In this sense, the course of emerging technologies will be socially beneficial only if public engagement is taken seriously; and debates expand beyond those relating to scientific risk. A more democratic governance of technology would see power drawn from 'the experts' and restored to the people. Finally, given each country's differing needs and levels of scientific capacity, technologies will need to be produced, used and adapted with the specific, local socioeconomic context in mind.

Whilst these two positions appear polar opposites, as I will show in the coming sections there is sometimes common ground when it comes to engaging with nanotechnology. Furthermore, it is not easy, nor perhaps useful, to always classify authors as fitting into one of these worldviews, given some draw from aspects of each (see, for example, Meridian Institute, 2004; UNESCO, 2006; Schummer, 2007).

In the remainder of this chapter, I will explore how the two interpretive frameworks play out in terms of debates about nanotechnology and development. The discussion is organised around my three themes for technological assessment: innovative capacity, technological appropriateness and approaches to governance. For each category I will review the common ground between perspectives before exploring the distinct sides of each argument. As contextualist positions are often a direct counter to instrumentalist positions, I will present the instrumentalist positions first.

4.4 Innovative Capacity

From the outset, there is mutual concern for some kind of 'nanodivide' (for the most comprehensive commentaries, see Barker *et al.*, 2005; Hassan, 2005). Most authors speak of a divide between the North and South, largely referring to differences in technological capacity (see Choi, 2002; Court *et al.*, 2004; Leahy, 2004; Sawahel, 2008). Yet for Hassan (2005), given the rise of certain countries from the South, such a dichotomy is less helpful:

> Today, the environment for research and development in nanoscience and nanotechnology in Brazil, China, India and South Africa bears closer resemblance to the research environment in Europe, Japan, and the United States than it does, for instance, to the research environment in the Dominican Republic, Laos, or Rwanda (p. 66).

In this light, others (Hassan, 2005; Meridian Institute, 2007) stress the divide between Southern countries. Here Hassan (2005) notes:

> ...there is a disturbing emergence of a South-South gap in capabilities between scientifically proficient countries (Brazil, China, India, and Mexico, for example) and scientifically lagging countries, many of which are located in sub-Saharan Africa and in the Islamic world (p. 65).

Some also acknowledge a divide, albeit of a different nature, within each 'nanotechnology-proficient' Southern country (Meridian Institute, 2007), with the UNESCO (2006) suggesting that "inequalities of access to [nanotechnology] research may be greater within nations, than between them" (p. 13).

However, there is also some agreement about nanotechnology's potential to be a "profitable industry for countries in the South" (Daar in Leahy, 2004), stimulating the ability to add value to basic commodities and goods (ETC Group, 2005c; Meridian Institute, 2007). Here, the Meridian Institute (2007) sees potential for the South to "engage in a number of new markets for novel nano-enhanced materials and production processes" (p. 3), whilst others (Sakar in Coupe, 2004; Singh, 2005a) speak of supporting 'cottage industries[11]' by using nanotechnology to build business based on traditional knowledge.

There is also some agreement in the literature on the limited nature of Southern engagement in global nanotechnology dialogues (Court *et al.*, 2004; ETC Group, 2004a; Barker *et al.*, 2005).

4.4.1 Instrumentalist Approaches to Innovative Capacity

From the largely optimistic, instrumentalist viewpoint, underdevelopment, as represented by poverty, remains largely unsolved due to a lack of technical capabilities, particularly those in emerging areas such as nanotechnology (see, for example, Juma and Yee-Cheong, 2005; Salamanca-Buentello *et al.*, 2005). Given a belief in nanotechnology's inevitability, the primary modes of engagement with nanotechnology R&D, for example as a resource provider, manufacturer, producer or consumer, are therefore said to determine a country's ability to address its development prospects (Schummer in UNESCO, 2005).

To avoid dependence and prevent technological inequality, it is argued that strong Southern engagement with nanotechnology is an imperative (Dayrit and Enriquez, 2001; Yonas and Picraux, 2001; Mnyusiwalla *et al.*, 2003; Onah in Malsch, 2008), with Hassan (2005) firmly stating: "developing countries have no choice but to embrace nanoscience and nanotechnology if they hope to build successful economies in the long term". Similarly, the then Zimbabwean Deputy Minister of Science and Technology Development, Patrick Zhuwawo, in 2006 said::

> Zimbabwe, like the rest of the world, cannot afford to ignore the nanotechnology and nanoscience revolution and hope to succeed in this highly competitive global village (Zhuwawo in Grimshaw *et al.*, 2006, p. 21).

In support, Onah (in Malsch, 2008) believes that "...not jumping on the nanotechnology bandwagon will have negative effects for developing countries, such as technological poverty, and could increase the brain drain". In this light, many governments are said to be focussing efforts on nanotechnology out of recognition for lost opportunities at the dawn of earlier technologies, such as the Human Genome Project, information and communications technology and biotechnology (Roco, 2001). For example, the sentiment from the Vietnamese National Assembly is that, as the country has not developed a biotechnology industry, it therefore needs to launch a nanotechnology programme immediately (Vu Long, 2004). Similar fears are held in Argentina, India, Mexico, South Africa, Zimbabwe, Sri Lanka and Malaysia (see Puig de Stubrin in Sametband, 2005; Revaprasadu in Etkind, 2006; Jarjis in Malaysian National News Agency, 2006a; Ramachandran, 2006; Amaranthunga in Warushamana, 2007), best demonstrated by the incredible case of Mexico, where there has been parliamentary support for a "National Emergency Program for investment in research and teaching of nanotechnology" (Foladori, 2006). Other key players are not exempt from such concerns, with China declaring nanotechnology a "critical R&D priority" in 2001 (Zhou and Leydesdorff, 2006, p. 15). The necessity for engagement is compounded by the paradigmatic-shift to a global knowledge economy. Whilst the knowledge-based model is already central to China's nanotechnology strategy (*ibid.*), others, such as the Brazilian President, frame nanotechnology as critical to a country's ability to "export knowledge" (Lula in Lemie, 2005).

There is also a strong desire to reduce international dependencies (see Rao in Press Trust of India, 2006), such as those relating to international commodities (see, for example, the case of Russia in The Associated Press, 2007). As Waruingi and Njoroge (2008) summarise, with respect to Africa:

> The question is, are we forever going to rely on developed nations to always unlock our rich natural resources and then sell them back to Africa at elevated prices? If the answer to this question is no, then it is about time for educational institutions, governmental organisation and research and development institutes to step up and develop capacity for understanding the nanotechnology revolution.

In terms of Southern engagement more broadly, there is a strong belief that both globalisation and nanotechnology open up new opportunities for R&D engagement, presenting an opportunity to 'catch up', with "qualitatively different development paths for some of the developing economies, enabling some regions to 'leapfrog' their way to leadership" (Henderson, 2002, p. 1).

Reasons for this optimism include advantages inherent in nanotechnology, comparative advantages held by Southern countries and advantages stemming from globalisation. In terms of technological advantages, instrumentalist authors distinguish nanotechnology's traits from those of the large-scale production and energy-intensive technologies that emerged from the first industrial revolution as well as the biotechnological and digital revolutions (Coupe, 2004; Burgi and Pradeep, 2006). Such authors emphasise nanotechnology's minimal requirements with respect to land use and the limited need for specialist skills, given the extent to which nanotechnology complements whatever scientific competencies exist within a country (Burgi and Pradeep, 2006). Furthermore, nanotechnology is said to present an attractive field for Southern R&D, as it requires modest industrial infrastructure (Cascio, 2004) and material requirements (Salvarezza, 2003; Shahani cited in Corporate Bureau, 2004; El Naschie, 2006). Nanotechnology is seen as relying little on economies of scale (Coupe, 2004), therefore needing little start-up capital (Salvarezza, 2003; Coupe, 2004; El Naschie, 2006), with Coupe (2004) claiming: "the investment required to produce niche, high-value products that give added functionality, is relatively small" (p. 33). Others [Welland, cited in Mantell, 2003; Shahani, cited in Corporate Bureau, 2004; Burgi and Pradeep, 2006) highlight the field of nanobiotechnology as particularly promising in terms of its limited start-up demands.

However, the reported expenses associated with establishing national nanotechnology initiatives in the South present conflicting evidence. Whilst the cost of establishing nanotechnology institutes has been claimed at approximately $5 million in Mexico, Sri Lanka and Vietnam (see Rao, cited in Ministério Das Relaçõs Exteriores, 2003; Thao, 2004; Hong, 2006; Perera, 2006), the national nanotechnology facility in Costa Rica, including a 'clean room[12]', was reportedly built for "about $50,000[13]" (Vargas, 2004). Creating similar confusion are claims made about the cost of an AFM, with Rao (in Patil, 2005) believing it to be approximately $1.5 million, whilst the ETC Group (2003a) puts this figure at $175,000. Malsch (2008) sheds some light on these conflicts, saying that the investment, as well as the minimum critical mass of researchers required for entry into nanotechnology R&D, depends upon the area of nanotechnology seeking to be explored. However, in apparent contrast to the evidence from Costa Rica, Malsch elaborates by saying that "...laboratories with clean rooms and high-tech instrumentation are much more expensive than the basic laboratories needed for other nano-based materials and devices".

In terms of Southern comparative advantage, authors highlight the South's ability to access natural resources, critical to nanotechnology (Schummer in Dayrit and Enriquez, 2001; UNESCO, 2005; Waruingi and Njoroge, 2008). Also seen as useful is the ability to easily build a critical mass through the availability of low-waged, highly educated workers who are eager to innovate (Bai, 2008; Waruingi and Njoroge, 2008).

In terms of advantages stemming from globalisation, the UNESCO (2006) says that nanotechnology can build on the information and communications technology revolution, with researchers "much more likely to have ready access to publications via the Internet" (p. 13). Furthermore, unlike earlier megatrends in science and technology, the U.S. does not have a commanding lead in nanotechnology R&D (Roco, 2002; President's Council of Advisors on Science and Technology, 2005). Similarly, the vast range of nanotechnology applications means no one country or region has a monopoly on the cutting-edge research capabilities necessary to apply nanotechnology to every part of every industry (Haworth cited in Roco, 2002; Changsorn, 2004). Therefore, "...nanotechnology stands to be a much more international scientific project than, for instance, research into biotechnology was in the 1980s and 1990s" (UNESCO, 2006, p. 13), with countries "...competing on a more equal basis for a slice of the action" (Watanabe, 2003, p. 478).

Instrumentalists, such as Court *et al.* (2004), say that these claims are supported by evidence from the South of a "surprising amount of nanotechnology activity". Of particular note is the emergence of China, a country "...poised to play a major — and in some cases, world-leading role in the development and implementation of nanotechnology" (Asian Technology Information Program, 2006, p. 3). As of 2005, more than 50 universities and 20 institutes within the Chinese Academy of Sciences had engaged in nanoscience and nanotechnology R&D, involving over 3000 researchers from around the country (Bai, 2005). Furthermore, there are reports that China "...appears to be leading the world in sheer numbers of new nanotechnology companies" (Choi, 2002, p. 345), with over 300 industry enterprises focussed on nanoscience and nanotechnology R&D in 2003 (Bai, 2005), growing to over 600 by 2008 (Bai, 2008). Similarly, India is touted as "...likely to become a leader in nanotechnology within the next five to ten years" (Pillai cited in Staff Reporter, 2005), with more than 30 institutions involved in nanotechnology research and training programs, as of 2004 (Dwivedi, 2004) and the Indian government hoping to capture 5% of the global nanotechnology market share by 2016 (Kalam, 2006). As another significant player from the South, Brazil has approximately 300 Ph.D.-level researchers working in nanotechnology (Leite, 2004) and, whilst "there were only 10 SPMs in Brazil in 1995...there are over 80 of those instruments operating in the country today" (Andrade in UNESCO, 2005, p. 14). For South Africa, reports from 2005 claim approximately 12 universities, 4 science councils and several companies active in nanotechnology R&D (Maruping, cited in Barker *et al.*, 2005).

Whilst nanotechnology engagement in China, India, Brazil and South Africa is to be expected, what Hassan (2005) finds extremely encouraging is that "smaller and poorer developing countries have also decided that this represents a strategic investment in future economic and social well-being that they cannot afford to ignore" (p. 65). For example, the Vietnamese Ministry of Science and Technology launched a nanoscience and nanotechnology infrastructure building program from 2004–2006 (Liu, 2004). At least six groups were working on nanotechnology in the Philippines in 2003 (Lee-Chua, 2003) whilst, as of 2005, Malaysia had six existing centres engaging in nanotechnology research (Hamdan, 2005).

Even more surprising are claims that nanotechnology has existed in the South for some time. For example, it is believed that African "... nanotechnology projects led to the discovery of chemistry during the Egyptian civilisation" (Waruingi and Njoroge, 2008). Similarly, many traditional Chinese medicines are now known to have contained metal nanoparticles (Huaizhi and Yuantao, 2001). Used for millennia in India, bhasmas — ayurvedic traditional medicines resulting from the combination of metals with herbal extracts — are nanoparticulate (Express News Service, 2005; Kumar, Nair, Reddy and Garg, 2006). Likewise, kajal — a cosmetic with which Indian women adorn their eyes — actually consists of inexpensively created carbon nanotubes (Singh, 2005a). But Ramachandran (2006) carefully notes that such examples are "...definitely not within the new paradigm of nanotechnology that has emerged in recent years..." (p. 113).

In terms of engagement around the new nanotechnology paradigm, with SPMs having stimulated initial Chinese interest in the 1980s (Bai, 2005), China claims to have had nationally run activities since 1990 (see Bai, 2001; Hsiao and Fong, 2004). Similarly, Vietnam is said to have commenced nanotechnology research in 1992 (Viet Nam News Agency, 2004), whilst nanotechnology research is believed to have been the focus of some African laboratories for up to 30 years (Peters and Page, 2003). The case of India provides some elucidation on the kind of nanotechnology with which Southern countries may have been engaged:

> In the late 1970s the Tata Institute of Fundamental Research (TIFR) of the Department of Atomic Energy (DAE) was carrying out studies in the application of fine-grained nano-crystalline materials in microwave and piezoelectric devices. In the 1980s, the Indian Institute of Technology (IIT) at Kharagpur was synthesising ceramic oxide nanoparticles. The researchers attempted industrial application of magnesium and aluminium oxide nanopowders in the cement industry (Ramachandran, 2006, p. 113).

Further evidence of the South acting as nano-innovators, albeit in an interdependent innovative framework, comes from early signs of bi- or tri-lateral international partnerships involving the U.S. and the E.U. and countries such as Vietnam, Costa Rica, India, Mexico, Argentina, Chile, China, and South Africa (see Waga, 2002; CORDIS, 2004; European Commission, 2004; Vargas, 2004; Foladori, 2006). Arrangements have included public-private partnerships, such as an agreement by U.S. company Lucent Technologies to develop products arising from state-funded nanotechnology R&D in Argentina (Sametband, 2005). The World Bank has also engaged in nanotechnology partnerships with the South, having provided $250,000 in 2003 for a 'nanoscience and technology observatory' as part of the Brazilian Millennium Institute in Nanotechnology (The Royal Society and Royal Academy of Engineering, 2004). Partnerships also cut across the South, such as a joint effort between scientists in Mexico and India to develop organic nanoherbicides (Roach, 2006), and tri-lateral efforts between science ministers from Brazil, India, and South Africa to identify areas for nanotechnology cooperation relating to HIV/AIDS prevention and treatment (Juma and Yee-Cheong, 2005). Virtual South-South partnerships have also emerged, such as the cooperative, online-running nanotechnology centre, launched by the Brazilian and Argentinean governments in 2005 in order to "develop joint projects, raise human resources capacity, create interchange grants for researchers and organise activities" (Almeida, 2005). Simultaneously, regional partnerships in the Asia Pacific, such as the Asia Nano Forum,[14] include China, India, Thailand, Indonesia, Malaysia and Vietnam, whilst the Asia Pacific Nanotechnology Forum[15] supports opportunities for Asian countries to collaborate and has already held international nanotechnology meetings on human resources development as well as environmental protection and pollution. Nanotechnology even appears to be a useful medium for breaking down international barriers, as in the case of North and South Korea that, in 2003, held a joint conference on nanotechnology to "exchange results of their research and discuss ways to cooperate in improving education on both sides...[anticipating] joint research opportunities in the near future" (Arirang News, 2003).

Reflecting upon all these advantages, Burgi and Pradeep (2006) state:

> Nano[technology] has the incomparable force to pervade all societies and economies, from the pre-industrial to knowledge societies, from ancestral to highly industrialized economies and is

not necessarily subjected to a nation's current development stage and/or geographical location (p. 647).

In this light, Hassan (2005) believes "nanoscience and nanotechnology may prove to be the first cutting-edge field to reflect the new realities of global science in the 21st century" (p. 65). He elaborates by saying that, over the last 20 years:

> ...[the numbers of] scientists and technologists from the developing world who choose to continue to work in their home countries are growing...[with many Southern countries, including LDCs] devising ever more sophisticated and effective science and technology policies (*ibid.* p. 65).

However, there is a wide disparity, in terms of the levels of Southern funding and support for nanotechnology, that has contributed to the belief that barriers to building and enhancing nanotechnology capabilities are largely endogenous (Court *et al.*, 2004; Singer *et al.*, 2005). In this light, Singer, Salamanca-Buentello and Daar (2005) say that "...the ultimate success of harnessing nanotechnology to improve global equity rests with developing countries themselves" (p. 62). Court *et al.* (2004) take a similar angle, saying that active participation on the part of Southern countries can deal with:

> ...displacement of traditional markets, the imposition of foreign values, the fear that technological advances will be extraneous to development needs, and the lack of resources to establish, monitor and enforce safety regulations.

Specific national barriers are said to include the capacity to house interdisciplinarity and cross-sectoral collaboration within research (Meridian Institute, 2007), as well as what some (Waga, 2002; Jarjis in Malaysian National News Agency, 2006a; UNESCO, 2006) see as inhibitive cost and infrastructural requirements for nanotechnology R&D, particularly when it comes to equipment (Dayrit and Enriquez, 2001; Meridian Institute, 2007). Furthermore, according to the Meridian Institute (2004), pre-existing infrastructural gaps in biotechnology will translate into infrastructural gaps in nanotechnology. In this light, some, such as Gold (in Sawahel, 2008), see nanotechnology as 'beyond' certain countries in the South; believing "few middle- and lower-income countries...have the capacity themselves to advance nanotechnology in a significant way". Tegart (2001) agrees, claiming that "the magnitude of investments will mean that all but the very largest economies will not be able to afford to have more than a handful of sites" (p. 20). Warungi and Njoroge (2008) add further support to this argument, believing that most African countries lack the physical infrastructure to support nanotechnology.

More generic barriers to innovation are said to include limited capacity to assemble and retain a critical mass[16] (Tegart, 2001; Salvarezza, 2003; Meridian Institute, 2004) and the difficulty of securing funding from risk-averse governments (Tegart, 2001; Meridian Institute, 2004). The latter is especially noted by Dayrit and Enriquez (2001) who say of nanotechnology:

> ...it faces the usual birth pains and expected comparisons with its very successful older siblings, biotechnology and ICT [information and communications technology]. In developing economies, securing support for the development of nanotechnology in competition with the more established fields can therefore become difficult (p. 1).

Fundamental questions also arise as to the ability of some countries to: support the creation of necessary government policy (Tegart, 2001; Meridian Institute, 2004); communicate new science to the private sector (Salvarezza, 2003); draft patents (Galembeck in UNESCO, 2003); and engage in public education (Meridian Institute, 2004). Endogenous challenges are seen as particularly heightened in those countries experiencing political instability (*ibid*.).

4.4.2   Contextualist Approaches to Innovative Capacity

To begin, the largely sceptical contextualist position rejects the developmentalist claim that the South must become nanotechnology-adapted, or remain underdeveloped (Shiva in CORDIS, 2003). Simultaneously, nanotechnology is seen as reinforcing the technology-inequality gap through a perpetuation and exacerbation of existing inequities and exploitation (Shiva in Thomas, 2003; Invernizzi and Foladori, 2005; ETC Group, 2008), masking ongoing "...oppression by industrialized nations" (Lovy, 2003). Here, some contextualists (Mooney, 1999; Shiva in Thomas, 2003) argue that nanotechnology's downfall is its emergence within an unjust system. Indeed, the ETC Group (2008) argues: "...if current trends continue, nanotech[nology] threatens to widen the gap between rich and poor and further consolidate economic power in the hands of multinational corporations" (p. 11).

Contextualists identify a number of different possibilities for Southern roles in global nano-innovation from those outlined by instrumentalists. Far from being exclusive, these possibilities may occur in tandem and can be closely linked. The South is firstly considered as a potential nanotechnology licensee, needing to pay fees for patented nanotechnology from the North, creating greater technological dependency (Salvarezza, 2003; Meridian Institute, 2006). Here the ETC Group (2005b) surmises:

> Researchers in the global South are likely to find that participation in the proprietary 'nanotech[nology] revolution' is highly restricted by patent tollbooths, obliging them to pay royalties and licensing fees to gain access (p. 19).

Foladori (2008) sees this as particularly likely for the LDCs:

> ...the reality is that LDCs will not own much of that technology: the structural challenges to ending poverty and inequality will remain, as the South works in service to the North, licensing patented technology from companies and governments from outside the region (p. 19).

The second possibility for Southern roles in global nano-innovation is as nanotechnology producers (Scott, 2003). In such a scenario, the exploitation of abundant labour combines with the extraction of raw materials for the development of nanotechnology-based products (Meridian Institute, 2004; Malsch, 2008). Taking the proposal that Latin American countries may act primarily as manufacturers for nanotechnology, Foladori *et al.* (2008) suggest that:

> ...the benefits will accrue to the south-north joint business partnerships, reproducing Latin America's long experience as the labour in service to multinational corporations and northern governments' brains (p. 18).

From this perspective it is argued that Northern companies will externalise production risks (Invernizzi *et al.*, 2008), with Foladori and Zayago Lau (2007) suggesting that MNCs are mainly attracted to Mexico's emerging high-tech parks due to "...the paucity of regulations and lax rules" (p. 224).

The third possibility for Southern roles in global nano-innovation sees the South as nanotechnology importers (Salvarezza, 2003).

Here, the South may be used as a 'spill-over' market and a dumping ground for unwanted products from the North, presenting dangerous potential to "...radically transform local economies" (ETC Group, 2003a, p. 9).

Closely linked, the fourth possibility for Southern roles in global nano-innovation is a fear that certain populations in the South will act as nanotechnology 'guinea pigs' (UNESCO, 2006; Schummer, 2007). Concern exists about the risks for trial populations as well as the transparency surrounding awareness of such risks (Meridian Institute, 2006), particularly in the development of nanotechnology-based antimicrobicides, aimed at preventing HIV/AIDS (Choi, 2002; ETC Group, 2008). In such situations, Schummer (2007) imagines exploitation could thrive "...because of lower wages, poorer regulations of human experiments, and less public attention to hazards" (p. 294).

The final possibility for Southern roles in global nano-innovation is that nanotechnology may threaten Southern commodity markets (see Shanahan, 2004; ETC Group, 2005c; Meridian Institute, 2006; Senjen, 2006; Foladori and Invernizzi, 2007). Here, the main concern is that new, nano-engineered materials provide industrial manufacturers with multiple raw material options, seriously challenging the existing markets of commodity exporting Southern countries (ETC Group, 2005c; Foladori and Invernizzi, 2007). According to the ETC Group (2005c):

> ...the very characteristics of nanotechnology that make it potentially suitable for developing countries also raise the possibility that is could displace commodities, labor and industries and worsen the position of developing countries (p. 38).

More specifically, Southern commodities may be under threat through circumstances in which "...copper wiring may be replaced by carbon nanotubes and platinum may be overtaken by a compound of nano-scale nickel and cobalt" (*ibid.* p. 5), affecting producing countries such as Chile, Indonesia, South Africa and Zimbabwe (*ibid.*). Further impacts are anticipated in areas including rubber in tyres, with aerogels and nanoparticles of clay both possible substitutes threatening the livelihood of workers in Thailand, Indonesia and Malaysia (*ibid.*), as well as the potential for tropical commodities, such as high quality cotton, to be replaced by cheaper raw materials including maize, oats and cotton leftovers (ETC Group, 2003a).[17] In terms of food, greater precision in design and lower production costs could enable the North to produce more food more economically, reducing dependence on cheap agricultural products from the South (ETC Group, 2004b). Such fears have been voiced in countries such as South Africa, whose government is worried that its country's national resources will be made redundant by "...cheaper, functionally rich and stronger materials" (ETC Group, 2005c, p. 27).

The elimination and migration of jobs, as well as decreasing future employment opportunities, are seen as key implications of nanotechnology-based commodity substitution (Galembeck in UNESCO, 2003; Grimshaw, 2004; ETC Group, 2005c; Foladori and Invernizzi, 2007; Invernizzi *et al.*, 2008). Impacts on labour are seen as particularly hurting the Southern poor and those most unable to adapt:

> Worker-displacement brought on by commodity-obsolescence or a drop in prices will hurt the poorest and most vulnerable, particularly those workers in the developing world who do not have the economic flexibility to respond to sudden demands for new skills or different raw materials (ETC Group, 2005c, p. 41).

In contrast to the instrumentalist position, contextualists argue that the major obstacles for Southern nanotechnology R&D, particularly for the LDCs, are exogenous (ETC Group, 2003a, 2005b; Invernizzi and Foladori, 2005). Of most prominence is the barrier posed by intellectual property rights. As I have already demonstrated in this chapter, research into global nanotechnology publication and patenting strength suggests the early concentration and privatisation of nanotechnology R&D, particularly for key nanomaterials and tools. Given the 'platform' nature of much nanoscience, there is fear such moves will block important Southern R&D (Mooney, 1999; ETC Group, 2005c; Meridian Institute, 2007; Invernizzi *et al.*, 2008). As I have already discussed, royalties and licence payments are seen as the residual obstacles arising from intellectual property rights held by the North.

Authors also identify trade barriers to the South developing nanotechnology R&D capacity (see Meridian Institute, 2004). Added to this is the threat of an exacerbated brain drain (Waga, 2002; Invernizzi and Foladori, 2005; Meridian Institute, 2007),[18] compounded by an existing and further projected global shortfall of 'nanotechnologists' (Ratner and Ratner, 2002; Harper, 2003a; Watanabe, 2003).[19]

On this note, North-South nanotechnology partnerships are also questioned in terms of whom they will really benefit (Meridian Institute, 2005), considering the limitations of the North's mainstream basis for measuring partnership outcomes (Stilgoe in Malsch, 2008) and the migratory lure from South to North historically associated with transnational public-private partnerships (Foladori *et al.*, 2008).

In summary, Foladori *et al.* (2008) believe that, for a number of countries in the South:

> ...[the] head-start by the developed countries, combined with LDCs challenges in infrastructure and workers with the appropriate skill base to support an emerging nanotechnology industry, appears nearly insurmountable over the long term (p. 18).

4.5   Technological Appropriateness

In terms of the appropriateness of nanotechnologies, both the instrumentalist and contextualist positions readily identify exciting technological potentials (see, for example, Juma and Yee-Cheong, 2005; Foladori *et al.*, 2008), with Foladori and Invernizzi (2007) noting that contextualists are "...not opposed to nanotechnologies in principle". Despite significant differences around questions of desirability, utility and outcomes, to be explored below, there is agreement that nanotechnologies could foreseeably contribute to efficiency gains across a number of sectors (Juma and Yee-Cheong, 2005; Department of Science and Technology, 2006; El Naschie, 2006; Meridian Institute, 2007; Scrinis and Lyons, 2007). In the words of the Director General of the National Agency for Science and Engineering Infrastructure in Nigeria, nanotechnologies offer:

> ...a vastly improved manufacturing process...greatly improved efficiency in almost every facet of life...better built, longer lasting, cleaner, safer and smarter products for the home, for communication, for medicine, for transportation, for agriculture, and for industry in general (Adewoye in Business Day Media, 2007).

From another angle, Scrinis and Lyons (2007) acknowledge the potential to integrate nanotechnologies into alternative agrifood practices and systems of production:

The organic agriculture and food industries, for example, may support the application of nanotechnologies, especially those that have the potential to enhance sustainable farming practices — for example by reducing chemical and water use (p. 34).

Even one of the most critical contextualist perspectives has this to say:

> ETC Group acknowledges that in a just and judicious context, nanotech[nologies] could bring useful advances that might benefit the poor (the fields of sustainable energy, clean water and clean production appear promising; applications to food and agriculture appear less so) (ETC Group, 2004b).

However, both instrumentalist and contextualist perspectives also recognise that nanotechnologies may present hazards[20] and exposure[21] risks to both humans and the environment (UNESCO, 2006). In terms of hazards, one of the most comprehensive literature reviews concludes:

> Recent studies examining the toxicity of engineered nanomaterials in cell cultures and animals have shown that size, surface area, surface chemistry, solubility and possibly shape all play a role in determining the potential for engineered nanomaterials to cause harm (Maynard, Aitken, Butz, Colvin, Donaldson, Oberdörster, Philbert, Ryan, Seaton, Stone, Tinkle, Tran, Walker and Warheit, 2006, p. 267).

In this light, some claim that there is greater toxicity when particles are scaled down (see, for example, Howard in Smith and Wakeford, 2003; The Royal Society and Royal Academy of Engineering, 2004). For instance, per given mass, Oberdörster (2004) reports an increase in pulmonary inflammatory response from ultrafine particles, as compared to larger particles, with such a phenomenon raising subsequent concerns for respiratory morbidity (see Feder, 2003). Consequently, comparisons are made to asbestosis,[22] given: "the physical characteristics of carbon and other nanotubes mean that they may have toxic properties similar to those of asbestos fibres" (The Royal Society and Royal Academy of Engineering, 2004, p. ix). Of particular concern are the potential health risks associated with nanoparticulate titanium dioxide (Sass, Simms and Negin, 2006),[23] cadmium-selenium[24] (Galembeck in UNESCO, 2003; Sass *et al.*,

2006) and carbon nanotubes (The Royal Society and Royal Academy of Engineering, 2004).

In terms of exposure, primary concerns surround the selectivity and generic ability for nanoparticles to enter certain parts of the body, such as the lungs and the foetus (Howard in Smith and Wakeford, 2003; Sass *et al.*, 2006), and across membranes, particularly those of the skin, cell and blood-brain barrier (see, for example, Haberzettl, 2002; Malsch, 2002a; Cass in Chung, 2003; Feder, 2003; Mnyusiwalla *et al.*, 2003; Howard in Smith and Wakeford, 2003; Oberdorster *et al.*, 2004). As the ETC Group (2008) notes:

> ...the very same properties that make engineering nanoparticles so attractive for the development of targeted drug delivery systems — namely, their mobility in the bloodstream and ability to penetrate cell membranes — could also be qualities that make them dangerous (p. 17).

Environmental impacts of exposure to nanoparticles are also a concern, with the Meridian Institute (2006) referring to potential dangers associated with "...nanoparticle release and exposure risks during production and use; spent waste and used cartridge disposal and nanoparticle concentrations in output water" (p. 2).

Of final note in the debates about both hazards and exposure are the critical findings of the comprehensive report written by the U.K. RS&RAE (2004) in which it was stated:

> Many nanotechnologies pose no new risks to health and almost all the concerns relate to the potential impacts of deliberately manufactured nanoparticles and nanotubes that are free rather than fixed to or within a material (p. ix).

As will become evident in the coming sections of this chapter, there is limited literature addressing the ethical issues nanotechnologies might raise for Southern populations. However, both instrumentalists and contextualists acknowledge the importance of considering such matters (see Tegart, 2001; Choi, 2002; Mnyusiwalla *et al.*, 2003; UNESCO, 2006). There is also agreement that, given nanotechnology is a platform technology, most risks and ELSI have an unusual potential to cut across multiple areas of application (Meridian Institute, 2007).

In this light, views about the novelty of issues raised by nanotechnologies are not easily distinguishable across instrumentalist and contextualist thinking. However, from my review of the literature, a common pattern emerges. Most perspectives (see, for example, Choi, 2002; Galembeck, 2003; Shiva, 2003; Meridian Institute, 2004; UNESCO, 2006; Bowman, 2007) generally agree that the typology of issues is largely common to other technologies, such as biotechnologies. However, with respect to these issues, nanotechnologies are also said to present a number of unique opportunities and challenges (Meridian Institute, 2007), with the properties that make nanotechnology novel seen to have the reciprocal ability to cause the most concern. In this light, Hodge and Bowman (2004) claim that:

> From an international perspective, nanotechnology appears to be significantly different to past technologies such as biotechnology and pharmaceuticals to warrant separate consideration of these dimensions (p. 30).

The Meridian Institute (2007) explains this more nuanced view of nanotechnology's implications as follows:

> While these issues may be generally applicable to technologies, the unique characteristics of nanotechnology may result in different considerations regarding each cross-cutting issue...[that] could, in turn, require new and different strategies for addressing these issues (p. 7).

4.5.1 Instrumentalist Approaches to Technological Appropriateness

From the instrumentalist perspective, nanotechnology offers a revolutionary paradigm-shift for both science and manufacturing, that may well prove disruptive[25] (Tegart, 2001; Yonas and Picraux, 2001; Meridian Institute, 2007), presenting new avenues to strengthen and supersede older technologies (Juma and Yee-Cheong, 2005).[26] In this light, Crow and Sarawitz (2001) note that governments, futurists and techno-pundits all promise nanotechnology's ability to "remake our world" (p. 61).

Nanotechnologies are therefore seen as able to 'solve' social problems within the South (see, for example, Juma and Yee-Cheong, 2005; Kalam, 2006; Malsch, 2008; Potoaçnik and Ezin, 2008). According to Ratner and Ratner (2002), some even go so far as to refer to nanotechnology as a 'panacea', with Singh believing it is "... the one-stop solution for everything" (2005b) and Swaminathan (2002) talking of nanotechnologies "...opening up uncommon opportunities for converting the goals of food, health, literacy and work for all into reality". Already, the UNESCO (2001) has identified nanotechnology as one of the areas presenting the greatest potential for Africa. For Mnyusiwalla (2003), this makes sense, given the belief that nanotechnology, "...were it to develop in the way it ought, might ultimately be of most value to the poor and sick in the developing world" (p. R11).

The certainty with which instrumentalist arguments are presented covers the potential for nanotechnologies to promote both social and economic development within a sustainable development paradigm (see Court *et al.*, 2004; Barker *et al.*, 2005; Juma and Yee-Cheong, 2005; El Naschie, 2006). Within nanotechnology, applications are said to be distinguishable in terms of their 'social' or 'industrial' focus (Department of Science and Technology, 2006).[27] In this respect, instrumentalist perspectives on nanotechnology's relevance for the South focus on social applications, with the South African Department of Science and Technology (2006) claiming nanotechnology 'solutions' as necessary in order to "...successfully assuage unemployment, poverty and underdevelopment, racial and gender inequities, among others" (p. 3).

However, many social applications are also said to hold promise for economic development (see Court *et al.*, 2004). Salamanca-Buentello (in Small Times, 2005), for example, is sure that "if you encourage the use of nanotechnology and education at all levels of science and technology, you are bound to increase the income of a certain country". Here, Invernizzi *et al.* (2008) believe that "the extreme instrumental positions substitute technology policy for social policy...[and] the traditional 'linear model of innovation' prevails..." (pp. 134–135). In this sense, improving competitiveness through innovation is seen as translating into "...greater development in both economic and social terms...[whereby] development and satisfaction of social needs emerge automatically from competitiveness improvement" (Invernizzi and Foladori, 2006, p. 115).

Pre-empting criticism, some say that nanotechnology is misconceived as entirely high-tech and engaging only high-profile scientists in complicated projects unrelated to the common man (Harper, 2003a; Mathur in Singh, 2005b). The Sri Lankan

National Science Foundation (2002) continues by arguing that, of the nanotechnology that is high-tech, much of it actually offers applications appropriate for use throughout the world. Similarly, Singer *et al.* (2005) claim that the call to focus on 'more pressing needs' lacks foresight:

> Some will argue that the focus on cutting-edge developments in nanotechnology is misplaced when developing countries have yet to acquire more mature technologies and are still struggling to meet basic needs such as food and water availability. This is a short-sighted view (p. 64).

However, the caution is added that nanotechnology should be part of a comprehensive technological approach, whereby "all available strategies, from the simplest to the most complex, should be pursued simultaneously" (*ibid.* p. 63).

Across the instrumentalist approaches to appropriateness, nanotechnology is viewed as beneficial: holding inherent, built-in advantages such as the ability to create significant employment (Sakar in Singh, 2005a; ASSOCHAM cited in Ramachandran, 2006) and, at worst, as neutral: with a belief that "there is nothing intrinsically good or bad about nanotechnology, it all depends on how it is used" (Choi, 2002, p. 353). As Singer *et al.* (2005) note:

> Each new wave of science and technology innovation has the potential to expand or reduce the inequities between industrialized and developing countries in health, food, water, energy, and other development parameters (p. 64).

This utilitarian line of thinking is supported by Cascio (2005), who suggests that "the issue isn't whether nanoscience will be of value to development, it's whether it will be applied in a way to benefit the greatest number of people". Juma (2005) shows how such an argument is extended in the case of nanotechnology's potential risks:

> Desirable properties of nanomaterials, such as high surface reactivity and the ability to cross cell membranes, could potentially have negative consequences if these technologies were used inappropriately (p. 74).

In this sense, nanotechnology's transfer between different sociocultural settings is viewed as unproblematic, demonstrating the tendency by instrumentalists to "...homogenize poverty issues and contexts, offering the same 'one-size-fits-all' technical solution to very different ecological, social and cultural contexts" (Invernizzi *et al.*, 2008, p. 134).

   With respect to identifying applications and perceived benefits in greater detail, Singer *et al.*'s (2005) Delphi study[30] drew on 63 experts from both the North and South[29] to identify and rank the 10 nanotechnology applications seen as most likely to benefit[30] the South in the 2003–2013 period. These 10, chosen applications (see Table 4.6) can be synthesised into five areas,[31] each commonly explored in the literature (see, for example, Dayrit and Enriquez, 2001; Tegart, 2001; Court *et al.*, 2004; Barker *et al.*, 2005). In descending order of popularity, according to Singer *et al.*'s (2005) study, the areas include: energy and the environment; agriculture; water; healthcare; and, to a lesser degree, construction (for the full table, see *Appendix D*).

Table 4.6   Top 10 Applications of Nanotechnology for Developing Countries, Ranked by Score (Adapted from Salamanca-Buentello *et al.*, 2005, p. 385).

Rank	Area of Application	Score*
1	Energy storage, production and conversion	766
2	Agricultural productivity enhancement	706
3	Water treatment and remediation	682
4	Disease diagnosis and screening	606
5	Drug delivery systems	558
6	Food processing and storage	472
7	Air pollution and remediation	410
8	Construction	366
9	Health monitoring	321
10	Vector and pest detection and control	258

* The maximum total score an application could receive was 819.

   Hopes for nanotechnology's applications across these areas are commonly quantified by linking nanotechnology potentialities with the MDGs (see Court *et al.*, 2004; Barker *et al.*, 2005; Hassan, 2005). Singer *et al.*'s (2005) study also proposes direct correlations between the 'top 10 applications of nanotechnology for the

South' and various MDGs (see *Appendix D*). Here there are high expectations that nanotechnology can "...reduce the cost and increase the likelihood of attaining the Millennium Development Goals" (UNCTAD, 2004), contributing to "...our ability to achieve these goals in an unprecedented way" (Hanekom in Bello, 2007, p. 3).

I will now explore applications across the top five areas, as noted in Singer *et al.*'s (2005) study.

4.5.1.1 Energy and the environment

In terms of applications for energy and the environment, the literature focusses on three main areas. Firstly, in the area of energy storage, production and conversion, it is suggested that photo- and thermo-chemical nanocatalysts can be used to generate hydrogen from water at low cost (Juma and Yee-Cheong, 2005). Also envisaged are novel hydrogen storage systems based on carbon nanotubes and other lightweight nanomaterials (*ibid.*). Combined, these could lead to a new generation of batteries and supercapacitors, useful for things such as efficient hydrogen-powered vehicles (Esteban, Webersik, Leary and Thompson-Pomeroy, 2008). Additionally, flexible photovoltaic cells and organic light-emitting devices could be based on quantum dots and carbon nanotubes within composite film coatings (Singer *et al.*, 2005). The latter are also said to offer "...strong, flexible conduits for electricity distribution networks" (Juma and Yee-Cheong, 2005, p. 73).

Secondly, in terms of air pollution and remediation, nanocatalysts are believed to offer more efficient, cheaper and better-controlled catalytic converters (Singer *et al.*, 2005). Moreover, titanium dioxide nanoparticle-based photocatalysis offers to degrade air pollutants through self-cleaning systems (Juma and Yee-Cheong, 2005), whilst it is claimed that magnetic nanoparticles can be used to remove oil and other organic pollutants from aqueous environments (Singer *et al.*, 2005). Nanodevices are also seen as able to absorb and separate toxic gases, whilst nanosensors will be able to detect toxic materials and leaks (Juma and Yee-Cheong, 2005; Singer *et al.*, 2005).

Thirdly, in the area of biodiversity and ecosystem management, nanotechnology is seen as enriching variety via new strains of species that exhibit novel properties when interacting with their environs (Juma and Yee-Cheong, 2005). According to Singer (2005), the increasing ability to create a database of the information underlying the planet's biodiversity, via faster nanotechnology-based DNA sequencing, will also help facilitate sensible ecosystem management.

Beyond applications, nanotechnology's appropriateness for energy and environmental needs in the South is explained in a number of ways. Most dominant is the belief that nanotechnology can contribute significantly to mitigating climate change (Esteban *et al.*, 2008). According to Singer *et al.* (2005), nanostructured materials, as the basis for solar and fuels cells as well as novel hydrogen storage systems, "...promise to deliver clean energy solutions" (p. 58). For example, "...in 10 to 15 years, projections indicate that nanotechnology-based lighting advances have the potential to reduce worldwide consumption of energy by more than 10%" (Roco and Bainbridge, 2001, p. 4). Given the ability to dramatically increase energy efficiency (Roco and Bainbridge, 2001; Harper, 2002), Singer *et al.* (2005) suggest that there is scope to de-couple development from environmental degradation. They continue, suggesting the dual benefit inherent in moving to energy self-sufficiency is the 'safety-proofing' of countries against an energy crisis as well as "...simultaneously reducing dependence on non-renewable, contaminating energy sources such as fossil fuels" (p. 58).

Instrumentalists continue by arguing that energy and environmental applications of nanotechnology are highly relevant to local communities in the South, with Potoaçnik and Ezin (2008) venturing that "energy production and storage, along with the creation of alternative fuels, is one of the areas where nanotechnology applications are most likely to benefit Nepalese people". Improvements in rural quality of life are envisaged through applications such as nanotechnology-enabled light emitting diodes for illumination (Juma and Yee-Cheong, 2005; Rajvanshi, 2006). Added to utility benefits are the advantages of significantly decreased costs, particularly in an area such as photovoltaics (Singer *et al.*, 2005; Esteban *et al.*, 2008), as well as the expectation that "applications will be robust and easily maintained and serviced" (Juma and Yee-Cheong, 2005, p. 73).

Finally, the instrumentalist perspective highlights flow-on effects from developing nanotechnologies for energy and the environment such as reduced air-pollution, leading to better human health (Roco and Bainbridge, 2001).

4.5.1.2   Agriculture

In terms of agricultural applications, the literature focusses on three main areas. Firstly, in the area of agricultural productivity, it is envisaged that zeolite nanoparticles, of differing pore size, can offer slow-release and thorough dosages of water and fertilisers for plants and efficient delivery of nutrients and drugs for livestock (Singer *et al.*, 2005). Similarly, nanocapsules are said to be able to release herbicides, in a slowly controlled manner, increasing payload efficacy (*ibid.*). Nanotechnology can also help develop a range of inexpensive applications to increase soil fertility and crop production (Roco and Bainbridge, 2001; Juma and Yee-Cheong, 2005). What is more, nanosensors are proposed as a useful tool to measure soil quality and monitor plant health, whilst 'nanomagnets' present new opportunities for the removal of soil contaminants (Singer *et al.*, 2005).

Secondly, in terms of food processing and storage, nanotechnology-based antigen-detecting biosensors could be used to identify pathogen contamination, whilst antimicrobial nanoemulsions could be used to decontaminate food equipment, packaging and food itself (*ibid.*).

Thirdly, in terms of vector and pest detection and control, nanosensors could be used for monitoring crop health and could also be applied to the skin of livestock or sprayed on crops to detect the presence of pathogens (Juma and Yee-Cheong, 2005). Furthermore, nanoparticles could be used for the creation of new pesticides, insecticides and insect repellents (Singer *et al.*, 2005).

Beyond applications, nanotechnology's appropriateness for agricultural needs in the South is explained in a number of ways. By increasing soil fertility and crop productivity, it is claimed that nanotechnology could "help eliminate malnutrition" (*ibid.* p. 58), thereby reducing childhood mortality (Juma and Yee-Cheong, 2005). In this light, the Indian President[32] sees nanotechnology ushering in a "second green revolution" (Kalam in Economy Bureau, 2006).

Finally, the Meridian Institute (2005) adds that it sees particular appropriateness for the South in the potential for nanotechnologies to "...make food products cheaper and production more efficient and more sustainable through using less water and chemicals" (p. 7).

4.5.1.3   Water

In terms of applications for water, the literature focusses on nanotechnology's potential to assist treatment and remediation. Here it is believed that "nanotechnology promises new or improved solutions to challenging obstacles to providing clean water" (Meridian Institute, 2006, p. 2), with nanosensors able to detect contaminants and pathogens such as arsenic, fluoride and nitrates in water supplies (Juma and Yee-Cheong, 2005). Furthermore, 'intelligent' nanomembranes, nanoclays, nanoporous zeolites and polymers, magnetic nanoparticles and attapulgite clays are said to be able to purify, detoxify and desalinate water more efficiently than conventional bacterial and viral filters (Juma and Yee-Cheong, 2005; Singer *et al.*, 2005). Similarly, titanium dioxide and iron nanoparticles can catalytically degrade water pollutants, with the by-products of remediation, such as toxic metal ions, able to be transformed into useful inorganic nanomaterials (Juma and Yee-Cheong, 2005).

The appropriateness of nanotechnologies for water needs in the South is largely explained in terms of technological utility. Advances in water technology are seen as leading to improved human health through the maintenance of a safe water supply, with subsequent reductions expected in water-related diseases, such as diarrhoea, cholera, typhoid and schistosomiasis (*ibid.*).[33] The ability to remove toxins, such as arsenic, from groundwater (Potoaçnik and Ezin, 2008) facilitates "...the use of heavily polluted and heavily salinated water for drinking, sanitation, and irrigation" (Juma and Yee-Cheong, 2005, p. 73). With this in mind, Sagman (in Wootliff, 2003) boldly claims that nanotechnology will lead to the global alleviation of water scarcity.

Finally, in terms of appropriateness to community settings, nanotechnology-based systems for water treatment and remediation are seen as likely to be inexpensive, portable and easily cleaned (Juma and Yee-Cheong, 2005; Singer *et al.*, 2005).

4.5.1.4   Healthcare

In terms of health-related applications, the literature focusses on the potential for nanotechnologies to offer new rapid, accurate and timely methods for disease diagnosis and screening as well as drug delivery (Salvarezza, 2003; Juma and Yee-Cheong, 2005). In terms of disease diagnosis and screening, Juma and Yee-Cheong (2005) present possibilities for nanotechnology-based, microfluidic lab-on-a-chip systems and nanosensors using quantum dots, magnetic nanoparticles, nanowires, antibody-dendrimer conjugates and carbon nanotube-based arrays. They add that atomic wires and 'nanobelts' can also be used to detect diseases such as cancer, since these nanomaterials are capable of revealing specific malignant agents, through changes in their electronic transport characteristics (*ibid.*). Fluorescent semiconductor nanoparticles are also useful as medical image enhancers, offering significant diagnostic advantages over conventional fluorescent dyes (*ibid.*).

In terms of drug delivery, some (Barker *et al.*, 2005; Juma and Yee-Cheong, 2005) envisage improved treatments with existing drugs, via nanocapsules,[34] liposomes, dendrimers, buckyballs, 'nanobiomagnets'[35] and attapulgite clays for slow, selective sustained-release systems. Using encapsulation methods, drugs can be protected, whilst ensuring delivery and release when needed (Juma and Yee-Cheong, 2005). According to others (Mnyusiwalla *et al.*, 2003; Salvarezza, 2003) such methods of delivery have the added benefit of being safer.

Other health-related opportunities for nanotechnologies are said to lie with regenerative medicine, nanoscale surgery and more durable medical prosthetics, utilising 'nanoceramics' (Juma and Yee-Cheong, 2005).[36]

Beyond applications, the appropriateness of nanotechnologies for healthcare needs in the South is explained in a number of ways. In terms of utility, health-related nanotechnologies are seen as "...especially promising, particularly for diagnostic tools, drug and vaccine delivery, surgical devices, and prosthetics" (*ibid.* p. 72). Prominently mentioned are HIV diagnosis and slow-release mechanisms considered "... especially useful for drug regimens that are long and complex, such as those used to treat tuberculosis" (*ibid.* p. 72).

Instrumentalists consider health-related nanotechnologies particularly appropriate for non-technical use in rural areas. In this respect, Harper (2003a) argues that nanotechnology-based pulmonary or epidermal drug delivery applications "...have the potential to free up the large numbers of trained medical personnel who are currently engaged in administering drugs via hypodermic needles". Furthermore, in countries without adequate drug storage capabilities and distribution networks, slow-release drug delivery mechanisms are highly valuable, obviating the need for regulated administration of medication (Barker *et al.*, 2005; Juma and Yee-Cheong, 2005). Areas with limited refrigeration will also benefit from nanotechnology "...improving shelf-life, thermo-stability and resistance to changes in humidity of existing medications" (Juma and Yee-Cheong, 2005, p. 72).

Finally, affordability resurfaces as a measure of appropriateness, with Juma and Yee-Cheong (2005) suggesting that nanotechnologies offer "...affordable methods of diagnosis and prevention...[such as] relatively inexpensive sensors in local clinics, using diagnostic kits" (p. 72). Ratner and Ratner (2002) go so far as to propose that disease screening could be inexpensive enough to be "...comprehensive even in low-income countries...as soon as the next two to three years" (p. 106). One reason provided for such optimism is the potential to reduce transportation costs through previously mentioned enhancements to the shelf-life and thermo-stability of medications (Juma and Yee-Cheong, 2005).

4.5.1.5 Construction

Although less discussed in the instrumentalist literature, 'construction' is the final area proposed by Singer *et al.*'s (2005) study. Here it is believed that nanostructures could make asphalt and concrete more robust to water seepage, nanomaterials could block ultraviolet and infrared radiation, whilst bioactive coatings could offer 'self-cleaning' surfaces (Singer *et al.*, 2005).

Beyond applications, the appropriateness of nanotechnologies for construction needs in the South is explained in a number of ways. In addition to improving environmental sustainability (Juma and Yee-Cheong, 2005) and the potential for increased production capacity (Meridian Institute, 2007), 'cost' is again prominent, with inexpensive nanomaterials seen to support cost-effective building and construction (Wiltzius and Klabunde, 1999; Singer *et al.*, 2005).

Finally, according to Juma and Yee-Cheong (2005), the development of low-cost, durable building materials could mean the provision of better living and educational environments, potentially reducing child mortality, improving maternal health and helping to meet the MDG of universal primary education.

Overall, within the instrumentalist perspective one can recognise the mechanical view that, if a problem can be identified correctly, only a form of nanotechnology need be applied for the problem to be solved (Invernizzi and Foladori, 2005).[37] A case in point is Potoačnik and Ezin's (2008) argument that nanotechnologies will help farmers to "...increase the agricultural production, thereby increase their income levels and improve the health of the people by decreasing malnutrition". Foladori (2006) notes that a similar logic exists amongst almost all of the Latin American countries that believe, "...by quickening the pace of nanoscience and nanotechnology, there will be an improvement in a country's competitiveness". Zimbabwe's Deputy Minister of Science and Technology Development[38] (Zhuwawo in Grimshaw *et al.*, 2006) shows a comparable train of thought, stating: "...this technology will now usher in a new era of economic growth through enhanced innovation and increased productivity..." (pp. 20–21).

4.5.2 Contextualist Approaches to Technological Appropriateness

In response to instrumentalist claims about the benefits and appropriateness of nanotechnologies, contextualist authors present various critiques, with some (Gould, 2005; Miller and Senjen, 2008) suggesting that the detriments stemming from nanotechnologies will likely outweigh their benefits. As Gould (2005) succinctly notes:

> Given the context of the current global political economy (which promotes corporate profitability as a central value, economic growth as the overriding social goal, and competitive advantage over appropriate caution), and the lack of understanding of (and thus, ability to mitigate) the ecological consequences of nanotechnological developments, the social and environmental costs of such developments are likely to outweigh the benefits promised by the individuals and institutions involved in their production (p. 16).

According to many (Mulvaney in Thomas, 2003; The Royal Society and Royal Academy of Engineering, 2004; Invernizzi and Foladori, 2006; Grimshaw, 2008) nanotechnologies are the latest to be presented as a 'technical fix' or 'silver bullet'. Here the ETC Group (2003a) is concerned that, "rather than confront the underlying problems of over-consumption and waste, industry could see Atomtechnology as a means to 'medicate' a solution for the earth" (p. 30). It is argued that nanotechnology does little to effect change in socioeconomic structures or redress base causes of existing inequities (Invernizzi and Foladori 2005; Scrinis and Lyons, 2007). In this light, the ETC Group (2003a) declare that the promises of nanotechnologies sound familiar:

> The hype surrounding nano-scale technologies today is eerily reminiscent of early promises in biotech[nologies]. This time we're told that nano[technologies] will eradicate poverty by providing material goods (pollution free!) to all the world's people, cure disease, reverse global warming, extend life spans and solve the energy crisis (p. 8).

Scrinis and Lyons (2007) elucidate, in relation to agricultural biotechnology, saying that there are:

> ...strong similarities and continuities between genetic engineering and nanotechnology in regard to the types of agricultural practices, farming styles, patenting regimes, and corporate structures these technologies are being used to support and transform (p. 36).

Furthermore, the entry of Southern countries into global markets, and simultaneous growth in a country's 'nanotechnology competitiveness', can mask a lack of social development, with no guarantee of reductions in internal national inequalities (Invernizzi and Foladori, 2006). As Invernizzi and Foladori (2005) explain:

> ...even if large developing countries that could join the nanotechnology wave (such as China, India, Brazil, etc.) can produce nanoproducts that could eventually result in clean and cheap energy options, in clean drinking water or in greater agricultural yields, this does not mean that the poor majority will benefit... (p. 110).

In this light, contextualists conclude that nanotechnologies cannot solve essentially political and social problems, such as poverty and unsustainable growth, "...despite rosy predictions that nanotech[nologies] will provide a technical fix for hunger, disease and environmental security in the South" (ETC Group, 2005b, p. 5).

Moreover, many instrumentalist perspectives are said to be insufficiently aware of "...the broader problems of development" (Foladori *et al.*, 2008, p. 4), failing to consider historical trends and current barriers to technological access in the process of identifying nanotechnologies (Invernizzi and Foladori, 2005). As a general critique, those pushing the 'nanotechnologies for the poor' or 'competitiveness' approaches are said to:

...overlook that the market is a barrier for wide sectors of the impoverished population to access the potential benefits of nanotechnology. They do not analyse the socio-economic starting point of this technological revolution, which is an extreme concentration of wealth and a huge gap between haves and have-nots. They do not understand that the main impulse for innovation is not the satisfaction of social needs but the drive for profit (*ibid.*, p. 116).

Commenting on Salamanca-Buentello *et al.*'s (2005) proposed 'top 10 nanotechnologies for developing countries', Invernizzi and Foladori (2005) say that "most of the examples used do not take into account the reality that the relationship between science and society is much more complex than identifying a technology and its potential benefits" (p. 105). Subsequently, Invernizzi and Foladori (*ibid.*) critique the belief that quantum dots could reduce the prevalence of HIV/AIDS through early-stage detection, by highlighting the challenges presented by the South African pharmaceutical 'patent wars'.[39]

Against instrumentalist claims, contextualists argue that Southern engagement with nanotechnology R&D appears largely geared towards inappropriateness with the trajectories and applications developed or described, to date, not those that will address the immediate problems facing 'the poor' in the South (Meridian Institute, 2005; Invernizzi *et al.*, 2008). Rather, most Southern countries[40] are adopting strategies entirely focussed on driving consumerism by creating a 'market push' for affluent purchasers, particularly those in the North (Barker *et al.*, 2005; Invernizzi *et al.*, 2008). Baya-Laffite and Joly (2008) shed the following light on this phenomenon: "as stated in most national initiatives for nanotechnology, the stated goal of participatory governance is to create a propitious environment for the successful development of nanotechnology" (p. 2). Similarly, as Malsch (2008) astutely notes, "so far, nanotechnology has been an area of 'technology push'...still mainly a solution looking for problems to solve".

Early nanotechnologies emerging from the South have therefore been export-oriented, building on existing international markets. In Thailand, for example, the initial focus has been on developing: waterproof, more durable silks; 'smart packaging' to monitor and maintain the state of food; more productive wine fermentation; 'self-sterilising' rubber gloves; and new car body materials (Changsorn, 2004). Early products in the Philippines have included semiconductors, diodes, lasers, computing and optical and high-speed electronics (Lee-Chua, 2003). In Vietnam, the planned research areas for the country's first two nanotechnology laboratories include applications for coal, solar batteries and optical electronics, all of which are expected to have large market prospects (Thao, 2004). And for Sri Lanka, initial nanotechnology research will focus on: "...industries such as apparel, rubber, ceramic, chemical products such as paints, activated carbon, mineral and herbal products" (Amaranthunga in Warushamana, 2007). Speaking of the Indian scenario, Rajvanshi[41] (in Padma, 2007) makes the bold claim that "there has been no effort to link the technology's potential with development in agriculture and addressing the needs of people in rural areas".

It follows, therefore, that the early focus of Southern governments is upon improving national corporate competitiveness, rather than developing local applications (Barker *et al.*, 2005). As an example, the head of the National Nanotechnology Centre in Thailand[42] (Tanthapanichakoon, 2005) says that his government: "...is determined to promote and accelerate nano science and technology as a crucial instrument of sustainable economic growth and international competitiveness" (p. 64). Almost identical sentiment is found in documents outlining the objectives of governments from Brazil, Colombia and Kenya (see Foladori, 2006; Waruingi and Njoroge, 2008). Furthermore, through an assessment of media news reports, Srivastava and Chowdhury (2008) show that nanotechnology debates and discussions in India have overwhelmingly focussed on nanotechnology's range of applications and promises for business.

Also key to early global developments has been a strong orientation towards military research, particularly in the U.S. (see Roco, 2003), where "many of the nanotech initiatives are being bankrolled by the U.S. Defence Advance Research Projects Agency..." (ETC Group, 2002, p. 3). Here, the ETC Group (2004b) remind us of nanotechnology's pervasiveness, suggesting agriculture may serve as "...the possible testing ground for technologies that can be adapted for surveillance and biowarfare" (p. 1). From the South, China is reportedly focussing on nanotechnology's potential military applications (Nemets, 2004). Moreover, 10% of Russian nanotechnologies are predicted to be developed for military purposes (Ivanov in The Associated Press, 2007), with new nanotechnology-based weapons expected to be "...designed in Russia within 15 years for combating radiation, chemical and biological terrorism..." (Fillipov in ITAR-TASS News Agency, 2007). Additionally, India and Malaysia have both expressed interest in nanotechnology's military applications, with prominent figureheads publicly noting nanotechnology's potential to revolutionise warfare technology (see Kalam in The Tribune, 2004; Malaysian National News Agency, 2006b).

As with endogenous Southern innovation, the promises for appropriate applications designed by the North, for and with the South, have proven largely hollow. Here, the Meridian Institute (2005) notes further similarities with biotechnology, in terms of the early orientation of applications:

> ...nanotechnology promises new cancer treatments, cheaper energy, and purer water, but the first products offered to the public have been more airtight tennis balls, transparent sunblock, and stain-resistant trousers (p. 13).

When it comes to social applications in an area such as healthcare, Howard (in Thomas, 2003) speaks of the 'cost' barrier as a challenge to a more equitable orientation of research:

> The vast cost of undertaking nanotech research will necessitate most of the effort going into profitable medicines — i.e. medicines for the lifestyle conditions of the rich over the life threatening illnesses afflicting much of the world's poor (p. 37).

Similarly, Invernizzi *et al.* (2008) claim that:

> Products such as personalized medicine, intelligent materials, human enhancement devices, supercomputers and other areas of nanotechnology research will be completely out of the reach of the poor (p. 136).

Indeed, from a contextualist perspective, the development of nanotechnologies is actively set to harm many in the South through the extension of large-scale, capital-intensive industrialisation and mass production (Scrinis and Lyons, 2007; Miller and Senjen, 2008). In the field of agriculture, Scrinis and Lyons (2007) imagine that farmers will be locked into new dependencies relating to seeds and chemical inputs. They envisage that "the 'efficiencies' and productivity gains of remote sensor farming, for example, may only be realised on large-sized, capital-intensive farms" (p. 33), with the resulting expectation that nanotechnology drives "...cost-effective mass-production of cheap and standardised food products" (p. 33). As Miller and Senjen (2008) summarise:

Nanotechnology appears likely to result in new pressures to globalise each sector of the agriculture and food system and to transport agricultural chemicals, seeds and farm inputs, unprocessed agricultural commodities and processed foods over even further distances at each stage in the production chain (p. 32).

Such developments are seen as leading to further appropriation and loss of local and traditional knowledge, skills and practices in areas such as food and farming, whilst greater automation, tied with the increasing capital costs associated with the development of nanotechnologies, threaten to reduce and displace farm labour and challenge the economic viability of small-scale agriculture (Scrinis and Lyons, 2007; Miller and Senjen, 2008). In this sense, the ETC Group (2003a) claim: "it is myopic and naive for Atomtech advocates to claim that a technology that the poor cannot control will somehow be used for their benefit" (p. 54). As an extension of modernisation's efforts, Scrinis and Lyons (2007) see these approaches falling within a 'nanocorporate' paradigm:

> To refer to a 'nano-corporate' paradigm is to both emphasise the dominance of the corporate economic form *per se* in the contemporary period, as well as the close interconnection between these respective technological and economic forms... there is a very strong sense in which nanotechnology — and other recent techno-scientific forms, such as genetic engineering — are *corporate technologies*...in the sense that corporations are using these technologies as one of their primary strategies for restructuring and extending their control of the agri-food system (p. 35).

The ETC Group (2003a) come to similar conclusions, pronouncing a 'new nano-economic order':

> Just as biotech[nologies] came to dominate the life sciences over the past two decades, ETC Group believes that nano-scale convergence will become the operative strategy for corporate control of commercial food, agriculture and health in the 21st Century (p. 9).

The market-based, nanocorporate paradigm is said to be further entrenched via broad-brush patenting (Barker *et al.*, 2005; Bawa, 2007),[43] leading to multisector techno-commodification through patents that span numerous industry sectors (ETC Group, 2005c). In this respect, nanotechnologies could accelerate the existing corporate takeover of the life sciences and associated development of patent monopolies (Mooney, 1999; UNESCO, 2006; Invernizzi *et al.*, 2008). As Becker (2001) notes:

> While some may question whether a company can patent an atomic structure, most legal experts agree that there are enough precedents involving biotechnology and genetic material to cover nanotechnology.

According to the ETC Group (2003a), nanotechnologies therefore contribute to the "...unprecedented potential for sweeping monopoly control of elements and processes that are fundamental to biological function and material resources" (p. 8). Furthermore, the lifespan of some existing patents may be extended just by scaling down from the micro- to the nanoscale (Foladori and Invernizzi, 2007), with Pandey (2005) claiming that this will be likely in India, where size considerations are enough to distinguish the technology from the prior art. Whilst Bowman and Hodge (2007b) carefully note that "not all nanotechnology applications may be protected, as Article 27(1) [of the TRIPs Agreement] provides patent protection only for inventions and not mere discoveries" (p. 31), they claim there has also been a "...blurring of the invention/discover interface" (p. 31).

Of considerable alarm to contextualists (Mooney, 1999; Invernizzi and Foladori, 2005; Senjen, 2006; Scrinis and Lyons, 2007) is that, in contrast to the early days of biotechnology, the big MNCs are engaging with patenting nanotechnologies from the outset. Furthermore, the market is consolidating, with nanomaterials produced by the global chemistry industry and dominated by multinationals (Cientifica, 2007). As Senjen (2006) explains:

> Nanotechnology will be the first 'platform' (or enabling) technology that is almost wholly owned by private interests...this is unusual as the basic building blocks of other major technologies (e.g. biotechnology, computer software, hardware) were initially all in the public domain (p. 34).

In fact, early MNC engagement has come across all industry sectors, with active companies including IBM, Canon, L'Oreal, Kodak, Procter and Gamble, Syngenta, BASF, DuPont and Dow Chemicals (*ibid.*). According to Kokini (cited in ETC Group, 2003a), "every major food corporation has a program in nanotech[nology] or is looking to develop one" (p. 39), with Kraft Foods having established its industry's first nanotechnology laboratory in 1999 (ETC Group, 2003a). The extent of corporate engagement is staggering, with Lux Research (2003] claiming that companies such as 3M, IBM and Hewlett Packard are allocating approximately one-third of their R&D budgets to nanotechnology.

Not surprisingly, says the ETC Group (2003a), the U.S. Navy and Army are at the forefront of nanotechnology patenting. Whilst the world's most prestigious universities are also staking a claim, the ETC Group believe that these developments, occurring 'on the back' of public research, are ultimately controlled by private spin-offs:

> ...also keeping with the biotech[nology] model, Atomtech is travelling on the backs of taxpayers...and public science...And, as ever, the profits will accrue to the elite academic entrepreneurs and the industrial giants that ultimately absorb the most promising start-ups (p. 41).

Following another contemporary trend, the ETC Group (2004a) find that certain companies, such as U.S.-based 'Nanosys', do not have any products but, rather, a portfolio of over 200 nanotechnology patents. Another example, 'C Sixty Inc' — a Canadian proprietary holder in fullerine and dendrimer technology — highlights the potential roadblocks created by such companies, with its Chief Executive Officer (Sagman cited in ETC Group, 2002) stating, "if people want to get in this game they have to deal with us" (pp. 2–3). Evident here is the 'patent land-grab', also witnessed with biotechnology (UNESCO, 2006). Critically for the South, this is "...a syndrome that will limit the number and types of products that may become public goods" (*ibid*. p. 13). Moreover, the TRIPs agreement obligates even the LDCs to enforce nanotechnology patents by 2006 (ETC Group, 2005b). As Bowman and Hodge (2007b) note:

> If a nanotechnology patent application satisfies the criteria of novelty, inventive step (or non-obviousness within the United States), utility and public disclosure, members of the WTO are prohibited from excluding it from patent protection under their domestic legal framework... (p. 31).

Beyond patenting, the appropriateness of nanotechnologies for Southern situations is questioned more deeply, with Invernizzi *et al.* (2008) claiming:

It remains uncertain as to whether a technology will work well in different contexts...critics recall past experiences with failed, universal technological 'solutions' in developing countries (pp. 136–137).

Here, issues of accessibility, particularly in terms of cost, remain central (Healy cited in The Royal Society and Royal Academy of Engineering, 2004; Grimshaw *et al.*, 2006; Invernizzi *et al.*, 2008), at least in the short-term:

> ...in the countries like India and China, where people have difficulty with basic daily needs, advanced high-tech medical diagnosis and treatment technologies with nano-devices will be a long time coming into popular use (Choi, 2002, p. 357).

As arose from one Zimbabwean dialogue on nanotechnology and water,[44] there is a belief that "Zimbabwe's water needs are unlikely to be met with nanotechnology until we can prove it is cost effective and sustainable" (Grimshaw in Majoni, 2006). On this point, Schummer (2007) notes that "...filters based on zeolites and ceramics, which are nowadays subsumed under nanotechnology, have been produced since many decades, without meeting the needs of developing countries" (p. 296).

Tied into questions surrounding costs and efficiency is a concern for job losses, via automation (*ibid.*). Here, Invernizzi and Foladori (2007) provide the example of lab-on-a-chip devices that, they claim, could make health workers obsolete.

The contextualist view also critiques the lack of cultural sensitivity embedded in the development of nanotechnologies. As noted by two South African scientists (Hillie and Hlophe, 2007) working on nanotechnologies for clean water:

> ...developing countries are at different levels of scientific advancement and have different priorities, so they cannot be subjected to a general prescription for solving water-related and other developmental problems (p. 664).

In this sense, Invernizzi and Foladori (2006) are concerned that nanotechnology-related rhetoric, debate and hype privileges a single, technological trajectory. Such an approach threatens to divert scarce resources and political will from context-friendly alternative trajectories, often involving less-expensive, more sustainable, 'low-tech' approaches (Scott, 2003; Mulvaney cited in Smith and Wakeford,

2003; The Royal Society and Royal Academy of Engineering, 2004; Invernizzi and Foladori, 2005). According to Grimshaw (2004), such exclusion would extend to "...proven technologies where there is capability in developing countries" (p. 13). In the case of healthcare, this could potentially translate into a situation in which:

> ...research into nanomedicine is not aimed at developing traditional, alternative or complementary medicines that are the basis of the health systems of possibly most of the world's population (Invernizzi and Foladori, 2006, p. 117).

Furthermore, as the dialogues on nanotechnology and water in Zimbabwe have shown, there remains ambiguity as to whether or not new technologies are needed to address existing problems (Grimshaw *et al.*, 2006). In support, Invernizzi and Foladori (2005) provide the example of malaria amelioration in China:

> Nanotechnology is not necessary to reduce malaria radically... in the Hunan Province of China, malaria was reduced by 99% between 1965 and 1990 as a result of social mobilization backed up by fumigation, the use mosquito nets and traditional medicine (p. 108).

Additionally, the grounding of many nanotechnologies in sophisticated processes is said to create an unhelpful knowledge barrier between various communities (Invernizzi and Foladori, 2006). Schummer (2007) elaborates:

> Apart from considerable cultural barriers, people need to build up and learn how to use basic electric facilities, including cables, switches, fuses, transformers, and rechargeable batteries, in addition to the electric devices for which the whole setting is built up. Nanotechnologies cannot contribute to that. They can perhaps improve the efficiency and price of solar cells by a few percentage points, or make solar cells smaller, more flexible, and transportable, which are humble contributions to the real problems...the real challenges are very basic and largely of educational and cultural nature (p. 297).

Overall, contextualists argue that nanotechnologies are not neutral, but socially conditioned, embodying social relations, political power and cultural values, with the overall trajectory associated with the development of nanotechnologies seen as "...

intrinsically limited in its ability to improve the living conditions of the underprivileged..." (Invernizzi *et al.*, 2008, p. 136).

Finally, in terms of uncertainties and risks, contextualists (Montague, 2004; UNESCO, 2006) reiterate the dangers of nanotechnologies, given the existence of many 'unknowns' for environmental and human health. Of particular concern are the unknowns surrounding ecological impacts[45] and, more generally, nanoparticle penetration, toxicity, accumulation and release (Malsch, 2002a; ETC Group, 2003a; The Royal Society and Royal Academy of Engineering, 2004; Sass et al., 2006).[46] Compounding these problems is a paucity of data (Malsch, 2002a; ETC Group, 2003a; Sass *et al.*, 2006), as well as the fact that "the state of research concerning... the behaviour of nano-particles is actually rather limited..." (Haum, Petschow and Steinfeldt, 2004, p. 2). The situation has been further compromised by "contradictory" evidence (*ibid.* p. 2), resulting in:

> ...no consensus[47] on whether nanoparticles or nanomaterials should be treated as something entirely new, or as a subset of existing materials, for the purposes of regulation or labelling (UNESCO, 2006, p. 16).

With these points in mind, researchers say it will be at least 10 years before they can give answers in an area such as the potential health and environmental toxicological effects of nanotechnologies (Feder, 2003).

In reflecting upon the novelty of potential risks, contextualists argue that the appropriateness of nanotechnologies is misrepresented by manipulative proponents who have adopted an advocacy stratagem of "same but different" (Paull and Lyons, 2008, p. 4). In this way, nanotechnologies are presented to patent offices and investors as different, whilst they are presented as the same as previous technologies to regulators (*ibid.*).[48] In response, contextualists argue that there "...may be inherent unpredictability and unmanageability associated with atomic and molecular level manipulations of nature" (Dupuy and Grinbaum cited in Scrinis and Lyons, 2007, p. 32).

As with risks to environmental and human health, a great deal of ambiguity exists about the ethical, legal and social implications arising from nanotechnologies. Of particular concern to the ETC Group (2004e), is the shift to 'atomically modified organisms' (AMOs). In 2004, the group reported the activities of researchers in Thailand who claimed to have atomically modified local rice varieties by "...drilling a nano-sized hole...through the wall and membrane of a rice cell in order to insert a nitrogen atom...to stimulate rearrangement of the rice's DNA" (*ibid.*), this 'mutation breeding' bypassing the process of genetic engineering. The reported aim of this research was to "...develop Jasmine varieties that can be grown all year long, with shorter stems and improved grain colour (*ibid.*). The ETC Group continues, saying that this case heralds an unknown, new frontier, whereby "nanobiotech[nology] takes agriculture from the battleground of GMOs to the brave new world of Atomically Modified Organisms". The response from local civil society groups to this research has been unfavourable, with the head of Biodiversity Action Thailand[49] (Lianchamroon in ETC Group, 2004e) saying:

> We don't consider atomically modified rice any safer or more socially acceptable than genetically modified rice...it sounds like the same high-tech approach that does not address our needs and could cause severe hardships for Thai rice farmers.

On the topic of nanotechnologies and their uses, whilst there is little Southern consideration for privacy concerns (Choi, 2002), there are serious concerns raised about nanotechnologies for military purposes (see, for example, Altmann and Gubrud, 2002; Altmann, 2004; UNESCO, 2006), and the potential for further uneven power relationships between the North and South (Choi, 2002). Already an alliance between India, Israel and the U.S. has been confirmed in the area of nanomaterials for electronic warfare systems (Political Bureau, 2004).

Whilst the fusion of information technologies, biotechnologies and nanotechnologies[50] is more the focus of the North (Choi, 2002), development debates have likewise touched upon the role of nanotechnologies in human enhancement. Here, one concern is that expensive nanotechnologies, directed towards human enhancement and targeting people with disability as well as others in the general population, will create new divides between the North and South (Gordijn in Meridian Institute, 2005; UNESCO, 2005), improving the lives of a few but making the lives of many others worse (Wolbring, 2006). The ETC group (2008) further cautions:

> ...in the world where 'enhancement' becomes an imperative, the rights of the disabled will be further eroded if disability is perceived as one more technological challenge rather than an issue of social justice (p. 56).

In this sense, others worry about privileging the 'medical' model ahead of the 'social' model of disability.[51] Here, Invernizzi and Foladori (2006) claim that the approach of nanomedicine is a "...reductionist and top-down medical approach..." outside the reach and worldview of 80% of the African population that uses traditional medicine for healthcare (p. 117). They add that "individualized medical treatment may be an advance in technical terms, but it is doubtful whether it is an advance in social terms" (*ibid.* p. 117).

Rounding-out the contextualist critique of the appropriateness of nanotechnologies, there is a suggested need for 'on-the-ground' engagement and awareness to consider potential risks and other implications, across cultures, given "differences in risks may be affected by differences in environments (natural and social) in developed and developing countries" (Meridian Institute, 2005, p. 10). Choi's (2002) study on nanotechnology in Asia also cautions that "people perceive ethics with different weight based on cultural background, people's belief systems, traditions...ethics has a lot to do with one's and the nation's moral philosophy" (p. 354).

4.6 Approaches to Governance

In terms of the commonality between instrumentalist and contextualist perspectives relating to approaches to nanotechnology's governance, there is agreement about the significant lag[52] in considering nanotechnology's ELSI (Mnyusiwalla *et al.*, 2003; Foladori and Invernizzi, 2007). As Choi's (2002) study of Asian countries suggests, this lag is even greater in the South where, despite annual increases in nanotechnology funding, many nations have not allocated significant portions of their budgets to assess ELSI, although some important developments have occurred.[53] Similarly, Foladori's (2006) research has shown that national nanotechnology initiatives in Latin America[54] are commonly characterised by a failure to consider nanotechnology's socioeconomic impacts or conduct studies into environmental health and safety (Foladori *et al.*, 2008). Following a review of the Indian nanotechnology situation, Srivastava and Chowdhury (2008) highlight the implications of this lag:

> There is such an emphasis on the production of nanotechnology that other aspects involved in commercial application — social-economic issues, risks, potential environmental and health

hazards, occupational safety etc. are being ignored. A small amount of risk and toxicity research is also being funded but they are confined to select institutes...and not yet part of mainstream nano[technology] related research...there is little possibility that the current regulatory mechanism will be equipped to address the worries associated with the application of emerging technologies like nanotechnology... (pp. 21–22).

Scott (2003) adds that there are few people in the South equipped to ask broad questions about nanotechnology's trajectory, "...let alone begin to answer them from their own perspective".

In this respect, the Meridian Institute (2005) claims that nanotechnology's risks "...are less well understood among developing world publics" (p. 10), with Southern countries having "...a hard time communicating technology risks in a way that facilitates public dialogue and decision-making" (*ibid.* p. 10). Similarly, Foladori's (2006) research has shown that the experience of nanotechnology in Latin America is commonly characterised by a failure to generate a process for widespread participation, with preference for discussions by a select group of scientists. According to some (Court *et al.*, 2004; UNESCO, 2006), this absence of widespread and cross-sectoral engagement creates fertile ground for a 'GMO-style' situation, with Court *et al.* (2004) explaining:

...as in the case of GM foods, lack of knowledge about the health and safety effects of nanotechnology can result in restrictions, outright bans and complex international conflict over production and transport of such materials (p. 12).

In this light, Scott (2003) claims that there is a need for:

...the UK and other industrialised countries to support awareness raising about nanotechnology amongst Southern policy makers, and to build capability in developing countries to engage in the debate about nanotechnologies and how they should be regulated internationally.

Furthermore, new and inclusive international approaches to nanotechnology's development are seen as necessary (Court *et al.*, 2004; ETC Group, 2004c). According to the ETC Group (2008), "the South, especially, needs a coherent U.N. approach to nanotechnology" (p. 61). Across instrumentalist and contextualist perspectives, an international body for governing nanotechnology's emergence is seen as essential to improving global outcomes. Court *et al.* (2004) envisage this as:

> ...[a] global focal point to commission and collect research results, promote awareness of the potential applications of NT [nanotechnology] for development, create new regulatory regimes (or build upon existing ones) for managing NT's [nanotechnology's] associated risks and promoting global public goods, provide a platform for constructive dialogue among all stakeholders — including representatives from government, industry, academe and citizens groups — and engage the voices of people in developing countries...and advocate for the interests of those in developing countries.

With somewhat more cautionary intent, the ETC Group (2005c) speaks of:

> ...an independent body that is dedicated to assessing major new technologies and providing an early warning/early listening system...to create a sociopolitical and scientific environment for the sound and timely evaluation of new technologies in a participatory and transparent process that supports societal understanding, encourages social and scientific innovation, and facilitates equitable benefit-sharing...[to] ensure the conservation of useful, conventional or culturally distinct technologies and, in particular, promote technological diversification and decentralization (p. 46).

Here, the UNESCO (2006) suggests itself as a body to facilitate the development of relevant international standards.[55]

4.6.1 Instrumentalist Approaches to Governance

With respect to nanotechnology's governance, instrumentalists focus on how to develop innovative capacity and manage risks. Leapfrogging aside, to best develop innovative capabilities, instrumentalists suggest the South should "...emulate the science and technology development model that allowed industrialized nations to become wealthy in the first place" (Salamanca-Buentello *et al.*, 2005).[56] The fundamental attributes of a successful nanotechnology strategy, say Hsiao and Fong (2004), should be the same in the U.S. as in China. Similarly, El Naschie (2006) presents the case of Israel as

"...a magnificent model for what a serious nanotechnology program should look like in all of the Middle East" (p. 772). In terms of Southern research, the President of the Chinese Academy of Science, Chunli Bai (2008), says that emulation of the North is standard practice, given "the research agenda in the developing world has largely taken its cues from research initially done in the developed world" (p. 37). Roco [2001) adds that an extra benefit of replicating models is an increasing homogeneity amongst the global workforce, thereby assisting the unifying power of global science.

Critical to the development of Southern capabilities in nano-innovation is said to be the development of long-term strategic plans, typically 10 years in nature. As examples, instrumentalists point to China's decade-long 'climbing up' project on nanomaterial science during the 1990s (Bai, 2005), the Chinese 'Compendium of National Nanotechnology Development (2001–2010)' (Zhou and Leydesdorff, 2006), the Iranian 'ten-year program for nanotechnology' (Islamic Republic News Agency, 2006), and the Argentinean ten-year bill proposing a national strategic plan for the development of micro- and nanotechnologies (Sametband, 2005).

Within the instrumentalist approach, diverse views are held as to what should be considered key outcomes of a country's engagement with nanotechnology. Some speak strongly of a need to focus on applied research,[57] and the ability to ensure absorptive capacity for recent worldwide developments (Asgar, 2003; Paterson, cited in Ministério Das Relaçõs Exteriores, 2003). Yet, Desai (cited in Scott, 2002) believes Southern countries must first develop basic, domestic science capacity that can then lead to engagement with R&D.

As an extension to the discussions on applied research, others (Dayrit and Enriquez, 2001; Schummer, 2007) focus on Southern countries developing niche export markets, based on strategic importance and comparative advantage, through value-adding to contemporary and traditional knowledge. Such an approach is reportedly desired or underway in Brazil, Colombia, India, South Africa, Sri Lanka and Thailand (see Revaprasadu, 2003; Changsorn, 2004; Almeida, 2006; Foladori, 2006; Rao in Press Trust of India, 2006; Financial Times, 2008). In this respect, Dayrit and Enriquez (2001) recommend that niche areas be based on the strategic importance of the technology in terms of: its science and technology characteristics; what it offers to industry and agriculture, society, the environment and national security; the presence of strategic advantage or national strengths and; current and potential capacity of the science and technology community, including the ability to harness expatriate and foreign capabilities.

Finally, many (see, for example, Hassan, 2005; Schummer in UNESCO, 2005) argue that Southern countries must focus on, and support, national policies and development plans that address critical, endogenous social and environmental concerns. Whilst, for these authors, a 'dual-track' approach — seeking to ensure applications are geared towards both market and social concerns — is the logical compromise. Others, such as Bai (2005), suggest this is a somewhat false dichotomy, arguing it is possible to directly couple social and economic development in national nanotechnology strategies. As an example, Barker, Lesnick, Mealy, Raimond, Walker, Rejeski and Timberlake (2005) claim the likely existence of Southern markets for Southern-developed nanobased water filters or photovoltaic devices. Salamanca-Buentello (in Small Times, 2005) concurs, finding that industries in Brazil, China and India have already responded to this market opportunity, harnessing the chance to link with "...local scientists and engineers and develop nanotechnologies that can solve important problems".

To overcome barriers to developing Southern R&D capabilities, instrumentalists (see, for example, Roco, 2001; Juma and Yee-Cheong, 2005) argue that the State must provide the lead in supporting nano-innovation. This fits with nanotechnology's historical introduction to the South via state-led efforts in countries such as China, India, Malaysia and the Philippines (see Lee-Chua, 2003; Nemets, 2004; Najib in Malaysian National News Agency, 2006b; Srivastava and Chowdhury, 2008), whilst in Iran, Malaysia and Thailand, the government has established either a national centre or a national nanotechnology 'headquarters' (see Lin-Liu, 2003; Islamic Republic News Agency, 2006; Malaysian National News Agency, 2006d). Similarly, to oversee national planning and policy, there are national councils or steering committees for nanoscience and nanotechnology in China and Colombia (see Bai, 2005; Foladori, 2006). Political leaders are seen as particularly important for ensuring initial government support for nano-innovation, with prominent examples having included the Presidents of Iran[58] and India (see Islamic Republic News Agency, 2006; Ramachandran, 2006).

In terms of government support, instrumentalists (see, for example, Roco and Bainbridge, 2003; Kalam, 2006) are particularly interested in regulation that supports nano-innovation and commercialisation, ensuring the market is nurtured. Already, the beneficial outcomes of national support have been noted in China, where growth in business expenditure for nanotechnology has outpaced government expenditure since 1997 (Zhou and Leydesdorff, 2006), and Sri Lanka, where for "...the first time the corporate private sector will collaborate with the government in research and development" (Amarathunga in Warushamana, 2007).

A key method for market facilitation is to actively engage the private sector by offering incentives, as with Vietnam and Iran's 'high-tech zones' that include tax breaks for nanotechnology businesses (Vu Long, 2004; Islamic Republic News Agency, 2006). Such zones often take the form of a networked technology park that can "...facilitate engagement with private sector companies and other research institutions" (Warushamana, 2007). Also popular, and evident in Argentina, Costa Rica and Malaysia, is the creation of nanotechnology-focussed 'centres of excellence', to facilitate scientific and resource mobility (see Vargas, 2004; Foladori, 2006; Tun Razak in Malaysian National News Agency, 2006c). According to Hassan (2005), all countries should consider such a strategy, as evidenced by his belief that it would be valuable to:

> establish nanotechnology centers of excellence in sub-Saharan Africa and other least-developed regions within existing competent institutions capable of partnering with other centers both in the South and North on joint projects (p. 66).

As outlined in the South African national nanotechnology strategy (see Department of Science and Technology, 2006), another mechanism for scientific exchange is the fostering of national systems of innovation. Clustering, as adopted in Thailand (Tanthapanichakoon, 2005), and other forms of facilitated nanotechnology collaboration are also believed to "...strengthen networks and stimulate cooperation between universities, research institutions and technology-based companies..." (Lemie, 2005). Furthermore, decentralised online research networks are seen as important for geographically large countries with relatively small populations, such as Brazil (The Royal Society and Royal Academy of Engineering, 2004; Andrade in UNESCO, 2005).

Whilst Tegart (2001) says that these approaches to research help facilitate a necessary cross-disciplinarity, in order to build the requisite critical mass for "...smaller economies and particularly in less developed economies" (p. 12) he says that there must be equally "...major changes in teaching" (*ibid.* p. 20). Fundamentally, this would involve changing traditional mindsets (Ramachandran, 2006) and developing 'nanotechnology experts', with interdisciplinary skills (Tegart, 2002). Waruingi and Njorge (2008) add that for a country such as Kenya to position itself as a global player, nanotechnology must be introduced early into the school curriculum, with Alpert (cited in Invernizzi and Foladori, 2005) proposing that, in the South, "changes in study plans would have to take place starting at primary education" (p. 109).

Another crucial way for Southern countries to overcome infrastructural barriers, is to share resources, knowledge, experiences and applications across international boundaries (Najib in Malaysian National News Agency, 2006b). As necessary and useful strategies, authors suggest drawing on diasporas (Singer *et al.*, 2005) and North-South partnerships between universities and research centres (Harper, 2003a; Hassan, 2005), particularly in the pre-commercialisation R&D phase (Roco, 2002, 2003). For India's President (see Kalam, 2006), such partnerships must be geared towards applied research. Yet, despite few governments having made the connection between nanotechnology and their aid programs, and "...no examples of pro-poor business projects that contain nanotechnology" (Meridian Institute, 2005, p. 13), Salamanca-Buentello, Persad, Court, Martin, Daar and Singer (2005) argue that there is fertile ground for multilateral approaches to nano-innovation, focussed on Southern needs. They therefore propose an initiative to address global challenges using nanotechnology, modelled on the Bill and Melinda Gates Foundation's '14 Grand Challenges for Global Disease'[59] and funded by national and international foundations, as well as collaborations between nanotechnology initiatives in the North and South. Others, such as Tegart (2001), focus on what can be done regionally amongst Southern countries, with Hassan (2005) proposing that "developing countries should devise broad-based strategies that include ample investments in South-South cooperation" (p. 66).

In terms of governance frameworks to manage risk, from the instrumentalist perspective, nanotechnology is already inherently regulated.[60] Despite claiming there is limited capacity in the South to address risks (Singer *et al.*, 2005; Meridian Institute, 2007;

Schummer, 2007), instrumentalists argue that risks should still be 'managed' (see, for example, Court *et al.*, 2004; Juma and Yee-Cheong, 2005; Meridian Institute, 2005; UNESCO, 2006). In the current environment, this means the ethics of nanotechnology "... must catch up to the science in order for the technology to progress in a socially responsible manner" (Court *et al.*, 2004). This means conducting further research (Asgar, 2003), with issues of safety, toxicity and environmental impact best dealt with through "... sophisticated techniques of risk analysis, scientific experimentation, and the legal re-evaluation of existing regulatory systems" (p. 17). Given evidence that current testing is adequate for identifying hazardous nanoparticles (Hoet, Brüske-Hohlfeld and Salata, 2004), such approaches are seen as able to occur in tandem with the technology's development (Asgar, 2003). In this light, the UNESCO (2006) argues that the appropriate question for regulators and policy makers is not whether nanotechnology is safe, but 'how can it be made safer?' In this sense, others (Juma and Yee-Cheong, 2005) speak of the need to ensure that nanomaterials are contained and disposed of appropriately, and potential risks are avoided by way of a 'safety by design' principle (Colvin in Choi, Kaplan, Mody and Roberts, 2008).[61]

Given a belief in inherent regulation, as well as the State's role in providing supportive legislation for nano-innovation and commercialisation, it is no surprise that the instrumentalist view favours 'soft' or voluntary regulation for nanotechnology (ETC Group, 2008). Examples include an international 'code of conduct' on the responsible development of nanotechnology,[62] as well as the U.S. '21st Century Nanotechnology Research and Development Act',[63] which mandates the establishment of research programs to address "...ethical, legal, environmental and other appropriate societal concerns", along with bi-annual reporting to the U.S. President on these issues (Meridian Institute, 2004, p. 34).

In a similar vein, some instrumentalists (Court *et al.*, 2004) reject considering the precautionary principle that, they say, currently equates to 'scaremongering'; threatening to draw attention away from identifying and applying nanotechnology for the South. As these authors note:

> ...the emerging tendency to raise fears about NT [nanotechnology] before there is much scientific evidence has the potential to stall progress in less advanced nations if only the interests of the

wealthy — and not the distinct needs of the poor — are allowed
to dominate discussion and influence decision-making...

The same authors subsequently argue that any moratorium on
the new production of nanomaterials would marginalise the South
further, given:

> Wealthy countries have the resources to continue to invest in NT
> [nanotechnology] despite a moratorium on public funding...They
> also have great incentive to invest in NT [nanotechnology], as they
> will reap great economic rewards when a ban is lifted and they
> are technologically far ahead of other countries (ibid.).

Rather, according to the ETC Group (2004c), instrumentalists
argue that nanotechnology should be considered safe, until proven
otherwise, and that the onus should be on the public to demonstrate
nanotechnology's risks when and if they arise.

Consequently, and to varying degrees, instrumentalist
perspectives are said to reproduce technological determinist
approaches, "...since they stress the beneficial impacts of a given
artefact on society" (Invernizzi and Foladori, 2007, p. 125).
Furthermore, there is evidence that suggests public engagement
is viewed by instrumentalists (see, for example Court *et al.*, 2004;
Department of Science and Technology, 2006) as a key mechanism
to ensure nanotechnology's acceptance, with nanotechnology's
development "...taken for granted as inexorable..." (Invernizzi *et
al.*, 2008, p. 134). Here there is a tendency towards an instructive
approach, whereby experts 'educate' their fellow citizens. In Kenya,
for example, Waruingi and Njoroge (2008) envisage the primary
purpose of public outreach as enhancing 'nanoliteracy'.

4.6.2 Contextualist Approaches to Governance

Fundamental to the contextualist position is the rejection of the
claim that the South must develop nanotechnology capabilities by
replicating Northern models of innovation. Here it is believed that, in
addition to the inappropriateness of such a proposal (Pratap, 2005),
the instrumentalist perspective fails to acknowledge the historical
grounding of Northern innovation in the exploitation of the South's
natural resources, thereby making effective replication impossible
(ETC Group, 2004c).

In terms of national regulation, contextualists (see, for example ETC Group, 2004b; Bowman and Hodge, 2007b) argue that a lack of nanotechnology-specific regulation[64] creates a vacuum that demands a unique, government-led regulatory response (ETC Group, 2003a, 2005c). Fundamental to such an approach is the need to ensure consumer rights through actions such as nanotechnology labelling (Miller, 2008). However, with governments both funding and regulating nano-innovation, there is concern that they are too compromised to approach nanotechnology assessment and regulation objectively (ETC Group, 2003a; see participant views in Meridian Institute, 2006), with the ETC Group (2008) claiming that "governments are so far acting as cheerleaders — not regulators" (p. 60).

Whilst risks can be country-specific, new international regulation is needed, given that national, regulatory fissures will be magnified at the international level (Bowman and Hodge, 2007b). Particular consideration will need to be made here given that the contemporary, international IP framework, primarily governed by the TRIPs Agreement, "...was realised at a time when nanotechnology was simply a futuristic aspiration" (*ibid.* p. 308). Overall, from this perspective the parallels between nanotechnology and biotechnology are profound (Scrinis, 2004), with nanotechnology remaining: "...unregulated, untested, unlabelled, prematurely released and commercialised, and developed and patented by large corporations in cosy partnerships with public research institutes" (*ibid.*).

Given the vast uncertainties I have already outlined, a belief that nanotechnology is advancing faster than previous 'revolutions' (Foladori and Invernizzi, 2007), and that its speed of advancement is too fast for regulatory structures to keep up (Mooney cited in Mantell, 2003), contextualists such as Montague (2004) argue that nanotechnology needs to slow down through the "proactive introduction of protective measures" (p. 19). Hard regulation, such as a moratorium, is therefore favoured:

> To make wider evaluations of nano-scale science and technology, including the impacts of intellectual property, South[ern] governments may wish to consider establishing a moratorium on nanotechnology until regulations are in place to protect workers, consumers and the environment — and until wider social impacts are considered (ETC Group, 2005c, p. 47).

From a contextualist, precautionary perspective, the onus should be placed on nanotechnology researchers, globally, to demonstrate the safety of the materials with which they are dealing:

> ...producers of nanomaterials have a duty to provide relevant toxicity test results for any new material, according to prevailing international guidelines on risk assessment (Hoet *et al.*, 2004, p. 19).

To crystallise such sentiment, in 2007, the U.S. Centre for Technology Assessment coordinated the development — with 40 civil society organisations[65] from around the world — of a set of principles for 'adequate nanotechnology oversight', focussed on practical and holistic forms of accountability (see Table 4.7).

Table 4.7 Principles for Adequate Nanotechnology Oversight (Adapted from ETC Group, 2008, p. 20).

I.	**A Precautionary Foundation:** Product manufacturers and distributors must bear the burden of proof to demonstrate the safety of their products: if no independent health and safety data review, then no market approval.
II.	**Mandatory Nano-Specific Regulations:** Nanomaterials should be classified as new substances and subject to nano-specific oversight. Voluntary initiatives are not sufficient.
III.	**Health and Safety of the Public and Workers:** The prevention of exposure to nanomaterials that have not been proven safe must be undertaken to protect the public and workers.
IV.	**Environmental Protection:** A full lifecycle analysis of environmental impacts must be completed prior to commercialization.
V.	**Transparency:** All nano-products must be labelled and safety data made publicly available.
VI.	**Public Participation:** There must be open, meaningful and full public participation at every level.
VII.	**Inclusion of Broader Impacts:** Nanotechnology's wide-ranging effects, including ethical and social impacts, must be considered.
VIII.	**Manufacturer Liability:** Nano-industries must be accountable for liabilities incurred from their products.

However, according to the ETC Group (2005a), international bodies that have sought Southern involvement in the debates around nanotechnology's risks, ELSI and general development have, so far, been Northern-based, leading to tokenistic, inequitable and industry-biased constituencies, with the tacit presupposition of outcomes pre-empting effective regulation (see, for example, Powell, 2004; Service, 2004).

Overall, early signs within the nanotechnology discourse suggest "undeniable parallels" with the biotechnology experience (ETC Group, 2008, p. 60). Public information, debate and policies are said to be running at least 8–10 years behind developments in innovation (ETC Group, 2004b), and even more so in the South (Invernizzi *et al.*, 2008). Here it is noted that a limited framework for assessment is resulting from nanotechnology "...emerging in a situation of 'risk sensitisation'" (Kearnes *et al.*, 2005, p. 285). In such circumstances, as with biotechnology, Mooney (1999) argues that nanosafety protocols will be used to impose monopolies, under the pretext that the necessity to 'feed the world' or 'safeguard the environment' warrants risk-taking by trusted enterprises searching for a high-tech solution. There is therefore a need to widen debates beyond technical issues that were largely the focus of the early GMO debates (Johnson, Raybould, Hudson and Poppy, 2007). Invernizzi and Foladori (2005) agree, picturing: "...[a] much larger and important debate which seeks to challenge the dominant socio-economic hierarchies in which nanotechnology development and application actually occur" (p. 101). In this light, some (Kearnes *et al.*, 2005) worry about discussions hitting fundamental limits:

> New government commitments to 'upstream'[66] public dialogue with science — if taken seriously — may run rapidly into head-to-head conflict with concerns about global competitiveness and the economic potential of national science systems competing aggressively for global investment and trained personnel (p. 283).

Others, such as Gould (2005), worry that it is already too late to radically alter the nanotechnology agenda through mainstream means:

> Although nanotechnological innovation remains at a relatively early stage, and public input into its form and efficacy is beginning to emerge, that input is emerging at a stage that is still too late

to preclude significant lines of research, and only after powerful institutional proponents have staked out interests and developed campaigns to ensure that such interests are achieved (pp. 15– 16).

Assessing levels of public input, the ETC Group (2004b) claim that, to date, "...participants in the discussion have been largely limited to scientists, investors and industry executives, primarily in the OECD nations" (p. 3), or, at best, the 'big South': Argentina, Brazil, China, India, South Korea, Mexico and South Africa (2005a). Small economies, the U.N., civil society and social movements having been sidelined (ETC Group 2004b, 2005c; UNESCO, 2006). Coming under particular scrutiny from some contextualist authors (Invernizzi and Foladori, 2005) are scientists involved in nanotechnology who, even when from the South, are said to not always be the most appropriate spokespeople for the poor or their needs. Strand (2001) sees this as a more universal phenomenon:

> ...in the eyes of any reader well informed of the traditions of the humanities and the social sciences, deliberations upon ethical and social aspects of nanotechnology written by natural scientists... are often preciously naïve with respect to the cultural and political assumptions and underpinnings they make along the way (pp. 9–10).

In response, it is argued that society has a right to shape nanotechnology's trajectory by determining "...the goals and processes for the technologies they finance" (ETC Group, 2008, p. 60).

Yet, for many (Kearnes *et al.*, 2005), there remains hope for reflexivity and the ability to shape alternative, future worlds given current awareness of the challenges facing the nanotechnology discourse offers a chance to address nanotechnology's implications with "...more success than ever before" (UNESCO, 2006, p. 20). By recognising historical mistakes and realities, Invernizzi and Foladori (2005) believe that the nanotechnology community can "...help to avoid repeating the mistakes of the pharmaceutical and biotechnology industries... help[ing] nanotechnology become a tool which can alleviate disparity rather than widen it" (p. 101). Scrinis and Lyons (2007) agree, claiming:

> ...there is considerable potential for civil society groups, workers' unions, farmer and producer organizations, environmental

and consumer groups, to challenge and shape the development and implementation of this technology, and to thereby support alternative applications, regulatory regimes, and techno-economic paradigms of development (p. 38).

Promoting an ideal approach, Foladori *et al.* (2008) say they "... favour an analysis of the socioeconomic context parallel to influencing technology development" (p. 4). In this light, Fudano (in UNESCO, 2005) proposes that 'constructive technology assessment'[67] be part of each country's process of engaging with nanotechnology R&D, if at all. In a similar light, the UNESCO (2006) suggest that public input should happen earlier and more often, with others (Invernizzi and Foladori, 2005) saying that such an approach is particularly important for the poor:

> the later we choose to address their social and economic implications, the less chance there will be for the technology to help the poor because nanotechnology will put down roots within the mainstream hegemonic socioeconomic structure, characterized by worldwide inequality (p. 110).

In agreement, Gavelin, Wilson and Doubleday (2007) say that early-stage engagement will increase the chances for the public to set the nanotechnology agenda, but Wickson (2007) adds that this will only be possible if technological trajectories are, in fact, flexible.

Several authors (see Mooney, 1999; ETC Group, 2004b; Gerbert in CORDIS, 2006; Wolbring, 2006, 2007) say that definitions of 'the public' must be more inclusive, to actively engage marginalised populations, especially women, indigenous peoples, farmers' organisations, unions, the disability rights movement and consumer organisations. There is a particular need, say the ETC Group (2004b), to involve social movements, trans-nationally, from both South and North. Furthermore, despite currently limited interest (Meridian Institute, 2005),[68] NGOs are seen as critical actors for ensuring the shift to a 'bottom-up' nanotechnology discourse and the assurance that nanotechnology is steered towards social needs (ETC Group, 2003a; Invernizzi and Foladori, 2007). But the unhelpful belief that expert views are scientific and objective whilst public views are unscientific and value-driven must change, say the ETC Group (2004b), as part of a redistribution of power, given:

...[nanotechnology raises the troubling issue of] the very structure of science itself...the legitimacy of scientific results, as well as the public trust in those results and the use and abuse of them by governments, corporations or nonprofit entities (UNESCO, 2006, p. 17).

However, the UNESCO highlights the opportunity that emerges from such despair:

The case of nanotechnology may represent one of the first where scientists are no longer capable of autonomously directing scientific research due to the growth of external pressures, not only commercial, but from civil society and State actors as well (*ibid.* p. 17).

In terms of Southern involvement in global debates, the UNESCO further suggests that:

Even if nations are not actively pursuing research in nanotechnology, they should nonetheless have a stake in defining the proposed outcomes and actual course of research according to norms of equity, justice and fairness (*ibid.* p. 7).

Supporting these contextualist arguments, in 2002, civil society organisations from Benin, Ethiopia, Ghana, Kenya, Mali, South Africa, Tanzania and Zimbabwe signed the 'Cape Town Declaration',[69] calling for global participation in decisions about nanotechnology.

In concluding the contextualist position, authors (ETC Group, 2005c; Grimshaw *et al.*, 2006) reiterate a vision for nanotechnology research that is driven by human needs rather than market wants. In this sense, it is believed that nanotechnology must be governed with the specific socioeconomic context in mind:

...empowering people to identify their needs; providing access to information to assess the benefits and risks of specific technologies and select the most appropriate technologies; and building capacity for local production and control (Meridian Institute, 2006, p. 10).

4.7 Conclusion

When it comes to nanotechnology's potential to offer hope for a more equitable world, the literature is polarised. From the instrumentalist perspective, nanotechnology offers hope through a new scientific paradigm that, when combined with the potentials of globalisation, presents new avenues for and evidence of Southern leapfrogging, as well as the ability to solve development problems, en masse. Furthermore, risks can be managed, largely through inherent regulation and public education. From the contextualist perspective, nanotechnology offers false hope given that it emerges within, and offers little to change the unjust, global economic system. When it comes to matters of orientation, nanotechnology's promises do not seem to match early realities but, rather, reflect similar hype to that associated with the emergence of biotechnology. Moreover, the risks associated with nanotechnologies seem to emerge within a regulatory and participatory vacuum.

These perspectives highlight the perpetuation of common debates from the development discourse. Of the research that has been undertaken and commentary that has been made, what clearly emerges is that the instrumentalist perspective mirrors and extends modernisation and mainstream development approaches (Scrinis and Lyons, 2007). On the other hand, the contextualist perspective mirrors an alternative development perspective, with occasional consideration for post-development ideology.

The literature thus confirms the usefulness of investigating reflexivity around the three themes I identified in my review of the literature on technology, development and inequity, particularly given the ongoing emergence of questions similar to those I have posed for my study. In terms of innovative capacity, Kearnes *et al.* (2005), for example, ask of developments in nano-innovation:

> Might they allow developing nations to 'leapfrog' into a new technological paradigm, or might they reproduce inadvertent forms of epistemic exclusion, stratified industrialised knowledge-economy divisions of international labor? (p. 284).

In terms of technological appropriateness, the RS&RAE (2004) question: "can the future trajectories of nanotechnologies be steered towards wider social or environmental goals...rather than towards meeting short-term or developed world 'market' opportunities?" (p. 53).

And, regarding approaches to governance, Bowman (2008) asks:

...what approach or approaches to regulation should be taken by government?...And importantly, what role may other actors, specifically industry and non-government[al] organisations, play in promoting the responsible development of nanotechnologies? (p. 180).

However, the obvious confusion about how nanotechnology is understood by stakeholders within the South, as well as the noted importance of such understandings as the basis for all debates, suggests that there is value in expanding my analytical framework for assessing nanotechnology to include 'understandings' as my lead theme.

Whilst there may be general agreement that the typology of issues raised by nanotechnologies is applicable to most other technologies, there is acceptance that the unique characteristics of nanotechnology may result in different considerations regarding each cross-cutting issue. In this light, Hodge, Bowman and Ludlow (2007) highlight the need for further research, asking: "...from historical, philosophical and ethical perspectives, in what ways does nanotechnology differ from earlier technologies?" (p. 386).

In this chapter I have shown that a number of gaps exist with respect to both the existing knowledge and methods undertaken by research into nanotechnology and development, restricting my ability to adequately answer my research questions. Thus, to arrive at a balanced viewpoint of nanotechnology's potential implications for global inequity requires additional analysis of empirical data (Invernizzi and Foladori, 2005).

Quantitatively, there is no clear, contemporary picture of how many Southern countries are engaging with nanotechnology R&D and the level of that engagement. Similarly, there have been no studies investigating claims of Southern engagement in global nanotechnology dialogue. Moreover, analyses of patent data remain broad, often presenting conflicting results, with further studies seen as necessary (see ETC Group, 2005c). To address these gaps, quantitative research is needed to establish more clarity around the global 'state of play'.

In terms of qualitative research, beyond identifying potential benefits, consideration for Southern perspectives remains largely absent (Court *et al.*, 2004), as does consideration for these perspectives alongside views from the North. A great deal of the 'research' I have referenced is speculative, failing to draw on the knowledge of those with the experience that matters, whilst within some research there is a suggested bias towards scientific and industry perspectives. In this light, targeted qualitative research is said to be needed to draw out more in-depth perspectives on nanotechnology's potential and actual implications, with studies considering developments in specific countries able to contribute useful information to the global pool of knowledge (ETC Group, 2004c; Singer *et al.*, 2005).

In this chapter I have established important data about global engagement with nano-innovation and a broad base on which to analyse nanotechnology's foreseen implications for the South. My findings have also highlighted the need to conduct qualitative research in order to more fully understand how trends exposed in the quantitative research are playing out from the perspective of those 'on the ground', as well as the relevance of these perspectives, coming from both sides of the North-South divide. Finally, the research has informed the focus of my upcoming qualitative study, such as the need to consider nano-innovation as a global phenomenon and to be sensitive to the emerging divides around international participation in nanotechnology dialogue and scientific events as well as patenting.

Endnotes

1 Relevant, as of September 2008.

2 In obvious reference to Drexler's (1986) vision.

3 Except for the times when it is clearly near-term nanotechnology, rather than advanced nanotechnology, that is the focus.

4 Adding to this difficulty, the funding figures reported in my research are from varying years and do not take purchasing power parity into account.

5 The Lux Research data included U.S. state funding in the total for North America and incorporated figures from associated and acceding E.U. countries in the European estimate.

6 This excludes Taiwan, which, in Marinova and McAleer's (2002) study, is considered part of China.

7 Originally appearing as 'China (Taiwan)'.

8 For details about the USPTO's Classification 977 relating to nanotechnology, see: http:// uspto.gov/web/patents/classification/ uspc977/defs977.htm.

9   Using the 2004 classification for nanotechnology patents (Classification 977).

10 The synopses I present, regarding the instrumentalist and contextualist positions, are, largely, directly quoted from the work of Invernizzi *et al.* (2008). Yet, due to the partly synthesised nature of such quotes, it was easiest to avoid intermittent quotation marks.

11 Home- rather than factory-based industries.

12 A space where the amount of airborne particles is vastly reduced from everyday use and strictly monitored (ETC Group, 2003a).

13 Equipping the facility costs an additional several hundreds of thousands of dollars (Vargas, 2004).

14 See: http://asia-anf.org.

15 See: http://apnf.org.

16 With Peters and Page (2003) claiming that nanotechnology research in Africa "...has been largely academic and disparate".

17 All able to be sourced and manipulated closer to distribution centers in the North (ETC Group, 2003a).

18 According to Watanabe (2003), countries such as Japan are already actively trying to recruit post-doctoral nanoscientists.

19 Tegart (2002) points to figures that suggest the level of people needed for nanotechnology by 2010–2015 will range between 1.8–2.2 million, globally, including between 0.4–0.6 million in the Asia Pacific region, led by greatest demand in Japan.

20 Understood to mean: "...biological and chemical effects of nanoparticles on human beings or natural ecosystems" (UNESCO, 2006, p. 14).

21 Understood to mean: "...leakage, spillage, circulation and concentration of nanoparticles that would cause a hazard to bodies or ecosystems" (ibid. p. 14).

22 An inflammatory and fibrotic medical condition affecting the lungs, caused by chronic exposure to and inhalation of asbestos fibres that then remain lodged in the inner layers of the lungs.

23 The Scientific Committee on Cosmetic and Non-food Products "... considered the safety of nanoparticles of titanium dioxide when used as a UV [ultraviolet] filter and declared them safe for use at any size" (Scientific Committee on Cosmetic and Non-food Products cited in Meridian Institute, 2005, p. 9).

24 Used as a medium for quantum dots.

25 Here, 'disruptive' refers to innovations that improve products or service in unexpected ways, typically through lower prices (see Bower and Christensen, 1995).

26 Invernizzi and Foladori (2008) claim that the instrumentalist perspective proposes the assured technical superiority of nanotechnology applications.

27 The industrial focus includes areas such as chemical and bio-processing, mining and minerals, and advanced materials and manufacturing. The social focus includes areas such as energy, water and health. 'The environment' crosses both sectors (Department of Science and Technology, 2006).

28 A forecasting method in which anonymous experts answer an initial set of questions over a number of rounds, in between which the facilitator provides a summary of the group results, allowing for expert revision in subsequent rounds. The intention here is that the group's results will ultimately converge.

29 Sixty per cent of these experts are from the South.

30 Panelists were asked to consider technologies based on their potential impact, burden, appropriateness, feasibility, knowledge gaps and indirect benefits.

31 Although Grimshaw (2006) claims that "...views about the relevance of application areas for poor people converge on two sectors, namely water and energy" (p. 4).

32 At the time Dr A.P.J. Abdul Kalam.

33 A parasitic disease, spread by the bite of a sand fly and found in parts of the tropics, subtropics and Southern Europe.

34 A hollow, spherically-shaped nanoscale object, used to encapsulate certain materials.

35 Magnetically responsive nanoparticles.

36 Ceramic materials fabricated from nanoscale particles or structures.

37 See, for example, Schummer's (2007) writing on nanotechnology and electrification.

38 At that time, Mr Patrick Zhumawo.

39 As described by Dolmo (2001), in 1997 the South African government introduced parallel imports and compulsory licensing to rectify the domestic crisis of inequitable access to affordable and essential medicines for HIV/AIDS. According to Fisher and Rigamonti (2005), in 1997 the average annual income in South Africa was $2,600, whilst the cost for HIV/AIDS treatment with antiretroviral drugs was

approximately $12,000. In response to the South African government's measures, 40 major drug companies sued, supported by the U.S. government (Dolmo, 2001), claiming the move "...was tantamount to a complete abrogation of patent rights and that it violated the Agreement on Trade-Related Aspects of Intellectual Property Rights" (Fisher and Rigamonti, 2005, p. 5). However, over the following two years, a great number of people and movements rallied around the world against the pharmaceutical industry, resulting in the suit being dropped (Dolmo, 2001).

40 Exceptions include the South African national nanotechnology initiative that is exploring a dual strategy to "...enhance quality of life and increase economic growth...[particularly] the quality of life of previously marginalised sectors of the community such as...women and people with disabilities" (Department of Science and Technology, 2006, p. 13; 16), and Nigerian nanotechnology, that will focus on challenges in the energy, health and water sectors, such as steady power supply and high-quality medical care (Business Day Media, 2007).

41 Director of the Nimbkar Agricultural Research Institute in Phaltan, India.

42 At the time Professor Wiwut Tanthapanichakoon.

43 So much so that there is "...potential for overlapping and conflicting patent claims" (Bowman, 2007, p. 308), leading to "...unnavigable complexity" (UNESCO, 2006, p. 18).

44 In 2006, researchers from Practical Action, Demos and the University of Lancaster collaborated on a process designed to engage Zimbabwean community groups and scientists from both the North and South in debates about new nanotechnologies.

45 Such as "...the effect of nanoparticles on species other than humans or about how they behave in the air, water or soil..." (The Royal Society and Royal Academy of Engineering, 2004, p. x).

46 Particularly from products when they are incinerated, buried or degraded (Sass *et al.*, 2006).

47 Although the RS&RAE (2004) recommended that "...chemicals produced in the form of nanoparticles and nanotubes be treated as new chemicals" (p. xi).

48 For example, despite the recommendations of the RS&RAE's (2004) report, in countries such as the U.S., U.K., Australia and Japan, "...existing chemicals now being produced at the nanoscale are not considered to be 'new' for purposes of these regulatory frameworks" (Bowman and Hodge, 2007b, p. 36).

59 A civil society organization based in Bangkok, see: http://biothai.org.

50 Commonly referred to as the 'nanobiocognitive interface'.

51 The medical model of disability sees a person's functional limitations as the root cause of any disadvantages. From this perspective, such disadvantages can only be rectified by treatment or cure. The social model, on the other hand, looks at the disabling social, environmental and attitudinal barriers that limit a person's opportunities (Crow, 1996).

52 According to Foladori and Invernizzi (2007), less than 4% of global nanotechnology R&D funding goes towards assessing risks or ELSI.

53 Although, when it comes to environmental health and safety, some of these countries may present a different picture in recent years, with China, amongst other things, having established a Laboratory for Biological Effects of Nanomaterials and Nanosafety in 2006 (AzoNano, 2006) and a national data bank on nanosafety in 2007 (Chinese Academy of Sciences, 2007). Numerous toxicology studies are said to be occurring at multiple Chinese institutions (Bai, 2005). Furthermore, China has released seven nanotechnology standards since 2001, claiming these to be the world's "...first batch of national standards" (Reuters, 2005). India, Malaysia and China have also adopted national committees for formulating nanotechnology standards encompassing terminology, metrology and manufacturing norms (see Reuters, 2005; Kulshrestha, 2006; Ruddin in Malaysian National News Agency, 2007).

54 Although de Almeida (2003) agreed that there has been little engagement in Brazil, in terms of assessing safety and ethical impacts, she noted there are some exceptions, such as the University of Brasilia that is "...concerned with the need to identify the impact of new nanomaterials in health, and its research group is carrying out studies to better determine the biological behaviour of these new materials".

55 In 2005, the national standards institutes of 24 ISO member countries participated in the development of nanotechnology standards, for "...terminology and nomenclature; metrology and instrumentation, including specifications for reference materials; test methodologies; modelling and simulations; and science-based health, safety, and environmental practices" (Frost, 2005).

56 An example of such an approach is the planning for a 'nanopark' in Southern India, modelled along the lines of the Hsinchu Science Park in Taiwan (Chennai Interactive Business Services, 2007).

57 As is the case in Malaysia where the spending ratio is 2:1 in favour of applied, ahead of fundamental research (Malaysian National News Agency, 2006a).

58 At the time: Dr Mahmoud Ahmadinejad and Dr A.P.J. Abdul Kalam, respectively.

59 More details at: http://grandchallenges.org.

60 Huam, Petschow, and Steinfeldt (2004) say that early signs from the U.S. and U.K. Governments support this philosophy.

61 For example, technologies designed to biodegrade.

62 See: http://europa.eu/rapid/pressReleasesAction.do?reference=IP/0 8/193&format=HTML&aged=0&language=EN&guiLanguage=en.

63 A downloadable copy of the act is available at: http://nanotech-now. com/S189.pdf.

64 According to Choi (2002), "...no law deals adequately with the potential impacts of this new technology" (p. 358).

65 For a list of these organisations, see: http://icta.org/nanoaction/doc/ nano-02-18-08.pdf.

66 Early-stage input into the consideration of technological development.

67 An anticipatory process whereby a wide range of stakeholders consider technology's impacts at an early stage of technological development in order to provide constructive suggestions for adjusting the technology under development.

68 According to the Meridian Institute (2005), "...NGOs that have pioneered ways of getting appropriate technology to the developing world have not yet tended to focus on nanotechnology" (p. 18).

69 See: http:// activistmagazine.com/index.php?option=com_content&t ask=view&id=127&Itemid=143.

Chapter 5

The State of Play

In this chapter I will analyse my quantitative data relating to global engagement with nano-innovation in order to map the contemporary state of play and provide a foundation for further exploration of nanotechnology's foreseen implications for the South. I will particularly explore the extent to which nano-innovation and innovative capacity is globally decentralised.

5.1 Global Engagement

Just how globally widespread is nano-innovation? According to my research, the number of countries engaging with nanotechnology R&D on a national level has grown to 62, with 18 of these 'transitional' and 19 'developing'. A further 16 countries demonstrate either individual or group research in nanotechnology, three of which are 'transitional' and 12 'developing', including Bangladesh as the sole LDC. Fourteen countries have expressed interest in engaging in nanotechnology research. Of these, one is 'transitional' and 13 'developing', including three LDCs: Afghanistan, Senegal and Tanzania (for a graphical representation of this data see Fig. 5.1; for a full country breakdown see Table 5.1).

Nanotechnology and Global Equality
Donald Maclurcan
Copyright © 2012 by Pan Stanford Publishing Pte. Ltd.
www.panstanford.com

Table 5.1 Global Distribution of Nanotechnology Activity, by Country and Classification (2004); Data Sourced from Google.

'Least Developed'	Other: 'Developing'	'Transitional'	'Developed'
National Activity or Funding			
	Argentina; Armenia; Brazil; Chile; China; Costa Rica; Egypt; Georgia; India; Iran; Mexico; Malaysia; Philippines; Serbia and Montenegro197; South Africa; Thailand; Turkey; Uruguay; Vietnam	Belarus; Bulgaria; Cyprus; Czech Republic; Estonia; Hong Kong; Hungary; Israel; Latvia; Lithuania; Poland; Romania; Russia; Singapore; Slovakia; Slovenia; South Korea; Ukraine	Australia; Austria; Belgium; Canada; Denmark; Finland; France; Germany; Greece; Iceland; Ireland; Italy; Japan; Luxembourg; Netherlands; New Zealand; Norway; Portugal; Puerto Rico; Spain; Sweden; Switzerland; Taiwan; United Kingdom; United States of America
Individual or Group Research			
Bangladesh	Botswana; Colombia; Croatia; Cuba; Indonesia; Jordan; Kazakhstan; Moldova; Pakistan; Uzbekistan; Venezuela	Macau (China); Malta; United Arab Emirates	Liechtenstein
Country Interest			
Afghanistan; Senegal; Tanzania	Albania; Bosnia and Herzegovina; Ecuador; Ghana; Kenya; Lebanon; Macedonia; Sri Lanka; Swaziland; Zimbabwe	Brunei Darussalam	

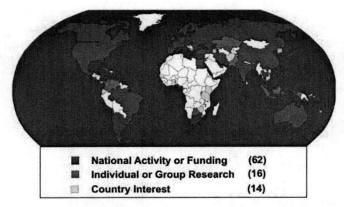

Figure 5.1 Graphical Representation of Global Nanotechnology, by Country and Level of Activity (2004). Data Source: Google.

A most prominent figure in this data is the number of countries engaging with nanotechnology on a national level at such an early stage in its global development. Although every country from the North, except Liechtenstein, is currently working at this level, the large number of Southern countries with national activity or funding reinforces Southern interest in building endogenous capabilities in nano-innovation. Of particular note, given their lack of mention in the literature, is the engagement of Serbia and Montenegro, and Uruguay at this national level.

However, it is clear from the data that the LDCs have not engaged with nano-innovation in any significant way. Particularly obvious is the lack of engagement with nanotechnology R&D from countries within Africa. A lack of engagement from large parts of the Middle East is also noticeable. These points raise the possibility that the divides within global engagement with nano-innovation could be as much within the South as between the South and North.

5.2 An Exclusive 'Global' Nanotechnology Dialogue

My research into the distribution of nanotechnology conferences or events in 2004, by host country, produced some interesting findings (see Table 5.2). As with engagement with nanotechnology R&D, a surprisingly diverse range of countries hosted nanotechnology events in 2004. Destinations included Bangladesh and Moldova, despite no national nanotechnology activity having been recorded in these countries. However, the data is highly centralised, towards certain Northern countries. The U.S. is shown to have hosted more than five times the number of nanotechnology conferences than any other country in 2004, with the U.K., Germany, Japan and France the only other countries who have hosted more than 10 events in 2004.

Table 5.2   Distribution of 2004 Nanotechnology Conferences or Events, by Host Country; Data Source: 'Nanotechnology Now'.

Country	No# Conferences	Country	No# Conferences
United States of America	164	India	2
United Kingdom	31	Ireland	2
Germany	30	Sweden	2
Japan	24	Vietnam	2
France	15	Bangladesh	1
Italy	10	Bulgaria	1
Canada	9	Czech Republic	1
Australia	7	Denmark	1
The Netherlands	6	Egypt	1
Russia	5	Greece	1
Singapore	5	Hong Kong	1
Belgium	4	Mexico	1
Spain	4	Moldova	1
Austria	4	Portugal	1
Israel	3	Slovenia	1
Poland	3	Slovakia	1
South Korea	3	South Africa	1
Switzerland	3	Taiwan	1
Brazil	2	Thailand	1

In building on this data, my analysis of country participation at three, key events held between 2003 and 2005: the International Dialogue on Responsible Research and Development in Nano-technology (IDRRDN); the International Nanotechnology Congress (INC); and the North-South Dialogue on Nanotechnology (NSDN), suggests high levels of inequity in international engagement relating to nanotechnology (see Table 5.3). Each of the meetings had between 18 and 25 countries represented, yet the numbers heavily favoured countries from the North, with over 50% of the speakers at the INC, for example, coming from the U.S. Furthermore, in a breakout group at the IDRRDN titled 'nanotechnology and developing countries', only

3 of the 13 representatives were from the South (Argentina, South Africa and Mexico). Moreover, participants in this group noted as insufficient the 'less than two hours' allocated for their discussions (Meridian Institute, 2004).

Table 5.3 Breakdown of Country Representation at Key Nanotechnology Conferences (2003–2005) by Presenter or Attendee*; Data Source: Various Conferences.[1]

Country	IDRRDN^ ('03)	INC^^ ('04)	NSDN^^^ ('05)	Country	IDRRDN ('03)	INC ('04)	NSDN ('05)
United States of America	7**	46**	10	New Zealand	1	1	—
Italy	1	1	72**	Russia	1	1	—
South Korea	2	8	—	Slovenia	—	—	2
India	1	5	2	Switzerland	1	—	1
Japan	5	1	1	Armenia	—	1	—
South Africa	2	—	5	Austria	1		—
Germany	1	4	1	China	—	1	—
Taiwan	3	2	1	Czech Republic	1	—	—
United Kingdom	1	2	3	Denmark	—	1	—
Canada	3	2	—	Egypt	—	—	1
Brazil	1	2	1	Georgia	—	1	—
Mexico	2	2	—	Ireland	1	—	—
Argentina	1	—	2	Malaysia	—	—	1
Australia	1	—	2	Nigeria	—	—	1
Belgium	1	.	1	Romania	1	.	.
Israel	1	1	.	Singapore	.	1	.
France	2	.	.	Uruguay	.	.	1
The Netherlands	1	1	.				

* The INC data refers to presenters whilst the IDRRDN and NSDN data refers to attendees

** Signifies conference host

^ The International Dialogue on Responsible Research and Development in Nanotechnology

^^ International Nanotechnology Congress

^^^ North–South Dialogue on Nanotechnology

Yet, Italian numbers at the NSDN aside, it is interesting to see that Asian representation at these events was greater than that of European representation, although Chinese delegates were notably absent from two of the three events.[2] Also of note are the high levels of Indian and South African participation in these events, as well as the inclusion of a Nigerian representative at the NSDN, despite Nigeria having no record of engagement with nanotechnology R&D (see Table 5.1).

5.3   Early Patent Control and Orientation

My assessment of health-related nanotechnology patenting showed that 34 countries have a share in the global distribution of 1256 patents (see Table 5.4).

Table 5.4   Total Number of Health-Related Nanotechnology Patents, by Country (1975–2004); Data Source: esp@cenet.

Country	No# Patents	Country	No# Patents
United States of America	420	Russia	6
China	260	Taiwan	4
Germany	166	Australia	2
France	161	Finland	2
South Korea	50	Hong Kong	2
Switzerland	39	Luxembourg	2
Japan	32	Poland	2
Ireland	29	Singapore	2
Canada	16	Virgin Islands	2
Israel	12	Austria	1
Spain	12	Bermuda	1
United Kingdom	11	Brazil	1
Netherlands	10	Greece	1
Sweden	10	Iceland	1
India	7	Norway	1
Italy	7	Serbia and Montenegro	1
Belgium	6	Slovenia	1

The three countries leading health-related patenting are the U.S. (32.8%), China (20.3%) and Germany (12.9%), with the top 7 countries holding 88% of the overall patent share (see Fig. 5.2) — a figure that is claimed to be reflected in the results relating to nanotechnology patents more generally.[3] However, unlike Compañó and Hullman's (2002) study, where no Southern countries ranked in the top 15 patent holders, my research showed that, in health-related nanotechnology patenting, in addition to China, holders from the South include India (0.5%), Brazil (0.1%) and Serbia and Montenegro (0.1%).

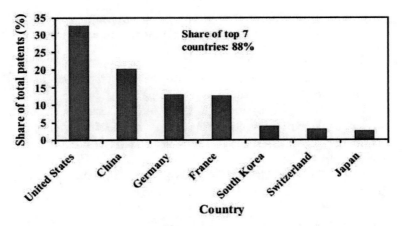

Figure 5.2 Distribution of 1975–2004 Health-Related Nanotechnology Patent Activity amongst the Top Seven Holders, by Country; Data Source: esp@cenet.

Whilst, overall, the U.S. has a very strong position in health-related nanotechnology, the 2004 data (see Fig. 5.3) is striking in showing China catching up, with 123 patents compared to 128 for the U.S. Furthermore, a considerable gap exists between China and third-placed Germany which produced 39 patents. This data also shows a higher concentration of patents amongst the top seven countries in 2004 than was the average for the period 1975–2004 (92%).

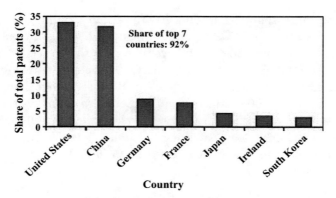

Figure 5.3 Distribution of 2004 Health-Related Nanotechnology Patent Activity amongst the Top Seven Holders, by Country; Data Source: esp@cenet.

When looking at the distribution of health-related patents, by continent (see Fig. 5.4), there is little separating Europe (36.7%), North America (34.2%) and Asia (28.8%). The large involvement of Asia suggests that nanotechnology may be the first widespread technology in which Asian countries have a foundational role. However, few or no patents are held in Oceania (0.2%), South America (0.1%) and Africa (0%). This furthers my earlier concerns about an emerging nanodivide within the South.

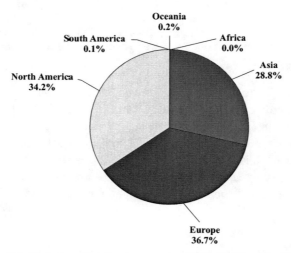

Figure 5.4 Global Distribution of 1975–2004 Health-Related Nanotechnology Patent Share, by Region; Data Source: esp@cenet.

With respect to sectoral ownership, the vast majority (77%) of health-related nanotechnology patents are held privately, with only 16% held by universities, 5% by government and 2% by independent/not-for-profit organisations (see Table 5.5). Compared with Rader's (1990) work in mapping the early stages of health-related biotechnology patenting, one can see that, whilst the share of patents held by the private sector has increased only slightly, there has been a strong shift from individual to company ownership. Additionally, gains in patent share by government have corresponded with slight and not so slight reductions for the university and independent/not-for-profit sectors respectively.

Table 5.5   Distribution of Health-Related Nanotechnology Patent Activity (1975–2004), by Sectors; Data Source: esp@cenet.

Sector	Patent Share (%)
Private:	
• Company	54
• Individual	23
University	16
Government	5
Independent/Not-For-Profit	2

The top 20 patent holders account for 28% of the patents, with the top 10 institutions holding 22% of the total (see Table 5.6). Interestingly, Sanofi-Aventis, GlaxoSmithKline, AstraZeneca and Merck, all top 10 pharmaceutical companies in the U.S.,[4] have engaged in nanotechnology patenting in the field of healthcare, whilst two further drug giants: Elan Pharma International and Novartis, hold strong patent positions in health-related nanotechnology. Yet, highlighted by a 2005 report (see Lux Research, 2005) claiming pharmaceutical giants are investing less money and people in nanotechnology than other industries, as of 2005, a number of big pharmaceutical companies were noticeably absent from health-related nanotechnology patenting. These include the top pharmaceutical manufacturer in the U.S., Pfizer, along with Johnson and Johnson, Bristol-Myers Squibb, Abbott Labs and Amgen.

Table 5.6 Top 20 Institutions with Health-Related Nanotechnology Patent Activity (1975–2004); Data Source: esp@cenet.

Rank	Assignee Name	No# Patents	Country*
1	L'Oreal	109	France
2	Elan Pharma International	38	Ireland
3	Nanosystems (ISRA Visions Systems Group)	31	U.S.
4	Henkel	28	Germany
=5	Cognis Deutschland	15	Germany
=5	Sanofi-Aventis	15	France
7	Amorepacific	14	South Korea
8	Vesifact	13	Switzerland
=9	GlaxoSmithKline	11	U.K.
=9	Japan Science and Technology Agency	11	Japan
11	Rohm and Haas	10	U.S.
=12	Centre National De La Recherche Scientifique	9	France
=12	Eastman Kodak Company	9	U.S.
=14	Ciba Specialty Chemical Holdings	8	Switzerland
=14	The Regents of The University of California	8	U.S.
=16	Diagnostikforschung Institute	7	Germany
=16	University of Texas	7	U.S.
=18	Alfatec Pharma	6	Germany
=18	Max Planck Gesellschaft	6	Germany
=18	Novartis	6	Switzerland

As confirmation that patents can cross over many industrial sectors, two of the top 20 institutions with health-related patents (Eastman Kodak and The Regents of the University of California) are also two of the greatest assignees for general nanotechnology patents (see Huang *et al.*, 2004).

My assessment of health-related patents by utility yielded some surprising results (see Table 5.7). In contrast to any belief that consumer health and diagnostic applications might be at the forefront of health-related patenting[5], the strongest emphasis is on therapeutic applications, such as drug delivery mechanisms. Of note here is the considerable number of applications that combine nanotechnology with traditional medicine for therapeutic benefit. Following therapeutic applications, however, almost a third of patents relate to consumer health, such as cosmetics and sunscreens.

Table 5.7   Categorisation of 1975-2004 Health-Related Nanotechnology Patents by General Utility; Data Source: esp@cenet.

Application	No# Patents	Patent Share (%)	Examples
Therapeutic	775	52	Drug delivery mechanisms, vaccines, nutraceuticals, bone scaffolds
Consumer Health	449	30	Cosmetics, sunscreens, antibacterial/antiseptic/antimicrobial coatings, water purification systems
Diagnostic	270	18	Sensors, biomarkers

An assessment of all patent titles and abstracts for cited health conditions showed a strange mix of results (see Fig. 5.5 for a list of the 10 most cited diseases). Predictably, lifestyle diseases feature prominently, with cancer receiving the greatest focus,[6] by far. Similarly predictably, cardio-vascular diseases are highly supported as a percentage of patented research, with some consideration also evident for nanotechnology research relating to diabetes,[7] osteoporosis and acne — the latter indicative of the many skin diseases highlighted in the overall results.[8] Whilst considering my overall results, I was struck by the very large number of citations linked to sexually transmitted infections,[9] including HIV/AIDS and vaginitis. Support for another communicable disease, influenza, was less surprising, given its high prevalence in the North, particularly in the U.S. However, somewhat unpredictably, two of the most cited conditions in my patent data: hepatitis (particularly hepatitis B); and beri-beri — an ailment of the nervous system caused by a dietary deficiency of vitamin B1, were areas in which there would seem to be less commercial opportunity in the global North.

Yet, overall, my research showed that there is very little commercial focus with respect to neglected diseases. Two of the world's greatest killers: malaria and tuberculosis, are noticeably absent from any significant level of nanotechnology patenting, with each recording just one patent.

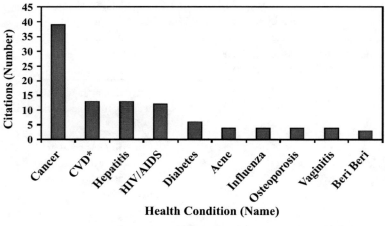

* CVD = cardiovascular disease

Figure 5.5 Categorisation of 1975–2004 Health-Related Nanotechnology Patents by Specific Utility (Health Condition); Data Source: esp@cenet.

5.4 Conclusion

In this chapter I have shown that more Southern countries are engaging with nano-innovation than previously documented, and that such engagement spans a number of different means. However, such figures would appear to shroud an increasing concentration of control over innovation in this field, albeit under the auspices of a slightly different mix of countries than witnessed with biotechnology's emergence.

In terms of nanotechnology R&D, a surprising number of Southern countries are already engaging with research and development on a national level. Although a number of these countries have already been flagged in my review of the literature, it is worth reiterating the different nature of a technological emergence in which Iran, Costa Rica, Georgia, Serbia and Montenegro, Turkey, Uruguay and Vietnam engage, in such a manner, so 'early on'. Yet, whilst engagement on the national level suggests some degree of coordination for nanotechnology R&D in parts of the South, there are also a great number of Southern countries for whom disparate research is the only form of noticeable engagement. This said, such research is occurring in unexpected places, such as Botswana. There is also a great deal of interest from other Southern countries, particularly in Africa, who are yet to engage with nanotechnology R&D. However, the noticeable absence of engagement by LDCs, in light of high levels of engagement from the 'big' and 'emerging' South, confirms that a research divide already exists just as strongly within the South as between the North and the South.

In terms of global engagement with nanotechnology conferencing and events, the wide range of countries acting as hosts suggests that there is a good deal of international hype surrounding nanotechnology, particularly given that some of the host countries are yet to engage with R&D in any significant way or, indeed, at all. Excluding the notable absence of China, as with biotechnology, participation in supposedly key 'global' events seems heavily biased towards 'wealthy' countries in the North and major trading nations in the South. In this light, my results suggest that considerable challenges may arise for reflexive engagement with nanotechnology at the level of international relations.

The overall picture of health-related nanotechnology patenting suggests control lies largely in the North, particularly the U.S. However, there is significant emergence of members of the 'big South'. Most notably, China looks set to play a considerable role in global nanotechnology R&D, as with biotechnology and information and communications technology. However, whilst the strength of China's patenting defies the general statistic that less than 2% of all the world's patents are granted to scientists in the South (Mooney, 1999), it obscures weak levels of patenting amongst the other Southern countries. Particularly outstanding are the low number of health-related nanotechnology patents emerging from India and Brazil, given the high levels of investment these countries have made into this field. Beyond the international analysis, ownership is shown to rest firmly with the private sector, following an earlier MNC engagement with nanotechnology than witnessed with biotechnology.

My results took an interesting turn when I looked at the orientation of patents in terms of utility and links to health conditions. Unexpectedly, I discovered that early commercial hopes lie largely with what nanotechnology can offer therapeutic applications, followed by consumer health and, to a much lesser degree, diagnostics. The results, with respect to health conditions, show that strongest patenting is for diseases of historical significance to the North, although it was surprising to see such strong patenting for HIV/AIDS, hepatitis and beri-beri. However, both Southern healthcare needs and global healthcare markets are changing. With different lifestyles accompanying market-led 'development', "... health differences between countries will be narrowed" (Maugh II, 1996). Thus, many of the conditions cited in the patents hold increasing relevance for the South. In terms of overall numbers, for example, cancer's burden is already greatest in the South (see WHO, 1997). Similarly, much of the projected doubling in cases of diabetes by 2025 will stem from increases in the South (WHO, 2005a). The centralisation of health-related nanotechnology patents is, therefore, of heightened importance to global equity, raising fears of even more restrictions to Southern R&D as well as the ongoing challenge of access to essential drugs that could be even further compromised by the emergence of a burgeoning pharmaceutical-consuming class in parts of the South. At the other end of the spectrum, the near void of patented research relating to neglected diseases highlights the ongoing challenge of addressing 'lack of interest' in these areas.

In this chapter I have established important data about global engagement with nano-innovation and a broad base on which to analyse nanotechnology's foreseen implications for the South. My findings have also highlighted the need to conduct qualitative research in order to more fully understand how trends exposed in the quantitative research are playing out from the perspective of those 'on the ground', as well as the relevance of these perspectives, coming from both sides of the North-South divide. Finally, the research has informed the focus of my upcoming qualitative study, such as the need to consider nano-innovation as a global phenomenon and to be sensitive to the emerging divides around international participation in nanotechnology dialogue and scientific events as well as patenting.

Endnotes

1 Serbia and Montenegro split into separate countries in 2006 but I will continue to refer to the two countries collectively, given my research occurred prior to 2006 and I am unable to know in what ways the results of this research would be applicable post-2006.

2 http://pharmabiz.com/article/detnews.asp?Arch=&articleid=24040 &sectionid=9; http://tinyurl.com/yfyu9mc; and http://ics.trieste.it/ Nanotechnology/.

3 Although a Chinese paper was distributed at the NSDN.

4 Compañó and Hullman's (2002) study of general nanotechnology patents from 1991–1999 shows the top 7 countries holding 92.1% of patents.

5 According to NDCHealth (2005) figures.

6 Resulting from expected lower costs and regulatory barriers.

7 In 2004 the U.S. Government set aside $144 million for nanotechnology-based cancer research in the U.S. (see National Cancer Institute, 2004).

8 My results do not distinguish between Type I and Type II diabetes, with only the latter being considered a lifestyle disease.

9 Examples include dandruff, eczema, psoriasis, rosacea and tinea.

10 Examples include gonorrhea, herpes, chlamydia and venereal disease.

Chapter 6

Understanding Nanotechnology

This is the first of four chapters in which I will seek interpretation of nanotechnology's foreseen implications for the South via the perspectives of 31 key informants from Thailand and Australia, supplemented by surveys of 24 Thai nanotechnology practitioners.

In this chapter I will examine how nanotechnology is understood across cultures in order to clarify the legitimacy and limitations of engaging differing groups of people in my assessment of nanotechnology. I will particularly explore the extent to which nanotechnology is understood in ways that allow common discussion about its implications for global inequity.

6.1 Characteristics

Given the definition is still evolving (Ford) and very broad (Tegart), some of my interviewees considered it difficult to define what nanotechnology 'is'. As Damrongchai, a Thai policy officer, noted, presently the "...definition has some diversity that can change according to the context". Others agreed that there will always be diversity of opinion (Kanok-Nukulchai), no matter what certain authorities might specify or claim (Yuthavong).

However, on the whole, interviewees from both Australia and Thailand presented surprisingly similar responses as to the characteristics that contribute to nanotechnology's definition. Six characteristics were seen as fundamental (see Table 6.1).

Nanotechnology and Global Equality
Donald Maclurcan
Copyright © 2012 by Pan Stanford Publishing Pte. Ltd.
www.panstanford.com

Table 6.1   The Six Fundamental Characteristics of Nanotechnology.

Number	Characteristic
1	It is based upon a size or length scale (the nanoscale)
2	It involves the ability to either 'control', 'manipulate' or 'engineer' on the nanoscale
3	It involves exploiting properties unique to the nanoscale
4	It is the practical application resulting from this exploitation
5	It is often the product of conducting 'old science' in a new way
6	It is the natural (but sometimes unconscious) progression for those working in cutting-edge areas of science and is, therefore, a new field rather than a new discipline

The most commonly defined feature of nanotechnology is that it relates to a size or length scale. Interviewees generally provided technical explanations, noting that there is a "...loose definition of nanotechnology to be between 1 and 100 nanometres..." (Berwick), with a nanometre being equal to '10^{-9}' metres (Weckert). Yuthavong, a prominent Thai scientist, highlighted other standard references, such as the 'nanoscale', and described this informally as "...mid-way between [the] atomic scale and the convention[al] scale that we are familiar with...[where one] would think of technology which deals with materials of a few atoms or a few molecules". Only Deutchmann, an Australian health practitioner, and Lynskey, head of an Australian NGO, provided non-scientific responses, referring respectively to nanotechnology as "miniaturisation" or "really tiny things".

Nearly half the interviewees referred to nanotechnology in terms of its command over the small scale. Berwick described nanotechnology as "...the control and ability to manipulate material at the atomic level". Tanthapanichakoon, Director of the National Nanotechnology Centre of Thailand (NANOTEC), talked in a similar manner of nanotechnology as "...the control of microstructure[s] or manipulation of the atoms or molecules or the clusters of molecules". Both Australian and Thai interviewees commonly referred to this trait as 'engineering' on the nanoscale.

A number of interviewees (Braach-Maksvytis; Tanthapanichakoon) highlighted that nanotechnology exploits unique properties not exhibited in bulk materials. The ability to utilise these unique properties was seen as the basis for enhanced research possibilities (Sawanpanyalert).

For many, it was important to make the distinction between nanoscience and nanotechnology. Interviewees distinguished that nanotechnology was the "practical application" of nanoscience (Chirachanchai), "...because it has got the word 'technology' in it" (Ford). This suggests an important distinction, particularly in terms of discussing a country's role in nanotechnology R&D, because it means a country's ability to produce the technology must be considered in addition to its ability to conduct research.

Ethicists and lawyers amongst the interviewees (Changthavorn; Ratanakul; Selgelid) presented nanotechnology as 'a new form of technology'. However, the majority of interviewees, particularly those with backgrounds in science and chemistry, claimed that nanotechnology was 'old' science done in a 'new way', or what Dutta referred to as "an old wine in a new bottle". Coyle lent her support to this argument, suggesting that, although "...the terms may be new...", it is, in fact, putting into effect much existing knowledge, with a new emphasis on the way to do things. In many instances, interviewees made the distinction between 'nanoscience' and 'nanotechnology', suggesting that nanotechnology builds on nanoscience knowledge that has "...been in existence for a long time..." (Ford), with Yuthavong presenting the example of liposome drug delivery as a nanotechnology process that has "...been going on for some time". Radt, an Australian scientist, explained his nanotechnology work in a similar manner:

> ...using particles loaded with a drug for drug delivery is very well established and old technology...the particles will become more sophisticated and will become more complex, but it will be the continuous change I see there that builds up from the brilliant work which is already published.

Chirachanchai made similar comments, noting that the contemporary term 'nanotechnology' can be used to classify previous work that occurred on the nanoscale "...even [if] we do not have the 'nano' wording...the way that people learn from experience and come to the molecules and start from molecules and go back, is already the nano[scale] work...".

Interviewees, such as Radt, highlighted a subsequent "...re-branding of old technologies" to fulfil an organisational objective. Chiranchanchai explained a common experience for many Thai scientists whereby their ongoing research was, all of a sudden, retermed 'nanotechnology'. Others noted surprise at discovering they had 'unconsciously' been working in nanotechnology. Coyle's first reactions highlight this point:

> When you talked about nanotechnology I thought 'what on earth is that?'...and then you sort of brought it down to atoms and molecules, and then, of course, I realised that the antigen/ antibody reactions which we have been dealing with for...lots of years, [are] at that scale.

Dutta saw positives in these points, suggesting that: "the attractive thing about nanotechnology is that everyone says 'hey, I am in it, I know it, I have been working on it but I have not been using that word'".

Furthermore, the shift to working in nanotechnology is seen as a "logical migration" (Cornell) for those at the forefront of various cutting-edge areas of science. As Warris noted: "if people are working in physics, chemistry and biology they are going to be working in nanotechnology because [it is at] the cutting edge of these topics". Interviewees suggested that this loose, and often unconscious, new grouping of research and its cross-fertilisation between disciplines and sectors, means that nanotechnology is a new field, as distinct to a new discipline or industry.

The sum of these findings suggests relatively common understandings in relation to nanotechnology's distinguishing features and gives credence to Turney's comments that there is no problem in interpretation and no need to get "...hung up on definitions". The commonality of understandings are made all the more surprising given 12 of the Australian and Thai interviewees had no background in nanotechnology, with some (Deutchmann; Selgelid) stating that their understandings were very limited.

6.2 Perceptions

Discussion of nanotechnology's scope was often prefaced by reference to its trait of "...organising present knowledge in various areas; in chemistry...in biology, physics, engineering and so on" (Cornell). Hence, many interviewees spoke of nanotechnology's wide-ranging nature. Tanthapanichakoon, for example, suggested that nanotechnology "...covers almost everything in all fields and at all levels". In this light, interviewees highlighted their support for the substitutability of the word 'nanotechnology' with that of 'nanotechnologies'. Yet, Ford believed the wide-ranging nature of nanotechnology means the boundaries of where nanotechnology starts and begins are unclear, with Tanthapanichakoon noting that this can create a tension between having a definition that is "comprehensive" yet "unifying". The wide-ranging nature of nanotechnology also means that there will be vastly different approaches to nanotechnology research undertaken by different groups. Kanok-Nukulchai, for example, recognised the possibility of employing either a 'top-down' or 'bottom-up' approach to research in an area such as material science.

Given its wide-ranging nature, I was interested in understanding how nanotechnology is generally perceived in terms of its technical complexity. Amongst interviewees, nanotechnology is commonly perceived as 'high-tech' (Arya; Chirachanchai; Deutchmann; Yuthavong) or "cutting-edge" (Selgelid). Interestingly, Australian interviewees saw nanotechnology as more 'high-tech' than their Thai counterparts. To some extent, the explanation for this difference comes from the differences in associated assumptions based on the nature of the word 'nanotechnology'. Deutchmann, an Australian international health practitioner, who noted the limitations of his understanding of nanotechnology, went on to state: "...it is all at the high-tech end". For most, however, nanotechnology's 'high-tech' label was justified by the demands it creates in terms of the level of human or technical resources required. Tanthapanichakoon, for example, spoke of the need for "well qualified technicians" holding advanced knowledge to operate or maintain nanotechnology equipment. He also highlighted a view held by many Thai scientists when he said that nanotechnology equipment is quite specialised and precise, with the ability to look at nanostructures requiring very high-resolution devices. Yet others, such as Radt, challenged the idea that nanotechnology relies on "highly sophisticated instruments", paving the way for a belief that, even if nanotechnology is perceived as high-tech, its scope includes a wide range of applications that vary with respect to the demands of required inputs.

A number of Australian and Thai interviewees (Kanok-Nukulchai; Tanthapanichakoon; Tegart) believed nanotechnology is not just high-tech, but spans low- through to high-tech. Tegart, for example, spoke of low-tech nanotechnology having "...existed for a long time in terms of micronised powders", with Damrongchai adding that such material can translate into everyday products like self-cleaning powders, or influence manufacturing aspects of textiles such as silk. Even Tanthapanichakoon, who had previously presented nanotechnology as "high-tech", spoke of its scope encompassing "very basic research", such as putting nanoparticles into wine or developing water-repellent surfaces for garments. Reinforcing that nanotechnology represents a spectrum of applications with varying input demands, Tegart and Tanthapanichakoon both highlighted, at the high-tech end of nanotechnology's spectrum, the example of quantum dots at the high-tech end of nanotechnology's spectrum that require complex knowledge and intense technical infrastructure. However, a number of interviewees believed that nanotechnology is often inaccurately perceived as purely high-tech, a mistake they believed will be clarified with deeper understandings. Tegart, for example, posited that people who have read about nanotechnology will see the "...'gee whiz' stuff...[but that] the people who know a bit about it may be a little bit more circumspect...".

6.3 Near-Term Nanotechnology or Molecular Manufacturing?

For some Australian interviewees there was a belief that governments in the South might engage with nanotechnology under the pretence of its potential for applications of a speculative, 'highly futuristic' nature. Ford, for example, was worried that Southern images of nanotechnology might include "nano-bots" ahead of examples such as "...energy efficient coatings for windows and paints". Some interviewees (Radt; Tegart) who saw these ideas driven by the media, thought the hype might be even more exaggerated in the South. In addition to futuristic claims, it was also believed this hype, as witnessed in the North, could lead to a polarisation within Southern discussions. Weckert saw this possible polarisation as similar to the phenomena witnessed with the emergence of Artificial Intelligence in the 1970s and 1980s, where groups of people thought it would "save the world" and others thought "...it was one of the worst things that could happen...". A recurrence of this approach was envisaged in terms of 'doom and gloom' scenarios relating to molecular manufacturing and uncontrolled atomic self-replication that would result in the 'grey goo' phenomenon.

Discussion of futuristic applications do play some part in the Thai public discourse on nanotechnology, as witnessed by the example from the Senior Researcher from Thailand's National Centre for Genetic Engineering and Biotechnology (BIOTEC) of his speaking about the film 'Fantastic Voyage'[1] to students at Sirinthorn International Institute. Interviewees also acknowledged that a discourse around futuristic threats exists, with Thajchayapong, saying that "...people are talking about the 'grey goo'". However, although a common belief was held that the bulk of nanotechnology's applications were some way off in terms of Thai actualisation (Panitchpakdi; Sriyabhaya), 'futuristic' applications were never central to interviewee responses about nanotechnology. Furthermore, descriptions about nanotechnology and its applications never implied an understanding of nanotechnology as molecular manufacturing. On the contrary, for the few times when molecular manufacturing was raised in conversation, Thai interviewees spoke extremely cynically of its feasibility, particularly in the coming 20 years (Dutta). Damrongchai claimed this cynicism is supported by most Thai scientists who dismiss the "...realisation of so-called 'self-replicat[ion]'", resulting in a general absence in Thailand of the Northern 'doom and gloom' polemic. In this light, the Thai situation in 2004 would seem to differ from that across the North, in which generally polarised views were appearing amongst the public, courtesy of the popular science media. As a speculative explanation, Weckert turned to the issue of culture:

> It might be sort of a cultural thing, too. The 'grey goo' is sort of a nice image that...our media can do a lot with...maybe it will not [be the same coverage] in some other countries, particularly...if they think that there are enormous benefits from other aspects of it.

In this respect, Yuthavong agreed that pressing issues, such as bird flu, mean that new technologies are presented in terms of their ameliorating capabilities, rather than the potentially dire future consequences.

However, Thajchayapong felt that, given the rapid, global nature of information dissemination in the 21st century, a uniform understanding about nanotechnology's overarching paradigm is not guaranteed and that hype and concerns relating to molecular manufacturing could capture the Thai public's mind and change the framework for the associated discourse. Yet, Damrongchai saw this only happening if Northern debates further infiltrated Southern settings, saying a shift in public debate could be prompted if more common reference was made to articles appearing in foreign new papers such as the *New York Times*.

6.4 Conclusion

There is agreement about nanotechnology's universal characteristics but, simultaneously, there are substantially different ways in which it is conceptualised. There is both common ground and critical difference in the way that nanotechnology is understood amongst the interviewees, raising both opportunities and challenges for common discussion about nanotechnology's implications for global inequity. For example, if nanotechnology's complexity is presented in opposing ways, is it ever worthwhile to compare arguments relating to the expected costs and infrastructure required for a Southern country to conduct nanotechnology R&D?

Although this research has only investigated the perspectives of a limited number of key informants from Australia and Thailand, the clear identification of six common characteristics, in terms of how nanotechnology is defined, raises the possibility that interactions between the South and North can be based upon shared foundations. These characteristics include: nanotechnology's length scale; its focus on the control of matter; its exploitation of novel scale-based phenomena; its practical nature; its rebranding and integration of existing practices; and its subsequent, natural emergence across a number of sectors resulting in a new field, rather than a new discipline or industry.

However, how nanotechnology is understood goes beyond its defining characteristics, as there appears to be large differences in nanotechnology's perceived scope and complexity, holding critical repercussions for policy debates in terms of the feasibility, relevance and limitations of nanotechnology R&D in the South. Yet, when comparing Thai and Australian interviewee perspectives, it becomes readily apparent that the distinctions in perception are less between countries than between interviewees with expertise in differing fields. This is particularly true in terms of nanotechnology's claimed novelty, its range of applications and its complexity, and may be explained by an individual's level of nanotechnology awareness or their self-interest in presenting nanotechnology in a certain way.

Overarching these debates is the paradigmatic framework encompassing nanotechnology. Contrary to popular belief amongst Australian interviewees, nanotechnology in Thailand is framed in terms of its near-term capabilities, rather than those attributed to the speculative paradigm of molecular manufacturing. Whilst one interviewee suggested this as a phenomenon grounded in cultural difference, the responses from Thai interviewees, as well as the findings in my literature review, suggest that the market guides the framing of nanotechnology in the South, thereby dictating a focus on the kind of nanotechnology that presents foreseeable outcomes.

Whilst these results differ from those presented by the literature, the literature may be illustrative of a broader definitional disjuncture that would understandably impede the usefulness of discussions about nanotechnology's consequences for global equity. In this light, this chapter has provided me with tentative confidence about the validity of my quantitative results, based on engagement around commonly identified phenomena. Furthermore, in exploring the questions at the heart of the coming three research chapters, I am mindful that, whilst discussions may focus on a technology considered to have common characteristics, the very same technology may be conceptualised in very different ways.

Such considerations may be most applicable to my investigation in the upcoming chapter, in which I will continue my qualitative research by exploring interviewee perspectives about Southern innovative capacity with respect to nanotechnology R&D.

Endnotes

1   A 1966 film in which humans venture in microscopic submarines into the human body to repair problems one cell at a time.

Chapter 7

Innovative Capacity

In order to further interpret my quantitative data and explore nanotechnology's foreseen implications for greater equity when it comes to global innovation, in this chapter I will examine interviewee perspectives about Southern engagement with nano-innovation. In particular, I will explore the extent to which nano-innovation and innovative capacity will be globally and locally decentralised and autonomous.

7.1 Understanding the Nanodivide and Its Constructs

Interviewees regularly referred to the term 'nanodivide' but assumed its meaning and knowledge of how it is constructed are commonly understood. Given the apparent difference in understandings that emerged, in this section I will seek to piece together interviewee's comments to establish some relevant clarity. I will first consider how the term 'nanodivide' is interpreted and understood. I will then explore some early characteristics of the nano-innovation divide in terms of its 'leaders' and those considered 'left behind'. Addressing these two points will establish a context and framework for my assessment, in this chapter, of Southern roles in global nano-innovation.

Generally speaking, the literature does not clarify what is meant by a nanodivide. Yet it is clear that, for interviewees, the term

Nanotechnology and Global Equality
Donald Maclurcan
Copyright © 2012 by Pan Stanford Publishing Pte. Ltd.
www.panstanford.com

nanodivide can have two different meanings. The first, that I term the 'nano-innovation divide', refers to inequity based on where knowledge is generated and retained and a country's capacity to engage in these two processes. Tegart presented this divide as one between the "...information rich and information poor...". Those on the 'leading' side of this divide are seen as able to actively contribute to and direct nanotechnology's trajectory, whilst those not leading are seen as playing passive roles, unable to exert influence over any sphere of nanotechnology's global trajectory. The second meaning, that I term the 'nano-orientation divide', refers to inequity based on the areas in which nanotechnology research is targeted, as compared to the areas in which it would address basic human needs. In this sense, Arya spoke of a differentiation between nanotechnology addressing 'real' and 'felt' needs. For many, this translated into a belief that nanotechnology would be governed more by market push- than social pull-factors.

Whilst these two divides differ in their nature, where research is targeted is often initially dependent on where knowledge is being generated and retained. Given that most of the world's R&D for emerging technology occurs in the North, comments on the orientation divide generally related to global inequities in terms of limited Northern research focussed on Southern challenges. However, the prospect of the South as active agents in global nano-innovation prompts additional consideration for inequities in the orientation of Southern generated knowledge (to be explored in the next chapter).

According to the interviewees such as Damrongchai, there is an increasing concentration of nanotechnology R&D generation and ownership in the hands of "limited leading countries". Interviewees classified countries as 'leaders' largely because of the high levels of early nanotechnology investment (Charinpanitkul; Deutchmann; Tegart), they also happened to be some of the more wealthy countries in the world (Kanok-Nukulchai). Leaders in nano-innovation were said to include[1] the U.S., Japan, Taiwan, Germany, Australia, Sweden, the U.K., France, Switzerland and Hong Kong. Despite often being grouped with the North, in terms of nanotechnology capabilities, China was not mentioned by interviewees as a nanotechnology 'leader'.

Given certain countries are seen as nano-innovation leaders, it was not surprising that interviewees (Berwick; Kanok-Nukulchai;

Sawanpanyalert; Tanthapanichakoon; Warris) referred to countries, at the other end of the spectrum with respect to engagement with nanotechnology R&D, as 'left behind'. The previously outlined pressure to be at the forefront of nano-innovation is often driven by a belief that if a country neglects nanotechnology it will be in an unenviable position later on (Tanthapanichakoon; Turney), having to try to "catch-up" (Berwick). According to Tanthapanichakoon, even a country with endogenous nanotechnology capabilities could fall behind if it did not seek to constantly develop its research position. The insinuation here is that, rather than all countries gaining from nanotechnology, no matter what the nature of their engagement, those that do not develop and maintain competitive innovative capabilities will actually lose out. Furthermore, as I shall soon explore, there is a perceived potential for nanotechnology to reinforce the underdevelopment of some countries by creating greater technological dependency (Arya; Charinpanitkul; Yuthavong).

The proposition that countries will play different 'roles' in global nano-innovation prompts greater exploration of exactly what kind of roles are envisaged for the South. Will the South be left behind by nanotechnology or will the situation present new opportunities allowing Southern countries to become agents in global nano-innovation? As shall soon be seen, this question leads to an investigation of the barriers and possibilities for Southern nano-innovation.

Four key issues relating to barriers and possibilities were presented as largely determining Southern roles in global nano-innovation: understanding, commitment, resources and infrastructure (Sawanpanyalert). Whilst many perspectives, particularly Australian, went beyond considering the development 'problem' as one solely influenced by issues of domestic Southern capacity by also looking at global externalities and contexts, these four issues form the crux of my discussion in this chapter.

7.2   The South Left Behind

Interviewees presented a number of different explanations for why Southern countries might be left behind in nanotechnology's global development. In the first part of this section I will look at some of the envisaged scenarios for Southern countries, exploring varying levels of engagement and different kinds of roles in global nano-innovation. I will then progress to addressing the underlying assumption — that the South cannot play an active role in global nano-innovation — by exploring the perceived challenges to developing innovative capabilities.

For some Australian interviewees, the possibility for nanotechnology's development in the South was tenuous; seen as a "contradiction in terms" (Coyle) or "...[not] a direct link, by any means" (Cornell). Those who struggled to see any link suggested that nanotechnology was irrelevant to the South and that, not only was endogenous R&D unlikely, but that they may not even play the role of 'recipient', given that "...existing, basic, often very cheap, sometimes even free, technologies or medicines, are not available in developing countries..." (Selgelid).

However, many of the interviewees, led by the Australians, saw the link between nanotechnology and the South via some form of passive diffusion, where the role of the South was as 'recipient', rather than innovator, particularly in the "...very, very poor countries in Africa" (Tegart). The common implication was that nanotechnology will most likely reach the South as a result of Northern influence. Selgelid's response highlighted this mindset when he commented: "nanotechnology would be great if someone really made it and provided it to developing countries". Bryce's reasoning was along similar lines, as he saw diffusion coming via a "serendipitous process".

Northern-dictated aid was seen as a likely mechanism for Southern engagement with nano-innovation. Berwick, a policy officer with Invest Australia,[2] saw potential for nanotechnology to be incorporated in "...world aid programs and assistance programs to help developing nations just help themselves develop further". Similarly, Cornell referred to potential areas of application in U.N. aid packages or U.S. or European initiatives operating in the South. Australian interviewees also saw the potential for international aid organisations to invest in Southern nanotechnology, commonly citing organisations such as the Bill and Melinda Gates Foundation. Braach-Myksvitis noted that this kind of activity has already commenced, with the Global Research Alliance[3] having been approached by a number of Northern foundations wishing to ensure the benefits of nanotechnology reach the South in this way. Damrongchai, a Thai technology policy officer, agreed that there was potential for nanotechnology to enter the South via aid, citing potential applications such as single-life diagnostic kits and methods by which to increase food preservation. His comments were distinct, with most Thai interviewees ambivalent about nanotechnology's potential delivery through aid and development assistance, having given little consideration to the idea. Panitchpakdi was one who spoke about the potential for greater Southern dependency as a result of becoming recipients of nanotechnology-based aid whereby it would "...all depend on the countries that are advanced [if they] are willing to share". For Damrongchai, the idea of a Northern-controlled situation raised considerable concerns about donors exerting political influence and impressing conditionality upon Southern recipient countries.

As an extension of aid, technology transfer was also discussed as a mechanism by which the South's role would remain passive.[4] Some interviewees (Bryce; Tegart) suggested that, with the right education and training, nanotechnology could be transposed from the North to the South. Importing products and technologies was viewed as a possible means of engagement for those countries without nanotechnology R&D capabilities (Arya). Kanok-Nuckulchai believed that LDCs, although unable to "build technology themselves" if they were to attempt to engage with nano-innovation today, can still start accumulating the knowledge and, once they have sufficient human resources and infrastructure, "...can absorb and transfer some of technology...".

Others, such as Cooper, saw this form of engagement with technology as entrenching the passivity of the South in global nanotechnology processes via a continuation of Southern technological dependency. Some interviewees, including Ratanakul, envisaged this dependency in terms of a 'trickle-down' of nanotechnology from the North to countries without endogenous innovative capabilities. Cornell, for example, imagined the South would only benefit "as a consequence" of 'spin-offs' from Northern nanotechnology advances in areas such as water desalination, cheap nutritional foods and low-cost fuels. Weckert was similarly cynical about the way in which many Southern countries would engage with nanotechnology, suggesting that until Southern countries "...get a bit more economically advanced", their engagement may be limited to Northern companies who "...see some big economic advantage" to distributing nanotechnology in the South. Tegart added that a number of U.S. companies might already be viewing the South as a "potential market" for Northern products. In this respect, a number of Thai interviewees (Chirachanchai; Damrongchai) confirmed that 'nanoproducts' have already entered the Thai market, with a strong fear emerging amongst interviewees of 'import dependency'. Here there was a belief that Thailand could end up "...buying a lot of things" (Yuthavong) and 'losing a lot of currency' by importing high-cost technology through both products and services (Charinpanitkul; Kanok-Nukulchai; Yuthavong). Furthermore, Thai interviewees, such as Charinpanitkul, believe that there is a danger that the Thai population will ignore domestically produced products once they "... get used to those [international] products". For Damrongchai, it is this continual buying of products from the North that will perpetuate underdevelopment.

Another kind of passive role presented the South as 'nanomanufacturers'. Thai interviewees, such as Sawanpanyalert, saw a role for the South to "...partner in the manufacturing stage" with associates in the North. For Weckert, this would merely be a case of Northern companies outsourcing work. Dutta saw such moves by companies as a natural outcome of the world system, explaining:

> ...for niche products, [where the] investments are lower, they [developed countries] will have to transfer the technology where you need more labour, where you need larger space to manufacture. You cannot keep it in developed nations, it is too expensive...

Others (Ford; Tegart) agreed it was likely that countries with strong nanotechnology programs would exploit countries playing passive roles in its development. There was recognition that Northern nanotechnology R&D partnerships with the South would seek to benefit from reduced costs (Bryce) and lower levels of regulation in the South (Cornell). Ratanakul was sceptical that such partnerships would allow Southern countries to play an active role in global nano-innovation, saying:

> The problem is these Western scientists are doing research for their own benefit...when they finish the research they go back and then they create new technology, based on the research, and they sell it [to] us.

The suggestion here was that the North will value-add to nanotechnology products possibly manufactured in the South, and that those not absorbed by Northern markets or considered 'too risky', will be off-loaded to the South, as often happens with pharmaceuticals. In this light, Deutchmann spoke of his concern that "...junk products would be dumped at a cheap price on the developing world...", with Cornell responding that many countries from the South will take "...whatever is available at a reasonable price".

At the heart of nearly all of these scenarios is the assumption that nanotechnology R&D, and therefore a potentially active role in global nano-innovation, is beyond the realm of the South. Turney and Berwick, respectively, spoke of a "cultural perception" and "natural tendency" to expect that nanotechnology will be a Northern technology. Interviewees, such as Chirachanchai, provided support for this hypothesis by speaking of the "advanced countries" assuming leadership roles in global nano-innovation. Additionally, there was a fear that "...developing countries are going to miss the boat...[and not take] advantage of nanotechnology to exploit their local, competitive advantages..." (Turney). Kanok-Nukulchai, amongst others, noted that his initial impressions about developing nanotechnology in a country such as Thailand were "distant", with ventures into research feeling "contradictory". Others spoke of the barrier of internal cynicism, suggesting that even in a country, like Thailand, that is making efforts to become active in global nano-innovation, there is potential for people to think "...it is just too difficult" (Turney), or that the research is "too late" to catch the North, leading to the cessation of activity (Chirachanchai).

Driving these perceptions was a conviction that "in-house" development of nanotechnology is too difficult for many Southern countries, given the weak capacity for innovation (Yuthavong). Thus, I will now explore a range of factors presented as challenges to developing endogenous Southern capacity.

Preceding the issue of basic innovative infrastructure was a belief that a lack of awareness, understanding and commitment could inhibit the ability for Southern countries to enter global nano-innovation. In this light, a number of Australian interviewees (Ford; Tegart; Warris) said that the initial barriers to Southern innovation include awareness of what nanotechnology is actually about and recognition of its potential and importance. This lack of awareness is said to be compounded by poor understanding, particularly amongst Southern leaders (Tegart). In contrast, Thajchayapong suggested that there seems to be a great deal of public awareness around nanotechnology and a general acceptance of nanotechnology's merits amongst policymakers in the E.U., Japan and the U.S.

According to many interviewees (Cornell; Kanok-Nukulchai; Turney), lack of Southern awareness and understanding about nanotechnology fits within a bigger picture, whereby nano-innovation is 'prioritised out' by Southern governments due to more immediate, basic needs. In this respect, Kanok-Nukulchai explained that nanotechnology is perceived as a "luxurious" investment, particularly in light of its embryonic state of development and the long-term nature of 'returns'.

Following on from this, some interviewees (Panitchpakdi; Tegart) saw a challenge for the South in gaining political commitment for nanotechnology. This, it was suggested, could translate into inadequate resource allocation (Sriyabhaya; Tegart), with a particular concern that Southern nanotechnology R&D would be underfunded (Dutta; Panitchpakdi; Radt; Sriyabhaya; Tegart; Warris). With regard to the Thai situation, Tanthapanichakoon noted: "...we still do not have a very strong budget or input into nanotechnology".

In addition to awareness, understanding and commitment, many other challenges were seen as reducing the ability to develop nanotechnology capabilities (Lynskey). According to Warris: "...[Southern countries] have got to get certain things in order before they can get into more high-tech applications, such as nanotechnology". In this light, interviewees pointed to challenges with respect to both basic knowledge and capacity.

In terms of basic knowledge, there was a belief that Thailand lacks the fundamental knowledge to become nano-innovators (Sawanpanyalert). Chirachanchai elaborated by saying that, in the past, "...understanding at the molecular level has been neglected". In this sense, the fact that "...the science is not quite there" (Thajchayapong) was believed to place the country "...a bit far behind from the very beginning" (Chirachanchai). More generally, Cornell was sceptical about the ability for the South to lead innovation, saying: "...the idea of them [Southern countries] being at the frontier of any of these areas is somewhat difficult to perceive, given the fact that it is born, really, at the very cutting-edge of developed country science...".

With respect to basic capacity, interviewees specifically referred to 'human resources' as the "biggest concern" (Damrongchai) and the "greatest obstacle" (Charinpanitkul) to Southern nano-innovation.

In this light, Thailand was already said to be experiencing a shortfall of researchers (Dutta; Kanok-Nukulchai; Yuthavong), compacted by the belief that the country will face difficulties in finding people with an 'interest' in nanotechnology (Arya; Sriyabhaya). Dutta spoke of particular Thai shortfalls with respect to those with basic knowledge in quantum physics and chemistry, or those with an ability to shift into nanotechnology from other fields. For Lynskey, this lack of a "critical mass" of relevant human resources in a country such as Thailand makes the idea of a country such as Eritrea developing endogenous nanotechnology capabilities "completely unrealistic". These challenges were placed in the broader context of a general shortage of science and technology researchers in the South[5] (Tanthapanichakoon) and critical weaknesses in terms of educational capacity (Tegart; Turney). Cornell said that this would be a particular barrier for Southern nanotechnology, given the lack of "...long-standing commitment to education" in areas that form the basis for developing nanotechnology capabilities, such as the molecular sciences. In addition to researchers, technicians were seen to be "...critical people...in the exploitation of much of this technology" and a further area in which the South was viewed as being in a much weaker position than the North (Tegart). In response, Kanok-Nukulchai noted that it might take some time to cultivate the expertise nanotechnology demands.

Furthermore, interviewees (Ford; Tegart) spoke of a strong potential for Southern nanotechnology researchers to be drawn to the North via the brain drain phenomenon. Brain drain was seen as a threat to retaining workers in a country such as Thailand if it lacked critical nanotechnology infrastructure, with a belief that the "best brains" would start to look to Singapore, the U.S. or Europe (Damrongchai). These challenges were seen as presenting a significant barrier to developing and retaining a critical mass of researchers in the South (Damrongchai; Ford; Tegart). Ratanakul believed this was already a major challenge for Thailand, stating that "...the government has not been thinking of the measures to prevent the well trained Thai scientists from being lured away by affluent nations...".

The other threat posed by brain drain relates to the challenges of reintegrating returned researchers into Southern contexts. Charipanikul believed that it would be quite difficult for Thais with nanotechnology expertise to find employment upon return from overseas work or training. Dutta explained that, whilst "...the Thai government has spent a lot of funds to train people abroad...", their adaptation time upon return is too long "...because there are no active groups working here".

In addition to human resources, many saw infrastructure as a basic requirement, and thereby major challenge, for the Southern development of nanotechnology capabilities. Interviewees first discussed problems with the amount of infrastructure available, referring only to 'hard' infrastructure, such as equipment and instrumentation. Like a number of others, Dutta believed that "... nanotechnology needs quite a bit of infrastructure", given the scale on which the research occurs. But Cornell was more forthright, saying that if a country wants to seriously engage in nanotechnology R&D then "...you could fill telephone books with the kind of infrastructure that you need". In elaborating, he listed, as necessities:

> ...an ability to work with ceramics, with plastics, organic chemistry development, with fine metals, thin-film deposition, you must have electronic foundries...you need everything that currently supports a modern industrial economy and that goes from screen-printing, paints, chemistries, lubricants, polymers, waxes, solvents, all of the moulding industries, the etching industries, electrochemical industries.

In this respect, two interviewees (Tanthapanichakoon; Tegart) referred to the "limited capabilities" within the South. According to Cornell, Southern countries have "...not yet advanced to the point whereby this kind of equipment, this kind of capability, is naturally part of their world". This was partially explained, he said, by the Southern absence of the military industrial complex — said to be foundational in driving nano-innovation in the case of the U.S.

The second aspect of infrastructural requirements relates to the quality and cost of instrumentation. In this respect, Thailand was seen to lack some of the required equipment (Dutta), with Kanok-Nukulchai noting a common perception that the technology Thailand requires is beyond its present capabilities:

> ...when we talk about nanotechnology most people think...it is something we cannot see, something that need[s] a lot of high-tech equipment and when we look back at Thailand we are not that advanced in terms of technology.

Prohibitive costs were presented as the main barrier to the acquisition of such instrumentation (Berwick), with Cornell believing that nanotechnology requires "...a fairly large investment in fairly expensive equipment".

In addition to challenges with respect to basic capabilities, interviewees spoke of challenges relating to the development and maintenance of competitive, nanotechnology R&D capabilities. There was a strong belief, for example, that access to appropriate instrumentation was a key barrier to the development of Southern nanotechnology capabilities. Radt believed many of the Southern countries actually have the instruments required to undertake nanotechnology R&D but saw the barrier more as a matter of access to, and maintenance of, these instruments. Dutta partly agreed, saying of the existing instrumentation, much of it is "underused".

The challenges around developing basic nanotechnology capabilities suggest coordination and strategic planning is required. However, coordination is another area in which interviewees saw challenges for Thailand, with a genuine concern that research will be "...unfocused and resource[s] will be scattered" (Kanok-Nukulchai). Chirachanchai, for example, saw problems in ensuring that each research effort was part of an overall strategy for Thai nanotechnology. In this light, Tanthapanichakoon highlighted that Thailand "...does not have a national strategy and all the labs, or centres, are working on their own interests or on their own subjects, without coordination...". Furthermore, he said that Thailand suffers from a significant breakdown in communication between many of the government agencies that will need to work together when it comes to nano-innovation.

Accompanying coordination of research is the Southern challenge of strategic planning. Tegart believed the initial planning difficulty is in assessing capabilities and then selecting focussed areas for research. Dutta's concern, that people talk generally about nanotechnology without a concentrated focus in any particular direction, was seen as part of a bigger fear that Thai nanotechnology lacks a clear and comprehensive vision for the future (Chirachanchai).

Developing focussed nanotechnology research can be made even more difficult if the ability to build knowledge is restricted by nano-innovation's global leaders. Interviewees strongly argued that some of the greatest barriers to Southern innovation relate to "...who's involved and actually creating the technologies" (Cooper) and, stemming from this, the "big issue" of control over IPRs (Damrongchai).

Concerns about the inhibitive impact of Northern nanotechnology patenting upon Southern attempts to develop innovative capabilities were seen to be uniquely exacerbated given nanotechnology relates to the fundamental building blocks of all material things (Damrongchai). Added to this, the potential disappearance of the 'cost-barrier' for nanotechnology R&D (to be discussed in the coming section), makes the issue of control over patents even more important because "...the powerbroker will be the knowledge" (Braach-Maksvytis). Furthermore, a major concern held by interviewees was that a great deal of nanotechnology patenting would be speculative in order to claim future applications. Pothsiri was worried that "...a Western country, particularly in the private sector...may try to play around with this kind of thing without making any attempt to find a new innovation". Additionally, a number of interviewees (Changthavorn; Selgelid) said that, in light of a greater research focus on atomic self-assembly, their concerns lay with the increasing move towards, or ambiguous nature of, 'process' patents.[6] In these respects, nanotechnology was seen as encouraging corporate monopolies (Ratanakul), thereby blocking potential avenues for Southern R&D (Cooper; Braach-Maksvytis).

One Thai interviewee (Arya) spoke very strongly about how nanotechnology patenting will maintain and promote the technological divide through ongoing oppression of the South. He presented IPRs as the "...new economic power...[and] new instrument of domination", with patent holders often over-exploiting their position of strength. He went on to say that the control of proprietary knowledge is driving greater oppression through a divide that, in addition to being technological, includes an:

> ...economic, social and also political divide, because those who
> have the new technologies will also invest, not only for the
> products to serve mankind but the products which can be used
> for domination, for hegemony, weapons of new kinds and so on
> and so forth.

General concerns were also held for the ability for Southern countries to translate nanotechnology research into patented knowledge. Charinpanitkul said that nanotechnology patenting in Thailand "...will be a big obstacle", given patent understanding, even in the university, is insufficient. Added to this, Changthavorn pointed specifically to a lack of nanotechnology understanding amongst Thai lawyers. Charinpanitkul saw the lack of knowledge as severe and inhibitive, highlighting, with respect to nanotechnology patent applications, that "...we do not know even what style or what wording we should add...".

A contributing factor to weak Southern patenting and another major challenge to the South playing active roles in global nano-innovation is the potentially poor level of private sector engagement with nanotechnology R&D (Turney). Thajchayapong claimed that the science behind nanotechnology actually demands greater participation from industry, with private sector participation suggested as a crucial driver for early nanotechnology success in Japan and Taiwan (Charinpanitkul).

In addition to foreseeable financial 'return', available financing and other financial incentives were seen as the initial drivers of private sector engagement in nanotechnology. However, building on earlier concerns about a general lack of funding, there was a belief that risks, particularly those associated with IP, could make access to nanotechnology finance and capital a serious problem in some of the Southern countries (Turney). In this light, many, such as Tanthapanichakoon, saw the countries of the North in comparatively strong positions, suggesting that a country such as the U.S. is in "the best position" when it comes to nanotechnology R&D, largely because of the "...good system of venture capital" in place.

A second challenge facing the development of Southern-owned proprietary knowledge is the difficulty of technology transfer from academia to industry. Whilst my survey of Thai nanotechnology practitioners showed that slightly more than two-thirds of the research currently underway is 'applied', the responses also demonstrated weak professional links between academia and industry. Most professional collaborations were with academia (57%), with only a quarter (25.5%) of collaborations engaging the private sector (see *Appendix J*). Thajchayapong said these weak private sector links were most visible in the poor levels of communication about nanotechnology between industry, government policymakers, researchers and academia. Finally, there was a belief that Southern firms are limited in their ability to absorb nanotechnology R&D from academia (Pothsiri).

Even leveraging from international partnerships to overcome capacity issues encountered scepticism. Interviewees reiterated barriers such as the lack of Southern infrastructure (Cornell), as well as the intensely competitive nature of contemporary global engagement with nano-innovation. To elaborate, Tanthapanichakoon used the example of the New Energy and Industrial Technology Development Organisation of Japan whose policy, in terms of collaborative research, is "...no grants at all in nanotechnology field[s]".

It is important to note that most of the challenges raised for the South to play an active role in global nano-innovation were not considered specific to nanotechnology but, rather, generic to all emerging tech fields. Some of the examples presented include: the low levels of investment (Panitchpakdi); the Northern concentration of proprietary knowledge (Tanthapanichakoon) — with nanotechnology said, in this respect, to raise issues similar to those at the forefront of the biotechnology debate (Damrongchai); there were also shared problems in the hesitation of the private sector to engage in nanotechnology R&D and difficulties associated with technology transfer (Tegart).

If many of the challenges are generic, is there anything to suggest that global roles in emerging innovation will change at the hands of nanotechnology? Will this divide be any different to preceding technological divides? Selgelid thought not, saying he saw no reason why the general situation relating to inequality would be any different for nanotechnology:

> The North-South divide is really complex and I do not see why there should be anything special or unique about the North-South divide or rich-poor divide as far as nanotechnology [is concerned]. I would imagine the same kind of dynamics that are driving inequality in all kinds of other domains would just apply to this domain, as well...

Others were concerned that nanotechnology's innovation divide could be "exaggerated" (Deutchmann) and worse than the divide currently witnessed with information and communications technology (Weckert). This view was partially justified by a belief that nanotechnology enters on a platform of existing and widening divides (Cornell; Tegart; Weckert), particularly driven by developments in biotechnology and information and communications technology (Tanthapanichakoon). In this light, Yuthavong saw potential for an extreme shifting in the concentration of R&D away from Southern countries, saying that nanotechnology is "moving too fast" for many Southern countries to "...really capture the benefits fully", with others (Sawanpanyalert; Tegart) adding that Northern countries are in a much more favourable position to respond and adapt their capabilities. However, as I shall now explore, others saw nanotechnology presenting new opportunities for the South to play an active role in global innovation.

7.3 New Opportunities

Whilst previous arguments suggest that an active role in nano-innovation is beyond the South, others (Dutta; Turney) suggested that the barriers are more matters of perception[7] and that nanotechnology can also be viewed as an opportunity for the South to engage in global R&D. As mentioned in an earlier chapter, despite seemingly universal understandings, nanotechnology conjures up a range of perceptions, some of which fail to consider nanotechnology in some of its more simple forms. 'Further consideration' is suggested as leading to more 'circumspect' perspectives (Tegart), with Selgelid's responses highlighting this point: after considering the issues in greater detail, Selgelid stated "...there is the possibility that nanotechnology is not out of the reach, or should not be out of the reach of developing countries". In this respect, a number of interviewees, particularly those from Thailand, saw alternative paths that involved the South as 'nano-innovators', actively contributing to nanotechnology's global trajectory. Nearly one-third of interviewees from both Australia and Thailand (Berwick; Charinpanitkul; Chirachanchai; Damrongchai; Deutchmann; Dutta; Ford; Tegart; Yuthavong) specifically referred to nanotechnology providing 'opportunities' or holding 'potential' for Southern innovation. In fact, some even suggested that Southern countries actually "have the advantage" (Turney) in terms of nano-innovation and that, on the back of various technologies, the South would "...be at the same level" (Dutta) as the North at some stage in the future. In this section I will provide the supporting arguments presented for many of these claims.

There are two main elements behind the argument that the South can play an active role in global nano-innovation. The first is that early signs of Southern nanotechnology commitment, as highlighted in my review of the literature, could set a platform for more active engagement in global innovation, including the involvement of some of the LDCs. In this respect, interviewees also outlined constraints as to the kind of nanotechnology activity that might be possible. The second element is the suggestion that Southern countries might be able to overcome a number of the previously raised barriers and challenges to developing innovative capabilities.

Interviewees introduced the issue of Southern nanotechnology commitment by citing nine Southern countries active in nanotechnology R&D and commenting on the perceived strength of each country's commitment. China was the Southern country presented as playing the most active role in nano-innovation and was often grouped with the North, given it is "...moving so fast and putting so many resources into nanotechnology..." (Tanthapanichakoon). Warris noted that "...China had the highest ratio [of nanoscience compared to their total science] in the world", highlighting that the Chinese government had identified nanotechnology as an area of increasing importance. Following China, India, South Africa and South Korea were all seen as playing highly active roles, whilst Indonesia, Malaysia, the Philippines, Thailand and Vietnam were viewed as playing moderately active roles. Although Tegart saw the Philippines, Vietnam and Indonesia as "much further behind", he believed there was a strong chance that each would have a strong nanotechnology presence in the future. As Tanthapanitchakoon noted, there is a lot of enthusiasm in Vietnam, with the country's government "...very keen to promote their nanotechnology program... [in order to] catch up".

In terms of Thailand, both Tegart and Turney spoke about the willingness and drive from various quarters to ensure the development of innovative nanotechnology capabilities. Thai interviewees confirmed the strong desire and ambitions, with Pothsiri noting that the Thai government's policy is "...to promote this kind of innovation to be...something that we would be able to do by ourselves...". Arya said that the Thai government's hopes are actually targeted at ensuring nanotechnology contributes up to 1% of the Thai GDP in the coming 10 years. Strategically, given its central location amongst South East Asian nations, Thailand is seeking to be a "hub" for nanotechnology (Charinpanitkul), with Tegart certain that Thailand could be among the leaders in South East Asia if it receives strong government support.

Pothsiri, speaking about the Thai nanotechnology situation, said that "right now the chance is quite good [to build capacity] because there is a policy commitment from the government". Interviewees, such as Damrongchai, said that the policy commitment had already resulted in initial funding, with a belief that there is enough money available to make a substantial investment in nanotechnology R&D. In addition to early funding, interviewees noted the establishment of a national nanotechnology centre and the earmarking of specific agencies to drive nanotechnology forward. Interviewees also noted that this policy commitment was translating into support for nanotechnology across a number of sectors. Thajchayapong highlighted examples from his own experiences with the Ministries of Commerce and Defence:

> ...I was invited by [the] Ministry of Commerce Permanent Secretary and I was explaining to them about nanotechnology, in front of 150 or 200 people...[similarly, at] the military school, they had about 150 student[s] listening to nanotechnology...

However, many interviewees (Changthavorn; Damrongchai; Kanok-Nukulchai; Sriyabhaya) were quick to highlight that these developments all stemmed from the Thai Prime Minister,[8] who was pinpointed as the main driving force for nanotechnology in Thailand. More explicitly, Turney noted that "the Prime Minister is actually driving this, personally, as something he wants to see happen", with Tanthapanichakoon adding that the Thai Prime Minister individually realised the importance of nanotechnology for Thailand's future. Thajchayapong, highlighting the importance of political leadership in a country such as Thailand, outlined the circumstances in which the Prime Minister initiated Thailand's first, serious foray into nanotechnology:

> ...about two years ago he [the Prime Minister] went to the science park. He visited us and he was surprised. He used the words to the effect, 'I did not realise that you have done so much' and then he mentioned about nanotechnology. And that is how we say, 'o.k., if Prime Minister use the word nanotechnology we have to respond' and we set up the centre.

Thailand's high level of commitment, combined with the previously mentioned nanotechnology activity in other countries, suggests the existence of foundations upon which Southern countries could play active roles in global nano-innovation. A few of the Australian interviewees (Braach-Maksvytis; Tegart) admitted surprise at the early nanotechnology capabilities in some of the Southern countries. Braach-Myksvitis saw the "early start" from Southern countries as something new in the science and technology arena. In this respect, Cooper said that nano-innovation does not have to follow the same distributive pattern as innovation in biotechnology, believing that early widespread engagement "...totally changes the picture". In presenting hopeful visions of Southern countries playing active roles in global nano-innovation, Cooper referred to other Southern successes such as the development of the Indian pharmaceutical industry in the 1970s and its associated production of affordable drugs, geared at local needs, whilst Lynskey similarly pointed to South Africa's recent ability to build a critical mass of scientific researchers.

Although the examples presented thus far deal mainly with the more technologically advanced Southern countries, there was a belief that nano-innovation could extend to the LDCs (Dutta; Ratanakul). Charinpanitkul, who assessed Thailand's nanotechnology capabilities in 2004, said that countries like Laos could be conducting nanotechnology research within five years. However, interviewees also placed limitations on the scope and nature of nanotechnology activity that might be possible in some Southern countries. For instance, there was a belief that Southern countries might only be able to engage with nano-innovation at certain stages of the R&D cycle. Although Sawanpanyalert did not see Southern countries as necessarily able to work on 'early-stage' nanotechnology research, he said that there are opportunities for 'later-stage' research. In this respect, Southern innovation was seen as dependent on partnerships with the North (Sawanpanyalert; Turney; Yuthavong), particularly in an area such as drug development where Southern countries "...cannot do it alone..." (Yuthavong). Dutta pointed to his collaborations with laboratories in Switzerland and Sweden as the kind of partnerships required for components of the R&D phases that cannot be completed in Thailand.

However, harking back to the way nanotechnology is perceived, Braach-Maksvytis noted that the scope of a country's contribution to global nano-innovation "...depends on what end of the scale you are talking about". She said that low-tech scale production of nanopowders for cosmetics, plastics and the polishing of silicon chips could be globally widespread and developed entirely within the South, whilst "...niche products for which you need very high investments" might remain outputs from the North. In terms of the nature of nanotechnology research, the ability to conduct both fundamental and applied scientific research across all fields was seen as beyond many of the Southern countries. Arya explained this situation with respect to his own country's context:

> ...Thailand is not in a good position to compete at the fundamental research level. We do not have the capacity, we do not have the potential. So, whereas we can do some fundamental research for nanomaterials or nanotechnology, we have to look, more, at the applied research.

As I shall explore in Chapter 9, the need to focus on applied research was closely followed by a need for Southern countries to find their own niche areas of application, building on niche knowledge (Arya; Songsivilai).

As I have already discussed in this chapter, a range of capacity issues contribute to the determining of a country's level of engagement with nano-innovation (Charinpanitkul; Sawanpanyalert). Nonetheless, arguments were made in six areas for why Southern countries might be able to overcome some of the previously raised barriers and challenges to developing innovative capabilities.

7.3.1   Availability and Demand of Human Resources

The first argument for hope in the ability for the South to engage in nano-innovation related to the level of available and appropriate human resources. With respect to the Thai situation, Tanthapanichakoon noted that, from the outset, the Prime Minister demanded particular effort in the area of "...human resources development in nanotechnology". The general belief was that Thailand possesses an adequate workforce to commence nanotechnology R&D initiatives (Damrongchai), with an estimated number of "...not less than 100 researchers[9]...[and] nearly twenty laboratories that have been working on nanotechnology, scattered around [the] universities" (Thajchayapong).

These views are coupled with a belief that nanotechnology might not require a big labour force (Kanok-Nukulchai); that a small number of scientists, with differing backgrounds but unified in their focus on nanotechnology, "...can be quite a big force" (Yuthavong). The implication is that countries can consider nano-innovation even if they only have a few researchers in nanotechnology-related fields (Warris). Lynksey, although previously sceptical about the hopes for endogenous innovation, said that getting a 'critical mass' is possible for countries — even some of the LDCs — that have a credible political approach, although this would be "...a couple of generations away". Yuthavong agreed that a focussed political strategy can make all the difference, stating that Thailand can develop a critical mass of scientists "...if it gets the right policy and right directives".

Furthermore, discussing nanotechnology with some of the Thai interviewees unearthed a surprising number of people who claimed that they, or their colleagues, had been working in nanotechnology for many years. Charinpanitkul cited Thai quantum dot research before the 1990s and, more recently, carbon nanotube research from the late 1990s. Furthermore, as I discovered through my surveying, Thai nanotechnology practitioners claimed to have been working in nanotechnology for an average of six years, with some claiming as much as 14 years. To explain this, a few interviewees, such as Damrongchai, referred to support for 'nanotechnology' as a recent Thai phenomenon:

> ...back 10 years or more...I was studying in that field [nanotechnology] and researching the molecular assembly of molecules...but at that time the environment in Thailand was not very supportive of doing research into nanotechnology.

In this light, Damrongchai believed that Thailand has enough workers with the appropriate skills and experience to seriously engage in nano-innovation. Songsivilai added that the diverse backgrounds of the Thai scientific workforce are an advantage in terms of the skill-sets nanotechnology R&D requires. In further support, my assessment of Thai practitioners who claimed to be working in nanotechnology showed a highly qualified workforce, with nearly all (91%) having completed doctorates. Kanok-Nukulchai added that, within three to five years, someone in Thailand will receive a Masters or PhD degree in nanotechnology. In the broader context, relating to other Southern countries, Berwick suggested that "...[a] mastery of the basic principles of nanotechnology can be applied across the board...".

7.3.2   Infrastructure

In addition to the hopes of addressing human resource barriers, interviewees suggested that nanotechnology could bypass some of the infrastructural challenges commonly associated with much emerging technology. Dutta suggested that developing bio-nanotechnology[10] applications, for example, does not necessarily need an extensive biotechnology centre. Similarly, Radt claimed that nanotechnology "...could transform any developing country to catch up with the scientific communities" considering that, in an area such as drug development, research does not necessarily rely on "...very expensive and long-lasting studies".

Additionally, these two interviewees (Dutta; Radt) challenged the belief that nanotechnology R&D requires special instrumentation or world-class facilities, claiming, rather, that many Southern countries have much of the standard instrumentation that is needed. With these points in mind, Radt distinguished nanotechnology from the capital-intensive nature of the previously emergent computer chip industry that needed "...big clean rooms, big plant[s]...[and] a lot of extremely expensive instruments to start".

Furthermore, interviewees such as Radt believed that nanotechnology R&D can utilise existing infrastructure and previous approaches "...because nanotechnology is based on the basic sciences: chemistry, biology and engineering...". Bearing similar thoughts, two interviewees (Sawanpanyalert; Tanthapanichakoon) presented nanotechnology as a 'natural progression' for Thailand, given its background in biotechnology research. For many interviewees, including Pothsiri, nanotechnology R&D is merely a modification, or "upgrading", of existing activities and infrastructure. As Chirachanchai noted, "...what we have now, we can apply it but just change the point of view from 'trial-and-error' to understanding at the molecular level". In this respect, Chiranchanchai said it is actually more a matter of identifying people's roles, with a belief that the infrastructure will develop as a natural process of the relevant people coming together.

7.3.3   International Knowledge

Building on the argument that nanotechnology can utilise existing infrastructure, many interviewees (Arya; Berwick;

Braach-Maksvytis; Changthavorn; Dutta; Lynskey; Pothsiri; Radt; Tanthapanichakoon; Warris; Yuthavong) presented a strong belief that Southern countries can 'leapfrog' their R&D capabilities and status within global innovation. When interviewees referred to 'leapfrogging', they spoke of utilising existing knowledge and research efforts by scientists in other countries, learning "...from what has already been done" (Berwick) in order to adapt and extend the lessons learnt (Chirachanchai). Thai interviewees, in particular, saw opportunities for countries to "jumpstart" their R&D activities (Ratanakul), avoiding the need to start from "square one" (Ratanakul; Tanthapanichakoon). Importantly, Braach-Maksvytis said that the potential for Southern countries to leapfrog in nano-innovation also paves the way for new innovation trajectories, whereby countries could "...steer a new, emerging science area into a very real and very practical outcome". Braach-Maksvytis elaborated, simultaneously challenging all involved:

> ...rather than going down the track of implementing technologies and devices that developed countries have now and then just repeating that, to actually look beyond already and use the new worlds that new technologies open up.

The ability to leverage global nanotechnology developments was seen as a major positive for a country such as Thailand (Arya; Tegart; Warris; Weckert). Tanthapanichakoon expressed what he saw as Thai thinking in saying that there is no need to "...reinvent the wheel...we want to know what the world has and what kinds of existing technologies we might make use of in order not to start from scratch". This is made easier given "...a lot of papers can get translated, [and] put on the Internet a lot quicker now..." (Berwick). Furthermore, there was a belief that, because nanotechnology is 'new' and not yet well defined, it is not too difficult to catch up, with Dutta saying that Thailand's relatively 'late' start "...does not necessarily mean that we have lost anything...".

On the contrary, Kanok-Nuckulchai suggested that the very nature of 'backwardness' means that 'falling behind' in nanotechnology may not be such a big issue for some countries "...because they can wait until the leading country develop[s] these technologies and they can follow". In fact, in looking to leap ahead, the South's 'backwardness' in nanotechnology R&D was even viewed as potentially advantageous in terms of saving "time and money" (Dutta).

Despite the scepticism of some, many saw usefulness in gaining support from the North to assist Southern countries in catching up. In this respect, Turney noted early signs of Northern assistance for Southern nanotechnology programs, highlighting that, whilst certain sectors in Japan may be wary of collaborative nanotechnology research with Thailand, the Japanese government is funding various forms of Thai education in nanotechnology. In this instance, Turney saw the net benefit as a flexible, highly skilled workforce from which both countries can benefit.

7.3.4 Research and Development Costs

Whilst entry costs were previously raised as a barrier for Southern innovation, many saw nanotechnology offering "...a door for countries into research even if they do not have as large funding bodies as perhaps highly developed countries" (Radt). In addition to the benefits I have already outlined, Ford said that nanotechnology offers considerable potential for Southern innovation "...for the simple reason that a lot of nanotechnology is extremely cheap". Some interviewees, such as Chirachanchai, agreed that big budgets are not necessary to commence working in nanotechnology, as highlighted by the Thai situation where the initial budget of six million baht[11] was "...enough, at that stage" (Thajchayapong).

However, it is important to note that small nanotechnology budgets do not necessarily equate to unsophisticated R&D. Rather, many nanotechnology approaches allow "...very useful and high-impact experiments with relatively small budgets..." (Radt). Self-assembly, for example, was seen as a potentially inexpensive means by which the South could play an active role in cutting-edge nano-innovation, given it utilises natural chemical reactions (Chirachanchai; Songsivilai). As Radt noted with nanotechnology-based drug delivery systems:

> ...most of the nanotechnological systems are self-assembling systems. Therefore, often it is possible to simply put reactants together in the right order...it is possible to produce these systems in parallel in a large scale...self assembling systems are easy to scale-up and that is why they are potentially useful and cheaper.

In addition to the cost reductions for drug development said to arise from chemical self-assembly, Chirachanchai saw nanotherapeutics as potentially inexpensive because increased understanding at the molecular level, combined with advances in simulation, should minimise the need for 'trial-and-error' research. A number of these cost-cutting factors contributed to Dutta referring to what he termed "poor man's nanotechnology", whereby nanotechnology could offer entry points to cater for all sorts of residual Southern capacity:

> If I think about a country like Laos...it is lacking resources...I would suggest that, 'look I am doing nanotechnology the poor man's way'...I am using colloidal nanoparticles and trying to find out applications of each ones. Laos could possibly also concentrate on things like this.

7.3.5   Comparative Advantage

Low costs are also assisted by traditional and new means of comparative advantage. Advantages were said to revolve around traditionally proposed areas, such as the relatively low costs associated with Southern labour, easy access to clinical trial participants and natural resource abundance. However, a number of advantages were suggested as specifically relating to nanotechnology. Interviewees spoke of the ability to utilise local biological products and raw materials, such as herbal products with natural resource bases (Charinpanitkul; Tanthapanichakoon). This was seen as a potential cost-saving advantage for a country like Thailand, with Chiranchanchai suggesting that, if Thailand produces a nanoscale product such as a base for drug delivery, "...the products, compared to other countries, may be cheap if we use our natural abundance for that".

7.3.6   Private Sector Support

In terms of some of the capacity matters raised earlier in this chapter, Damrongchai challenged the belief that researcher retention will be an issue for a country such as Thailand, saying that there are enough researchers who want to remain in Thailand. There was also a belief that the early signs in Thailand suggest adequate private sector support for nanotechnology. According to some (Charinpanitkul; Thajchayapong), industry is showing early interest in nanotechnology, signalling a positive change in perceptions amongst Thai businesses that now see it as a "promising field" (Charinpanitkul). As a consequence, Turney notes the emergence of Thai 'nanotechnology entrepreneurs'.

In this light, although Tanthapanichakoon agreed that Thailand faces technology transfer challenges and, as of 2004, does not have any industrial products,[12] he noted that some product samples and prototypes are being produced in labs. Tanthapanichakoon added that the commercialisation stage of mass production was probably another two to three years away.

7.4   Conclusion

Whilst some point to significant barriers for Southern nano-innovation, many take the polar view that nanotechnology presents a feasible entry point for almost all Southern countries into varying forms of R&D. In this sense, there is large variation in opinion surrounding issues of technological capacity, such as the expected entry costs and infrastructure requirements of nanotechnology R&D. However, despite the possibilities, perspectives on innovative capacity support the likelihood that nanotechnology will further the global centralisation of R&D, with extremely limited consideration for local engagement highlighting a lack of faith that nano-innovation can be localised.

My initial research shows the existence of two, slightly different 'nanodivides'. The 'nano-innovation divide' relates to where nanotechnology knowledge is generated and retained, whilst the 'nano-orientation divide' relates to where nanotechnology research is targeted. The nano-innovation divide is the platform on which the nano-orientation divide is determined. Early levels of nanotechnology funding in the North appear to be driving the innovation divide, which is framed in a polar way by reference to countries leading R&D and those being 'left behind'. Equity is a central ethical issue in the international debates around nanotechnology. However, current thinking, as supported by this research, suggests that greater equity can only occur on a platform of more widespread nano-innovation, where nations become more globally competitive. Ultimately, this thinking limits subsequent debates to a narrative underpinned by the contested philosophy of 'sustainable development' and denies the many creative alternatives emerging worldwide.

In terms of new possibilities for Southern dependency, fears were held for an increasing concentration of proprietary knowledge in the North, reinforcing an inequitable IP regimen, accelerating the global division of labour and reducing the chances for autonomous innovation in the South. Compounding these issues are the proposed capacity challenges facing a country's efforts to develop endogenous nanotechnology capabilities. Although most of these challenges do not appear exclusive to nanotechnology, some key informants perceive them as inhibitive for Southern engagement in global innovation.

However, it would appear likely that nanotechnology also creates new potential for Southern innovation. There was a strong belief that nano-innovation is misperceived as being beyond the capacity of Southern countries. This was supported by reference to the early levels of Southern nanotechnology awareness and commitment, and subsequently raised the possibility that LDCs could become active in global nano-innovation. However, interviewees noted that the possibilities were constrained in terms of the scope and nature of nanotechnology activity that might be possible in some Southern countries. In particular, there are significant differences in potential 'scale', ranging from those countries where nano-innovation might occur on a relatively small scale and be targeted towards a certain phase in the R&D process, to those for which the scale of nano-innovation is more broadly encompassing and comparable to R&D efforts in the North. Although some are generic, most of the factors raised in arguments supporting the ability for Southern countries to engage in nano-innovation relate specifically to nanotechnology.

Overall, Australian interviewees were more likely to consider Southern nanotechnology within a global context of external challenges to building endogenous capacity, identifying market-dictated challenges, such as the orientation of nanotechnology R&D, its costs and its availability, as well as other structural barriers relating to the global patent system. From this perspective, the development 'problem' is seen to include global institutions in the North, implying that any positive, sustainable alteration of the divides' trajectory must address international imbalances in systems of trade as well as other structural inequities.

Thai interviewees were more likely to consider nanotechnology within a domestic context of internal challenges to building endogenous capacity. In this respect, many identified capacity issues, such as developing adequate human resources and technical infrastructure, as well as ensuring political support and appropriate financing. From this perspective, the development 'problem' is seen more as a result of Southern deficiencies, with Southern countries viewed as 'backward' or 'lagging' and needing to 'catch-up'.

These responses have also reinforced the critical influence of how nanotechnology is perceived in terms of debates on how it will unfold. Building on my findings in the previous chapter, interviewees commonly made arguments based on fluid definitions for nanotechnology and the possibilities it raises. This point explains a number of apparent contradictions, particularly in areas such as nanotechnology's infrastructural and cost requirements, where interviewees presented an unusually wide range of views. The conflicting views also raise the interesting question as to the role of hype in nanotechnology discourse and its ability to obfuscate important realities.

My findings in this chapter show the validity of a range of concerns raised in the literature surrounding Southern roles in nanotechnology R&D, and provide increasing evidence that nano-innovation will be centralised, albeit amongst a larger scientific elite. My research also confirms that views on these issues are highly polarised and therein highlights the importance of how nanotechnology is conceptualised in determining its foreseen implications for the South.

Combined, these findings provide strong grounds on which to explore what kind of technologies are actually being considered and the associated claims about the foreseen implications of such applications.

Endnotes

1 In descending order of interviewee level of citation.

2 http://investaustralia.gov.au.

3 An alliance of nine knowledge-intensive technology organisations from around the world (http://research-alliance.net).

4 The role of licensing will be explored as a strategy for developing endogenous innovation capabilities (see Chapter 9).

5 According to Tanthapanichakoon, for every 10,000 people in Thailand there are 2.7 science and technology researchers.

6 A process patent is "a claim or claims to a process for the manufacture of a product, which may itself be the subject of a patent though it does not necessarily have to be" (UNCTAD-International Centre for Trade and Sustainable Development, 2005, p. 496).

7 According to Dutta, the barrier of inaccurate perceptions may include some people thinking: "...what the hell can you do in nanotechnology in Thailand?".

8 Dr Thaksin Shinawatra was the Prime Minister of Thailand from 2001–2006.

9 Thajchayapong drew this statistic from a 2003 study on Thai nanotechnology capabilities (see Unisearch, 2004).

10 A somewhat ambiguous term, generally referring to a subset of nanotechnology that uses biological principles and/or starting materials that can result in biological applications.

11 The transactional equivalent of approximately $165,000, as of September, 2004.

12 Although Panitchpakdi claimed Thailand has already produced a nanoscale diagnostic kit.

Chapter 8

Technological Appropriateness

In order to be able to more truly judge how likely it is that nanotechnology offers hope for a more equitable world, in this chapter I will examine interviewee perspectives about the supposed benefits and risks for the South arising from nanotechnologies. I will particularly explore whether nanotechnologies offer appropriate technologies for the South, referring largely to 'technologies' in the plural because I want to focus on the aspects directly relating to nanotechnology's artefacts.

8.1 Benefits for the South

From my discussions with interviewees, three arguments emerged in relation to the foreseen benefits of nanotechnologies for Southern development. The first relates to the novelty of nanotechnologies arising from their emergence within a new paradigm. Second, subsequent argument relates to the specific areas of benefit foreseen for Southern development. The third argument relates to the ability for nanotechnologies to be considered appropriate technologies, geared to meeting social needs in 'poor' settings.

8.1.1 A New Paradigm

From the outset, claims were made that nanotechnologies emerge from a "revolutionary" paradigm (Tanthapanichakoon) that offers an

Nanotechnology and Global Equality
Donald Maclurcan
Copyright © 2012 by Pan Stanford Publishing Pte. Ltd.
www.panstanford.com

opportunity to come up with new solutions by thinking "outside the square" (Braach-Maksvytis). Even interviewees without a scientific focus, such as Lynskey, referred to nanotechnology as able to "... change the way all sorts of things can be done...".

Closely tied with such claims was a belief that nanotechnologies present a great deal of hope, in terms of the level of benefits they offer. With respect to healthcare, interviewees described the promised benefits as "ideal" (Coyle) and "the dream" (Sriyabhaya), whilst a number of Thai scientists and policy officials spoke of nanotechnologies as "...the next big thing" (Thajchayapong), almost offering technological panaceas (Charinpanitkul; Thajchayapong; Sriyabhaya). The benefits of the new paradigm were suggested as including the elimination of a number of limitations held by previous technologies (Coyle; Sawanpanyalert).

In order to discuss the prospective benefits and risks of nanotechnologies, there is a need to establish the basis on which interviewees suggested nanotechnologies might offer something new. Interviewees outlined five central benefits of working on the nanoscale that contribute to the development of a new paradigm. These relate to increased control, reduced material and energy consumption and manufacturing costs, universality and complimentarity, potential to unify the disparate sciences, and the potentially sustained nature of any associated 'revolution'.

8.1.1.1 Control

The fundamental benefit of working on the nanoscale relates to increased control. This control stems from an improved ability to 'see' and more carefully manipulate matter and interactions on this level, explaining the claims I raised in Chapter 6 — that nanotechnology is old science done in a new way. Charinpanitkul claimed that the paradigm-shift stems from the ability to image on the nanoscale, reminding us that it is only "...just recently, [that] we can see in the nanoscale thanks to the scanning electron microscope, [and] atomic force microscope, which allow us to be able to see the real shape of molecules".

Interviewees claimed that such examples have led to new approaches, understandings and abilities to exploit nanoscale-dependent phenomena. Turney, Director of the Australian Commonwealth and Scientific Industrial Research Organisation's Nanotechnology Centre, noted that working on the nanoscale,

"...enables [us] to do things we could not have done previously". Sawanpanyalert, a Thai health policy official, added that the benefits stem from the ability to manipulate materials, thereby changing physical properties such as colour.

8.1.1.2 Consumption and costs

Interviewees explored how working on the nanoscale might positively impact upon material and energy consumption, as well as manufacturing costs. In addition to proposed improvements in operational efficiency (Coyle), there was a belief that working on the nanoscale means using less resources (Braach-Maksvytis; Cornell; Damrongchai) and "smaller functional units" (Damrongchai), resulting in less required energy for both the production of materials or devices and the maintenance of their functions (Braach-Maksvytis; Songsivilai). Combined, these aspects could mean cheaper production costs when nanotechnology becomes more 'mainstream' (Damrongchai; Sriyabhaya). Songsivilai said that the impact of working on the nanoscale, with respect to something such as diagnostic testing, means "...it will become less expensive...per unit test, because it can look at several parameters, simultaneously". Supporting this was a belief that working on the nanoscale will contribute to improvements in manufacturing capabilities (Tegart).

8.1.1.3 Universality and complementarity

According to a number of interviewees, one of the strongest benefits of working on the nanoscale relates to the universality of resulting applications. Interviewees suggested the potential range of nanotechnologies is extremely wide, possibly influencing every sector of society (Braach-Maksvytis; Selgelid). The general explanation for such a wide range of applications focussed on the ability to deal with the 'building blocks' of all materials when working on the nanoscale, with nanotechnology-based sensing techniques presented as a common example. In this way, nanotechnologies are seen as 'platform technologies', able to be applied across sectors.

However, a small number of interviewees (Arya; Deutchmann; Radt) saw nanotechnologies as more 'niche' than unlimited, with Radt pointing to past experiences, such as lasers, that were considered universally applicable but are now "...well established in certain areas for very specific applications".

Whatever the range of applications, the nature of working on the nanoscale is also seen as complementing, rather than displacing, existing scientific disciplines (Kanok-Nukulchai; Lynskey; Tegart).

8.1.1.4 Potential to unify

Moreover, related to the universal scope of nanotechnologies is the potential to unify the disparate sciences, given the development of nanotechnologies demand facilitated interdisciplinary research (Charinpanitkul; Kanok-Nukulchai; Radt; Warris). Warris, reflecting on his experiences in Australia, described how this outcome might now be possible as scientists from differing disciplines come together:

> Physicists were starting to talk to biologists and starting to realise that there are all these great things that they can use if they can just start to understand what the biologists are talking about, and vice versa.

Dutta's anecdote from his experiences teaching students from the disciplines of electronics, communications, engineering and computer science reinforced the 'great things' that can happen at the boundaries of different thought:

> The first few months I taught them to do drab chemistry. Essentially, they have quite a lot of trouble, initially, but after a few months they come up with interesting results, interesting points of view...that is the whole beauty of nanotechnology: that you can have people from a subject totally outside your domain and they can come and teach you.

Ford took these points a little further, saying that the real reason for increasingly blurred distinctions between traditional disciplines, and therein one of the main strengths behind nanotechnology, is that the research is "...happening on the boundaries and...being brought about by people who understand more than one discipline". Speaking from experience, Charinpanitkul, a chemical engineer, said that he has to learn many things about sensors from both electrical engineers and chemists and that this kind of trend is only going to increase: "in four to five years, multidisciplinary research will become [the] majority of research work".

In identifying their educational backgrounds, Thai nanotechnology practitioners showed further evidence of the merging of scientific disciplines at the nanoscale (see Fig. 8.1).

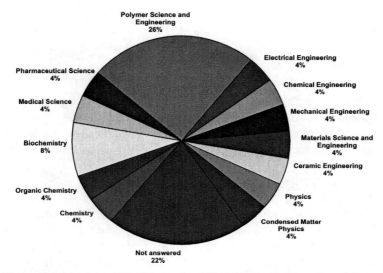

Figure 8.1 Educational Backgrounds of Thai Nanotechnology Practitioners.

Charinpanitkul gave a practical example of nano-innovation happening at the 'boundaries' by describing the necessary "... collaboration between the material science, electrical engineering and computer engineering [departments]..." in the development of Thai company Siam Cement's[1] nanotechnology-based, 'smart' sanitary wear. Tanthapanichakoon, having also witnessed the breakdown of traditional barriers between disciplines, believed working on the nanoscale is a very good opportunity to draw the sciences together. Yuthavong extended this point, saying that working on the nanoscale provides various disciplines and bodies of knowledge, such as chemistry, biology, physics and engineering, with a common goal and language.

8.1.1.5 A sustained revolution

The final key benefit presented for working on the nanoscale relates to the foreseen perpetuation of the 'revolution', given, as previously mentioned, it deals with the 'building blocks' of all matter and crosses so many industries and applications (Braach-Maksvytis; Radt). Moreover, some, such as Damrongchai, believed that, in terms of scientific necessity, working on the nanoscale heralds the "...end of the line". Cornell explained this in the context of medical diagnostics:

> ...one would have to say that you genuinely have, now, ultimately
> hit the wall...we now have the kit, which if we understand it to
> a sufficient level of sophistication, should provide answers to all
> problems.

Reflecting upon these claimed benefits, it would seem that, despite the finding in Chapter 6 — that nanotechnology is being conceptualised in different ways — the common understandings about nanotechnology's key characteristics mean general consensus regarding the nature of its paradigmatic novelty and resulting benefits.

8.1.2 Areas of Application

Having explored the beliefs that working on the nanoscale presents a fundamentally new approach for science, yet keeping in mind the tendency for divergent conceptualisations, it is useful to now explore how interviewees felt nanotechnology could be best applied in the South. This process included both an identification of relevant areas and applications as well as elaboration of the envisaged benefits associated with such applications. Although some interviewees (Damrongchai; Panitchpakdi) were unclear about prospective applications, most felt comfortable to either speculate or speak confidently about areas in which nanotechnology could be best applied in the South.

Despite some similarities, Thai and Australian interviewees largely suggested different areas in which nanotechnology holds its greatest Southern potential. Reflecting my interview sample, both Thai and Australian interviewees spoke, at length, about the benefits nanotechnology could bring for Southern healthcare. But whereas Thai interviewees focussed on the potential for nanotechnologies relating to food and agriculture, ahead of thin films and coatings and textiles, Australians referred more to applications in water- and energy-related areas.

In terms of healthcare, there was a general belief that nanotechnologies could be 'useful', with the responses of Thai interviewees suggesting slightly more optimism than those of their Australian counterparts. Interviewees explained the usefulness of nanotechnologies in two different ways. The first relates to philosophical reasoning that suggests that the nanoscale, given it is the level at which so much of the body 'functions' (Cornell; Radt), is the 'most appropriate' level on which to address many healthcare concerns such as combating viruses (Chirachanchai). Reference was also made to the highly sophisticated surfaces associated with work on the nanoscale as well as the usefulness of working on a scale at which biological interactions are so specific (Chirachanchai; Radt). Cornell added his belief: that we are entering a new age of biomimicry, with nanotechnology fundamentally opening up the ability to replicate nature, such that "...whatever nature can do, ultimately will be available for you".

The second explanation for usefulness relates to the practical applications of nanotechnologies across all components of healthcare administration (Sawanpanyalert). In this respect, nanotechnologies and associated benefits were identified in three main areas: diagnostics, therapeutics and drug discovery and design.

Whilst my quantitative research has shown that this is not the case in terms of global developments, diagnostics — such as 'biosensors' — were generally viewed as the forerunner for Southern, nanotechnology-based healthcare applications (Braach-Maksvytis; Cornell; Dutta; Tegart; Thajchayapong).

Interviewees explored a number of the technical benefits that nanotechnologies could potentially bring to the field of diagnostics. Multifunctionality — in the form of simultaneously conducting multiple assays — was seen as a key feature, meaning that the average diagnostic test would no longer be disease-specific (Songsivilai). However, multifunctionality was not seen as compromising accuracy. Rather, nanodiagnostics were viewed as potentially more targeted and specific (Coyle), particularly considering the reduced material requirements at both the detection and sample levels (Braach-Maksvytis; Songsivilai). Also suggested as an advantage was Cornell's belief that with nanotechnology-based diagnostics "...you do not have to separate out the various components of a biological sample to allow you now to make the assay...".

In this light, interviewees (Cornell; Songsivilai) suggested that nanotechnologies are driving the shift to 'point-of-care' diagnostics, whereby sample results can be rapidly and conveniently gained with small sample sizes, from any type of body fluid. Braach-Maksvytis added that the ability to quickly determine disease progression is something that is "just not available" in most Southern countries. Songsivilai further added that, overall, nanotechnology-based diagnostics would make for more 'user-friendly' processes. He cited the example of HIV, where he believed testing for viral loads and cell counts would become much easier, allowing greater patient information on the status of their disease as well as the ability for practitioners to make more appropriate adjustments to therapeutic dosage requirements.

In terms of therapeutics, there was a consistent belief amongst proponents that nanotechnology can provide more effective drug delivery to target sites (Chirachanchai; Radt), utilising both new mechanisms of delivery as well as the existing nature of nanoscale target interactions. Initial benefits were seen in terms of the improved ability for drugs to reach target sites, either through novel ways of breaching membranes, by shifting the focus to alternative methods of drug delivery or by exploiting nanoscale phenomena at the point of delivery. In this sense, Radt described the ability to hide a drug within a nanoscale structure whilst also using "active transport mechanisms" to ensure that certain drugs remain undetected by the human immune system. Chirachanchai also noted other nanotechnology-based methods, such as dermal, oral or nasal delivery, that will ensure medicines reach their desired sites. Furthermore, working on the nanoscale helps facilitate the controlled-release of a drug from the molecule of the drug itself (Chirachanchai). Highly targeted and responsive shells, facilitated by the scale on which nanotechnologies are operating, could react to indicators such as pH, temperature changes, light irradiation or magnetic fields, in order to release a drug to a targeted area, at a desired time, over a desired time period (Radt). A by-product of increasing the effectiveness of drug delivery would be the reduction of internal side-effects (Radt). When combined with alternative forms of administration and new mechanisms for drug delivery, Chiranchanchai and Radt believed that nanotechnology can ensure more user-friendly therapeutics. In this light, speaking of the impediments to immunisation coverage, Deutchmann suggested that nanotechnologies could offer "real opportunities" to explore "a non-injectable vaccine".

As with nanotechnology-based diagnostics, the multifunctionality of nanotechnology-based therapeutics was considered an important attribute. Interviewees (Deutchmann; Radt) spoke about the ability for certain nanotechnologies to simultaneously respond to different therapeutic problems. For example, Chiranchanchai held high hopes that nanotechnologies could facilitate the ability to simultaneously terminate DNA-RNA[2] duplication systems whilst controlling the mutation of viruses. Furthermore, theranostics — the combination of therapeutics and diagnostics — was mentioned as a potential area for nanotechnologies to emerge. In this respect, Chirachanchai said that producing a nanotechnology-based drug that can also function as a traceable sensor is a dream held by many Thai scientists.

The impacts of nanotechnologies on drug design and discovery were also considered "very important" (Pothsiri). Impacts were largely seen in terms of the ability to increase the options and speed associated with problem solving (Radt), thereby reducing the time needed for clinical trials (Pothsiri; Tegart), whilst improving 'benefit-to-risk' ratios and providing subsequent reductions in costs (Pothsiri).

Placing the foreseen therapeutic, diagnostic, design and discovery benefits of health-related nanotechnologies in a Southern context, Arya believed nanotechnologies could offer "a lot" for domestic healthcare, with others speaking of the potential for nanotechnologies to assist in achieving the MDGs (Braach-Maksvytis; Damrongchai). In this light, many interviewees focussed their comments on communicable and neglected diseases, where it was believed nanotechnologies could make "really important" contributions (Selgelid), particularly in light of South Africa's early efforts to address HIV/AIDS through appropriately targeted, endogenous nanotechnology R&D (Weckert). Interviewees spoke most about the potential for nanotechnologies to assist in the amelioration of malaria (Dutta; Lynskey; Turney), HIV/AIDS (Damrongchai; Lynskey; Songsivilai), tuberculosis (Deutchmann; Lynskey), cholera (Dutta) and chagas disease (Selgelid).[3] Having outlined to me some of the needs relating to tuberculosis, Deutchmann proceeded to say: "...tuberculosis is an outstanding candidate for the contribution nanotechnology can make around vaccines, treatments and the issues pertaining to diagnosis and then the whole management thing[4]".

Others saw the greatest potential for nanotechnologies relating to new social needs, such as chronic lifestyle diseases, with Panitchpakdi raising diabetes as a prime candidate for consideration. Yet some, such as Pothsiri, saw the most potential for nanotechnologies lying with even "more immediate" social needs relating to Avian Flu and SARS epidemics. There would seem to be widespread support for this view, given early Thai planning is underway to develop nanotechnology-based biosensors for bird flu and other epidemics (Tanthapanichakoon).

Beyond health-related areas, interviewees identified a range of agro-industrial and other nanotechnologies they felt could be applied to the South. For Thai interviewees, food and agriculture was clearly identified as one of the more important areas for R&D. Respondents explored how nanotechnologies could potentially benefit a range of processes within the food production and output cycles. Tanthapanichakoon, for example, spoke of the ability for nanotechnologies to "...improve the productivity of our farms or rice production". Others (Braach-Maksvytis; Turney) suggested nanotechnologies could help facilitate the monitoring of crops or aquaculture environments. Interviewees also spoke about improvements in the nutritional content of produce (Cornell) and benefits for food preservation, such as ensuring coffee beans keep their flavour (Dutta) and extending the shelf-life of vegetables and tropical produce (Tanthapanichakoon; Turney). Interviewees (Tegart; Turney) said that preservative benefits will stem from 'active packaging', wherein nanotechnology-based sensors can facilitate constant, inexpensive monitoring and reactions to the internal conditions of packaged food. The most commonly raised examples, however, were in viticulture. The two main benefits envisaged were the reduction of free radicals in wine, via the introduction of gold nanoparticles and other similar matter (Charinpanitkul; Thajchayapong), and the acceleration of a wine's ageing process during fermentation, via catalysis (Charinpanitkul; Dutta; Thajchayapong).

Water was also highlighted, predominantly by Australian interviewees, as a key area for the application of nanotechnologies in the South. This was explained in terms of inexpensive filtration and desalination (Cornell; Tegart) and nanostructured sensors that could provide advanced detection methods for pathogens, bacteria and water purity (Braach-Maksvytis; Dutta; Songsivilai; Turney).

To a lesser extent, both Thai and Australian interviewees presented 'the environment' as an area in which the application of nanotechnologies could find purpose in the South, although Tanthapanichakoon spoke more broadly about the potential for nanotechnologies to foster "sustainable development". Interviewees envisaged technologies that would monitor the status of the environment (Songsivilai), assist with bioremediation (Tegart) and increase energy efficiency (Arya; Cornell).

Thin films and coatings as well as textiles, were also highlighted as "key areas" of potential benefit for Thailand (Charinpanitkul; Thajchayapong). Thin film and coating applications were pictured in terms of 'self-cleaning' films for windows (Damrongchai; Dutta) or coatings for automobiles (Thajchayapong), whilst proposed textile applications include stain- or wrinkle-free silks (Damrongchai; Thajchayapong).

In addition to the benefits raised in this section, interviewees returned to the notion that endogenous Southern nanotechnology R&D would mean less dependence on Northern technology (Damrongchai) and, therefore, a greater chance to steer nanotechnology towards "home-grown solutions" (Berwick). Yet, when assessing the nature of applications in their entirety, it is clear that there is an implicit distinction between nanotechnologies for social development and those designed to service a market. This distinction exists particularly between Australian and Thai interviewees but also between Thai interviewees themselves.

The argument that nanotechnologies can be developed for end-users in the South, keeping in mind the best interests of those most in need, raises the issue of the appropriateness of nanotechnologies to development settings. An assessment of interviewee perspectives on this topic produced some interesting results. Although perceptions on the appropriateness of nanotechnologies to Southern contexts in the area of healthcare were mixed, there was a general belief that nanotechnologies can be 'appropriate'.

8.1.3 Nanotechnologies and Appropriateness

A number of Australian and Thai interviewees who initially rejected there being a link between nanotechnologies and Southern healthcare, given high-tech appearances (Coyle; Deutchmann; Kanok-Nukulchai; Selgelid), said that their perceptions shifted towards being more supportive as they discovered the broad-ranging nature of nanotechnologies. For Damrongchai, this kind of initial misperception reinforces the 'image' problems faced by nanotechnology, which he said often stems from a lack of in-depth understanding. Kanok-Nukulchai agreed, saying: "...after a while when we think of nanotechnology, this is not only high-tech, we deal with appropriate technology as well...". In this light, Bryce suggested that "...whether a technology is modern, high tech or low tech does not have anything to do with its appropriateness...". Braach-Maksvytis agreed, going so far as to say that "...very high-tech devices certainly can [be appropriate], and that we probably should be targeting them for developing countries...". With these points in mind, Lynskey believed that 'level of quality' should replace 'level of complexity', in terms of the mainstream understandings of appropriateness as they relate to nanotechnologies.

Combined, these comments appear to be representative of a shift in attitudes and understandings relating to appropriate technology in the South. According to Deutchmann, the last 20 years has seen a significant redefining of the criteria for 'appropriate' technology:

> ...if you go back 10, 20 years, appropriate technology was pretty much defined as simple technology...some of that is still appropriate in the sense that it needs to be a technology which requires either: little maintenance or maintenance that can be managed on site, and that was always the problem with high-tech. But...increasingly, high-tech has an application, provided it is accessible and can be maintained at the end-user point, rather than dependent on highly skilled people who are often either inaccessible...not present in sufficient numbers, or locally available...

Such a shift holds important repercussions in the debate about the appropriateness of nanotechnologies, opening up pathways for discussion but also raising questions about the fundamental characteristics of 'appropriateness'.

In terms of how nanotechnologies might be considered appropriate for development settings, four key features were outlined. The first was a belief that nanotechnologies would produce straightforward and simple solutions (Coyle; Lynskey; Sawanpanyalert). Dutta provided the example of "...very simple, small sensors" he envisaged might be appropriate for a country such as Laos:

> Laos has a lot of fish farms. Many a time, fish die in the pond. Most of the time, the fish die in the pond because of the change in acidity of the pond water. When we do sol-gel chemistry[5] we often see that the sol turns into a gel or the gel turns into a sol. When the whole sol of the gel changes the pH, you could have a very simple sensor there.

Placing the potential benefits outlined earlier in this section in the context of resource-poor Southern healthcare, Deutchmann spoke positively of what nanotechnologies might offer:

> ...[anything that] can be brought into a very simple health care setting that...is the means of diagnosis that is both low-cost and immediate...is a fantastic asset...whether that is bench-top HIV testing kits that come in or whether it is some of the new technology that enables one [assay] to test all.

As the second key feature, 'affordability' was seen as accompanying simple designs (Deutchmann; Sawanpanyalert). For example, Radt saw real potential for the self-assembly process to result in "cheap devices", once an application's IP time expires.

The third key feature was a belief that nanotechnologies can improve accessibility to technologies in an area such as healthcare by facilitating the decentralisation of services and strengthening available support at the periphery (Cornell; Songsivilai). In this respect, Sawanpanyalert saw the possibility that nanotechnologies offer something different compared to past technologies:

> ...when it [technology] is introduced in Thailand either locally or from abroad it usually is diffused from central, big institutions and then gradually to the periphery, but hopefully nanotechnology does not have to follow suit, if it can be applied in the makings that it can be used at the peripheral level...

To support such claims, Cornell mentioned nanotechnology-based, point-of-care diagnostics which he said would reduce the infrastructure required for deployment. He also spoke of the potential for portable diagnostic devices, carried to remote regions "...in the pannier bags of bicycles". In China or parts of Africa he envisaged:

> ...you would have small medical centres which would be able to perform tests and provide services way beyond anything that one would imagine possible, at present, and that would be brought about as a consequence of having portable, convenient, easy to maintain equipment, which currently is not available.

Complementary to simple, affordable and distributed solutions, the final key feature was the ability for nanotechnologies to empower communities by reducing dependence on specialists. Interviewees often claimed that, in healthcare, nanotechnology-based diagnostics could assist by increasing the level of individual self-monitoring

(Sawanpanyalert; Songsivilai). Similarly, Panitchpakdi suggested that nanotechnology could build on developments in mobile information technology and telemedicine, allowing more things to be monitored in the home. Charinpanitkul elaborated through the following example:

> ...in the toilet you have some nano-sensor that will detect the amount of sugar in your urine...[sending] the signal to the medicine doctor to inform them about your physical situation and then the doctor will be able to take preventative action.

Additionally, there was a belief that nanotechnologies could lead to more self-administered, semi-automated drug delivery devices (Radt); an aspect that Sawanpanyalert saw as particularly relevant for the future management of chronic diseases. Highlighting an underlying advantage of these expectations, it was suggested that future, Southern diagnostic testing and procedures should not require a doctor and could be easily administered by a nurse, paramedic or some other kind of healthcare auxiliary (Cornell; Sawanpanyalert; Sriyabhaya). Using the example of tuberculosis, nanotechnology-based applications might allow a healthcare auxiliary to administer services such as sputum examination and patient screening (Sriyabhaya), thereby "...eliminating the need for skills which just are not available in certain locations" (Coyle).

8.2 Contextual Challenges: Old Rhetoric, Old Reality

As explored in my literature review of technology, development and inequity, new technologies are often lauded as providing appropriate 'solutions' for Southern development problems. Yet, history suggests the proponents of such expectations repeatedly ignore or underestimate barriers to overcoming various challenges, and that there exists significant divides between rhetoric and realities.

Given the claims I have made earlier in this chapter, the above tension is worth exploring for nanotechnologies that have been presented as beneficial and appropriate for Southern contexts. In this respect, interviewees referred to two different forms of hype and expectation that, in turn, constitute the areas of discussion in this section. The first relates to claims about what nanotechnologies can achieve technically, with interviewees exploring whether projections that have been made are confounded by old scientific bottlenecks and whether such hype diverts attention from problems of scientific capability. The second form of hype relates to what nanotechnologies can achieve for development, with interviewees exploring whether projections are confounded by existing problems that are not of a technical nature, and whether nanotechnologies could actually amplify development concerns, based on inappropriate orientations, the associated end-user and the potential to create greater inequality.

8.2.1 Technical Claims

There was recognition of a global over-inflation of expectations associated with nanotechnologies (Warris) that had seemingly spread to Thailand (Yuthavong). Interviewee responses indicated that most of this hype has been driven by the market, where the term 'nanotechnology' is set to be a "...best-seller catchphrase or key word" (Damrongchai). Fluid definitions and a lack of clearly defined parameters for the field were seen as self-serving, promoted by companies that were 'jumping on the bandwagon' and often misusing the term 'nanotechnology' to describe unrelated, rebranded work (Damrongchai; Dutta; Weckert). As Damrongchai elaborated:

> The so-called 'nano-products'...have the 'nanotech' label but they are not much different from conventional production or manufacturing technology...[these use the word] 'nano' because the particle size happens to be on the nanoscale...on the other hand [there are] people who are trying to argue that they are doing nanotech research when, in fact, nanotechnology content may be only 30 or 40 per cent.

As I have already shown near the beginning of this chapter, a great deal of the hype surrounding nanotechnologies centres on claims about technical advantages. However, there was a belief that these claims are reminiscent of those made for previous emerging technologies (Cooper; Yuthavong), with Cooper specifically referring to claims such as 'localised and targeted intervention' as "familiar rhetoric". Cooper added that the approach of working on the atomic level supports "...a reductionist vision that was very prevalent in the heyday of molecular biology and that really kind of came to the fore with the human genome project".

Furthermore, returning to an idea raised in my review of the literature: that nanotechnologies have been around for quite some time, there was a belief that hyped rhetoric was not new for nanotechnologies either (Yuthavong). Here Yuthavong claimed that nanotechnologies had already resulted in much disappointment after Thai work on liposome delivery in the 1980s, "...did not really come to very much".

The noticeable scepticism about the rhetoric surrounding the scientific benefits of nanotechnologies was largely explained by scientific bottlenecks that were said to present "...the same problems as before" (Radt). Sawanpanyalert, who previously noted his strong belief that nanotechnologies could overcome a great deal of scientific stumbling blocks, was also quick to caution that "...it still has limitations, so it is not a panacea". For Cooper, one of the main problems was with the vision that things can be "totally controllable" at the nanoscale and that there is a "...linear model of what can be expected". She and others, such as Yuthavong, added that localised drug delivery still faces a lot of complex hurdles, with Radt explaining some of the biological challenges:

> ...in most cases, the particles...cannot cross barriers within the body and it is very hard to get even a nanoparticle...to the right place in the body. There is the immune system scanning everything and doing its best to keep particles [out]...

In addition to identifying old rhetoric and old barriers associated with the creation of nanotechnologies, interviewees claimed that hype and expectations for what can be achieved technically are diverting attention from real problems of scientific capability. Much of the rebranding of minimally related work as 'nanotechnology' was seen as a front for a lack of capacity to conduct 'real' nanotechnology research. In the Thai context, for example, Damrongchai says that research claimed as nanotechnology remains largely a form of chemistry given the country "...still do[es] not have the real capability to manipulate the atom on the nanoscale directly".

Furthermore, interviewees saw the number of people 'jumping on the bandwagon' leading to misleading figures about the human resource capabilities of a country. Yuthavong said that Charinpanitkul's 2004 national assessment suggesting Thailand has one hundred nanotechnology researchers (see Unisearch, 2004), was "generous" and that those recorded nanotechnologists are people "... who have been involved with the nanotechnologies of the familiar varieties for a long, long time". He believed that people working in an area such as chemical synthesis are 'familiar' with nanoscale chemistry and that there is a tendency for those working in drug design and delivery to think:

> ...'hey this is also my field' because, of course, even a small molecule of a small molecule drug is also nano-size and you can do a lot of things with it, model it, design and make new formulations and so on...

Rather, Yuthavong claimed that Thailand does not really have any "new nanotechnologists". Furthermore, for those who might be considered 'existing nanotechnologists', Kanok-Nukulchai said it would be difficult to classify them as 'experts in nanotechnology', considering they have just merged from other fields. Changthavorn agreed with these sentiments, estimating the number of Thai nanotechnology experts to be closer to "a handful".

Finally, misleading human resource figures raised a concern that funding would be allocated inappropriately. Yuthavong said that, in certain cases, people in Thailand were "going overboard", relating the following story as evidence:

> I just heard, the other day, that Kohn Kaen University, a provincial university in the North-East of Thailand, decided that they are going to devote some 450 million baht,[6] for nanotechnology and I am trying to think, 'who are their nanotechnologists?', and I cannot think of one.

These comments pick up and reinforce the earlier message about the fluid nature of how nanotechnology is defined. In these cases, the use of the term 'nanotechnology' would seem heavily driven by self-serving commercial prospects.

8.2.2   Development Claims

The other identified form of hype relates to what nanotechnologies can achieve for the development process. Building on the hope associated with the identification of applications and 'appropriateness', interviewees identified a tendency for nanotechnologies to be presented as able to provide a 'magic bullet' (Lynskey; Turney).

Yet, in a similar fashion to the hype and expectations around the technical brilliance offered by nanotechnologies, there was a view that the positive projections for the impacts of nanotechnologies on development processes constitutes a rhetoric that is all too familiar. More specifically, interviewees (Bryce; Cooper; Lynskey; Panitchpakdi; Radt; Sawanpanyalert) referred to cases where biotechnology and information and communications technology were also presented as a 'techno-fix', able to 'solve all problems'. Cooper, Sriyabhaya and Turney all highlighted inequality, poverty, hunger, environmental degradation and governance as areas in which science has been falsely promoted as holding 'the solutions'. However, Cooper claimed that this wider tendency to "...want to look to high-tech as a panacea" is a deliberate, strategic move by governments and big business to obfuscate the 'real' problems, with the focus on 'social needs' a justification for the pursuit of profitable commercial niches.

In response, some (Cooper; Weckert; Sriyabhaya) saw the 'real' problems as old embedded inequities that are beyond the ability for nanotechnologies to rectify, considering they are not problems of a technical nature. In this light, Deutchmann questioned whether there would be appropriate infrastructure and resources to fix, replace and maintain nanotechnologies with ease and at low cost. Others (Coyle; Sriyabhaya) challenged the proposed simplicity and user-friendliness of nanotechnologies, questioning the ability for people to use applications themselves, rather than requiring highly skilled people to operate the technologies. Yet, the most identified problems of a systemic nature related to issues of inequitable access to the fruits of technology (Braach-Maksvytis; Bryce; Deutchmann; Lynskey; Selgelid; Sriyabhaya). More specifically, interviewees (Ford; Panitchpakdi; Sawanpanyalert) defined three, inter-linked components of the 'access' problem: the underlying motivations for Southern engagement with nanotechnology, the orientation of nanotechnology R&D and the likely 'end-user' of subsequent nanotechnologies.

Given the various possibilities for Southern engagement that I outlined in the previous chapter, it is understandable that countries will be motivated to engage with nanotechnology for different reasons, many of these dictated by the country's economic or industrial development status (Tegart). Here I will look specifically at the issue of access as it relates to the nature of Thai engagement with nanotechnologies.

The Thai example presents evidence of a diverse range of motivations for engaging with nanotechnology R&D, corresponding to the growing number of 'stakeholders' seen to be building hype and expectations around nanotechnology. For example, some (Radt; Tanthapanichakoon) said that the hype from academia and business has little to do with the science and more to do with funding and politics. On the other hand, there was concern that the high levels of support coming from certain members of government might mask political self-interest and "...ulterior motives for funding projects" (Tanthapanichakoon). However, the Thai government, led by the Prime Minister — seen as the main driving force for engagement with nanotechnology[7] — is said to have a very clear motivation. As Dutta explained: "the whole impetus came in about a year back [2003] when the Prime Minister of Thailand starting pushing scientists in Thailand to start working in nanotechnology". According to Thai interviewees (Damrongchai; Pothsiri), the clear motivation was to promote economic development and growth. Evidence can be seen in the pressure to develop applied rather than fundamental research (Panitchpakdi), leading to the creation of "...products which will make money" (Dutta). Despite some interest in a domestic consumer market, competitiveness in world markets is being promoted as the real driver of economic development and growth, and therein the rationale for engagement (Arya; Chirachanchai; Pothsiri; Tanthapanichakoon; Thajchayapong). More specifically, the interest is in harnessing the financial benefits of knowledge creation, with a "...desire and a drive...to do something for Thailand and push it into a knowledge economy" (Tegart).

Given these comments and the nature of applications I identified earlier in this chapter, there emerged a new tension between nanotechnology rhetoric and reality that Pantichpakdi saw as the struggle between community and business interests. Fundamental to the outcomes of such a struggle is the way in which research is orientated. In this respect, I surveyed current areas of research conducted by Thai nanotechnology practitioners (see Fig. 8.2). Whilst the results highlight the diversity of efforts in nanotechnology research, they also show an emphasis in the areas of advanced materials, pharmaceuticals, energy and electronics. For Chirachanchai, these areas of focus could be considered indicative of the Thai emphasis on research that will produce practical applications. The results emerging from my survey of Thai nanotechnology practitioners support this belief; with just over two-thirds of researchers focussing their work on applied outcomes (see *Appendix J*).

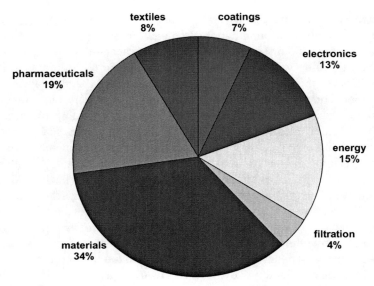

Figure 8.2 Areas of Research by Thai Nanotechnology Practitioners.

In this light, many (Braach-Maksvytis; Pothsiri; Sriyabhaya; Turney) suggested that the orientation of Thai nanotechnology research is largely representative of the global picture, whereby national orientation is foremostly towards the commercial, rather than public arena. In South Africa, for example, despite efforts to develop nanotechnologies to combat problems such as HIV/AIDS, Braach-Maksvytis said that "the focus is still that commercial industry focus…".

To gain a clearer picture of the output of Thai nanotechnology research, I have listed the kind of applications practitioners envisaged as most likely to result from their work (see Table 8.1). Research into advanced materials is said to translate into applications such as catalysts, composites and metal oxides. Foreseen pharmaceutical applications include diagnostics to screen for the Human Papilloma Virus[8] and others that use optical nanoscale phenomena for fluorescent labelling, as well as new forms of drug delivery and design. Potential energy applications are largely seen in terms of various membranes to enhance fuel cells as well as lithium-ion batteries. New thin films and coatings were also seen as an area of great potential, facilitating enhanced preservation for food products, automobile surfaces and pigment paints as well as environmentally useful applications, such as membranes for toxic filtration. Finally, nanotechnology-enhanced textiles were seen to hold great potential, presenting advantages in the quality and in-built functionality of materials such as silk and rubber.

Table 8.1   Potential Applications Resulting from Areas of Research by Thai Nanotechnology Practitioners.

Area of Research	Prospective Applications
Advanced Materials	Metal oxides, composites, polymeric biomaterials, catalysts, ceramic powders
Pharmaceutical	Diagnostics, for example: biosensors, fluorescent labelling, drug delivery mechanisms, pharmaceutical compositions
Energy	Fuel cells, batteries
Electronics	None mentioned
Thin Films and Coatings	Functional films, preservative coatings, membranes for toxic filtration
Textiles	Rubber, biodegradable polymers

These results support interviewee responses about Thai research directions in areas such as pharmaceutical, agricultural, textile and thin film nanotechnologies, in turn, contrasting the ideals of largely Australian interviewees with the realities of the Thai situation. In these realities the orientation of R&D is most commonly linked with existing exports or new industries such as the automobile, agricultural and construction sectors (Damrongchai).

A natural progression from discussions about research orientations was to talk about foreseen outcomes for Southern populations. Here interviewees first considered who would constitute 'the market' for nanotechnologies. For some (Damrongchai; Chirachanchai; Weckert) there was a belief that nanotechnologies could be focussed towards "the rich", with Damrongchai adding that if research is led by the U.S., then you will end up with a concentration on expensive areas of application, such as tissue repair. Here a number of interviewees (Arya; Coyle; Deutchmann; Ratanakul; Sriyabhaya) expressed their concern that inhibitive user-costs for nanotechnologies could be one of the main drivers of further inequity. In terms of healthcare, early costs associated with nanotechnology-based therapeutics, for example, could be particularly prohibitive for local hospitals (Panitchpakdi; Songsivilai). In speculating on these matters, Selgelid suggested that the focus of nanotechnologies towards benefiting the wealthy would be nothing out of the ordinary for emerging technologies. Others (Coyle; Radt) qualified their belief that nanotechnologies would be limited to the wealthy by saying this might only be a problem in the early stages of technological development. With respect to health-related nanotechnologies, Cornell believed that "...once the first threshold was reached of a really cheap and convenient way of molecular detection" there would be an "explosion" of consumer products such as 'simplistic diagnostics' for shampoos, soaps, or cloth-washing liquid.

Others argued that parts of the South present genuine consumer markets, particularly for 'high-tech' products that service urban Southern users (Braach-Myksvitis; Chirachanchai). Referring to the Thai situation, Dutta said nanotechnology R&D was mainly oriented towards a domestic consumer market and he did not see any potential export market in the immediate future. This argument was supported by the results of my surveys of Thai nanotechnology practitioners who reported 66% of their work was aimed at supporting domestic ahead of international markets.

Yet, interviewees, such as Tanthapanichakoon, agreed that, irrespective of who constitutes the global 'markets' for nano-technologies, applications would focus on consumer durables and luxury goods such as advanced cosmetics, rather than 'real' needs.

In terms of who determines the direction in which efforts are placed, Tanthapanichakoon saw nanotechnology 'convenience' or 'fashion' driving uptake, thereby creating consumer 'pull' factors. However, most saw applications, like the ones mentioned in this section, as part of a supply-driven techno-push, with Braach-Myksvitis saying she had witnessed a similar phenomenon in the development of Vietnam's nanotechnology programs. Arya said that the promises of nanotechnologies were driving 'felt' needs and actually creating "...more needs for consumption...more of the cheap product[s] that will pollute, not only our environment but our mind[s] as well...". Speaking of a vapid consumerism driven by nanotechnology, Arya imagined:

> ...[things such as nanotechnology-based drugs] can be completely luxurious, can be completely unnecessary...but people would say, 'oh, it is very important, we should have this for better social status...'.

Yet the main impact on the wider Southern population of the market orientation of nanotechnologies is the lack of applications addressing real Southern needs, given the inappropriateness of products produced for Northern markets to local development settings in the South (Bryce). For Cooper, the simple reason nanotechnologies are likely to maintain existing inequities is because "the vision of the way that these technologies are going to be used from the start is totally biased". According to Selgelid, such bias means that there is a real challenge in getting "...someone to make stuff that is aimed at meeting the needs of developing countries".

In Thailand's case, the lack of nanotechnologies aimed at human needs would appear to be part of a broader problem whereby "...social issues are secondary..." (Braach-Maksvytis). Despite the previous rhetoric about the potential for nanotechnologies to advance efforts towards the MDGs, interviewees, such as Tanthapanichakoon, were unaware of any Thai efforts to promote nanotechnologies in a way that specifically worked towards poverty alleviation or similar objectives. Interviewees provided a number of reasons for why this kind or orientation is unlikely in the near future. As Turney noted, examples such as arsenic in Bangladesh's groundwater and parasitic diseases such as malaria will probably not be addressed because the technology would need to be a "spin-off" from the North and these are not "major issues" for the North. In this light, the belief was that industry will be disinterested in areas that offer a low financial return (Yuthavong; Selgelid; Pothsiri; Cooper; Bryce; Deutchmann; Cornell). Moreover, Arya said that whilst nanotechnology presents "...the potential to do something for the benefit of humanity", it is compromised by the international IP framework, as well as "...the single, commercial mind[set]". Others, such as Braach-Maksvytis, spoke about problems within the 'culture' of the science community, with the first relating to training that is "...without any social context...[ensuring] end-user context is going to be bypassed". A further problem is that there is an 'acceptable culture' that academic researchers must "...go for international publications...", resulting in a focus on whatever is most intellectually popular or holds the most commercial potential (Thajchayapong). Finally, Tanthapanichakoon said that addressing needs in local healthcare, for example, is outside the scope of Thai national bodies such as NANOTEC:

> ...one limitation is that our centres or the universities will belong to the Ministry of Education, the Ministry of Science and

Technology, we are not in the umbrella of the Ministry of Health, or public health. So, we do not think it is our main mission to solve such a problem and even if we try to initiate such activities or program we might be viewed negatively, that we are encroaching on their territory...

The sum of these issues forms what can be considered the second kind of 'nanodivide': the 'orientation divide'. As I mentioned earlier, this divide relates to inequity based on where nanotechnology research is targeted and the predicted lack of consideration for Southern development needs, with Selgelid expecting a split similar to that of the 10/90 divide, irrespective of how much R&D occurs in the South.

Restricted access and inequitable distribution are seen as the by-products of the orientation divide combining with the innovation divide (Coyle; Lynskey; Ratanakul; Selgelid). In this situation, a great deal of the already limited research that is aimed at Southern needs is developed, and therefore controlled, by the North, raising concerns for Northern-generated barriers to access. For nanotechnologies, the main consideration here is that they are emerging in a period when IP is becoming increasingly competitive, resulting in growing concentration of ownership in the North that "restricts access" (Braach-Maksvytis; Bryce). Ford said this phenomenon, with respect to nanotechnologies, raises concerns of a repeat of the South African pharmaceutical debacle. Braach-Maksvytis agreed that a repeat of such drama is likely "...if we go down the same paths of what we have done in the past...", with Damrongchai adding: "...unless the patent holder realises not to repeat things like they did with biotech industry" it is hard to be optimistic. Changthavorn outlined one of these lessons, with respect to nanotechnology ownership and distribution:

> For rice, if these products happened to be patent[ed] by Thai authorities or the Thai government, it should be okay, and if this new product is to be distribute[d] to farmers that should be okay, but if this product happened to be in the hand of private enterprise, different story.

However, lack of access and inequitable levels of distribution were considered a problem as much within the South as between the North and the South. Thais, such as Ratanakul, constantly referred to internal issues of access, with corporate monopolies seen as driving an existing gap between Thailand's 'have' and 'have-nots'. In this light, Tanthapanichakoon suggested that, as a result of the introductions of nanotechnologies, "...maybe the majority of the Thai people might fall behind". Others saw it as specifically dividing the elites and the general public (Ratanakul) or "...those who can afford [it] and those who cannot" (Songsivilai). But Damrongchai saw local communities as the most marginalised, given:

> ...[they] are relatively far from technology. Most likely they will be the ones affected because they are in [an] agricultural culture and the wave of globalisation will affect those people in quite a hard way.

Others questioned whether nanotechnology is "...the best technology to decrease inequality" (Selgelid). Panitchpakdi, for example, challenged the rhetoric relating to empowering communities, claiming that nanotechnology will demand specialised people to administer some of its applications. Furthermore, many argued that feasible, proven interventions, including low-tech applications, are already available and that it is just a matter of "...putting political commitment and resources in[to] the area[s] we need it" (Sawanpanyalert). In this light, interviewees (Chirachanchai; Songsivilai) said that there are often cases where nanotechnologies are not needed. Referring to the MDGs and healthcare services in particular, Songsivilai claimed that: "...most of the technologies require[d] to achieve the goals [are] already there...existing point-of-care test[ing] is good enough to do most of the thing[s] that the nanoscale technology can do...".

By bringing in unnecessary technology, there was an associated concern of funding being drawn away from other important areas (Lynskey; Selgelid). Cooper took this argument one step further, claiming that high-tech could displace available, low-tech alternatives that often produce better outcomes, with an associated belief that nanotechnologies could destroy indigenous resources and technologies (Ratanakul). Returning to the area of healthcare, Cooper elaborated: "...quite 'low-tech' kind of responses, such as organising public health infrastructure on the ground, can be much more powerful in combating infectious disease than any kind of high-tech solution".

Interviewees such as Lynskey and Selgelid thus returned to the point that the rhetoric surrounding nanotechnologies appears even further from reality when considering that even low-tech, basic technologies often fail to reach those in need.

8.3 Societal Implications

In this section I will investigate the polarising issue in the literature I reviewed as to whether nanotechnologies raise unique societal implications. Amongst interviewees there was general agreement that nanotechnologies present a "two-sided coin" (Changthavorn), with both positive and negative outcomes (Turney). However, there was disagreement about the novelty of implications, with some suggesting that nanotechnologies are "...nothing which, necessarily, revolutionise anything..." (Radt), whilst others believed they could have a "disruptive[9] influence" (Turney).

Following the literature I reviewed, discussions about the implications of nanotechnologies can be grouped into four broad themes. The first surrounds the nature of human health and environmental risk. The second theme surrounds ethical implications. The third theme surrounds legal implications. The final theme surrounds remaining social implications that have emerged or may yet do.

8.3.1 Risk

The issues of health and environmental risks were important considerations for interviewees. However, both Thai and Australian interviewees focussed their attention on risks associated with human health where the wide range of responses explored questions of size-dependent, nanoscale phenomena. For many interviewees the first point of discussion was new forms of human exposure to potentially dangerous materials. Debates about exposure were seen as crucial, given the ability for nanoparticles to easily cross external and internal bodily membranes, such as the blood-brain barrier (Radt; Turney; Yuthavong). Another cited membrane was the lining of the pleural cavity, with Arya noting the ability for nanoparticles to be absorbed into the body via the lungs.

Creating particular controversy was the issue of toxicity as it relates to nanoscale matter. Interviewees raised concerns about the physical size of the particles (Weckert; Yuthavong), with Radt explaining that "...very small changes on the surface of the particle can result in a very different behaviour in the biological system, especially in a human body". Unease was raised for the possible internal effects of nanoparticles on humans, with Chirachanchai worried that, if commercial-grade polymers are used for nanotechnology-based therapeutics, they will have a carcinogenic effect at the nanoscale, whilst Panitchpakdi pointed to the unknown side-effects of nanotechnology-based drugs. However, although noting that higher surface areas might increase the likelihood that a particle will be toxic, Turney stated: "at this stage there is little, hard evidence that there are any severe toxicological dangers related to nanotechnology by the virtue of the fact you are working with nanostructures...".

The issue of potential toxicity means that exposure becomes a health and safety issue. For some, there were significant fears about human exposure to potentially toxic nanoparticles, with Yuthavong saying: "...we could have a case of 'nano asbestos'...that is really something that worries me a lot". Although he was not sure that nanoparticles should be compared with asbestos, Weckert said that there are "...obviously potential problems if these particles are 'free' and we are breathing them". But others challenged these suppositions. Turney and Damrongchai spoke of the lack of scientific concern shown regarding existing contact with ultrafine particles via things such as automobile combustion, with Turney noting that humans have been living with nanoparticles "...for millions of years..." and that "...every breath you take probably breathes in about 5 million nanoparticles". Furthermore, the suggested hazards from inhaling nanopowders or dealing with nanomaterials are seen as nothing new, in light of the concerns already raised within the chemical industry (Damrongchai; Radt).

Returning to concerns about the ability for nanotechnologies to cross internal bodily membranes, Yuthavong spoke of potential liabilities from the ability of nanoparticles to penetrate cells and "create havoc" in unknown areas. He added that "...nano-products going to where they should not be, or doing things they should not be doing, that is something new, something that we really have to be serious about...". But some Australian interviewees were less worried, with Turney saying that such matters need to be assessed on a case by case basis and Radt adding that cell permeability is nothing new, given that chemicals always "...cross these barriers without asking...".

There was also a belief that nanotechnologies, given their general distinction from biotechnologies due to the greater focus on industrial production, may "...not be as serious as biotechnology" (Thajchayapong). According to some, this means nanotechnologies will involve less of the health and safety concerns associated with biotechnologies (Yuthavong), leaving "...nothing to be concerned [with] at this stage" (Thajchayapong).

In addition to new considerations for human health, interviewees spoke, although to a much lesser degree and with little certainty, about environmental risks. Without elaboration, Songsivilai speculated that nanoscale production would be a lot more environmentally hazardous. Similarly, Arya said that, whilst nanomaterials may be harmless to humans, they could still be harmful to living creatures, introducing 'foreign' matter into various environs. In this respect, Arya was most concerned about the residual, environmental impact of something unnatural entering a natural system:

> ...[we are] creating something new and stranger to our ecology... It may change some balance in the delicate ecology and, if it is not recycled, or re-transmitted, in a certain way but remains artificially in our natural world then, in the long-term, what will be the effect?

Arya then spoke of the tough decisions needed to balance progress with the potential costs associated with such waste:

> ...we want to go ahead but, at the same time, if it will have a negative impact for the future generations then the cost that we are talking [about] now will later-on become very large if we have to invest to clear, or clean, the planet of nanotechnology material...

Also of note with respect to the environment was the disparity between perspectives regarding the immediacy of threats posed by nanotechnologies. Some spoke of this issue being a future concern (Arya; Charinpanitkul), with environmental risks seen as "less immediate" (Yuthavong), whilst others, such as Songsivilai, thought the issue was a "great concern now".

For both human and environmental health, the general comments reinforced Ford's belief that there is a lack of certainty about the impacts of nanotechnologies. But for Weckert the early stages of any new technologies are "...always a matter of the unknown", with Damrongchai adding that he witnessed a similar situation with the introductions of GMOs. Nonetheless, two broad reasons were presented for why a lack of certainty accompanies consideration of nanotechnologies, in terms of impacts on human and environmental health. The first is that it is too early to make reasoned judgments about technologies that are said, by Arya, to be in an "exploratory" phase. This problem commences with a lack of certainty about "...which applications are going to be most important or most commercialized" (Warris). As Songsivilai noted, Thai researchers are unsure if nanotechnology will be environmentally friendly because it is not clear what nanotechnology devices might emerge, or the materials that they would use. Even those who are focussed on researching certain specific nanotechnologies do not know all the implications (Arya). Furthermore, society is definitely unable to provide feedback on human and environmental health risks given they are yet to experience the impacts of nanotechnologies upon daily lives (Panitchpakdi). In this light, Yuthavong said that nanotechnologies are "...still very abstract because no real case of liability has come to light...". Partly as a result of the nascent state of nanotechnologies, there are a lack of coordinated, toxicological studies (Turney), creating a situation where "...the jury is still out" (Weckert).

The second broad reason for why a lack of certainty accompanies nanotechnologies is that there are some things that may never be able to be reasonably judged; nanotechnologies raise a number of imponderables (Arya). From a number of interviewees (Charinpanitkul; Tanthapanichakoon) came a shared sense that nanotechnologies produce unexpected and unintended implications and risks that industry and regulatory bodies may be unable to assess. Such imponderables are founded on an uncertainty about how to assess issues such as the safety of nanotechnologies (Cooper; Damrongchai; Sawanpanyalert). Presently, those involved in nanotechnology R&D would appear unable to provide assurances to consumers, particularly in terms of side- and long-term effects, with Panitchpakdi asking: "how will we know that nanotechnology... is safe for health issues like TB [tuberculosis]?". Added to this are concerns about the capacity for Southern countries to assess and "...make their own decisions" regarding nanoparticle safety and the level of required experimental testing, "...as opposed to accepting Western decisions and ideas about this" (Braach-Maksvytis). In support of this argument, Arya questioned whether Thailand has the capabilities to fully test nanotechnologies for potential harms.

8.3.2 Ethical

The increasing attention given to work at the nanoscale has also drawn light to a wide range of ethical considerations (Changthavorn; Songsivilai). As was the case with human and environmental health considerations, the main tension to be explored here is whether nanotechnologies actually raise any new ethical issues.

Interviewees presented a range of perspectives about the novelty of ethical concerns raised by nanotechnologies. Whilst some, such as Deutchmann, imagined that applications "...will provoke, case by case, new ethical or guideline concerns", others did not see ethical implications as a threat, with Yuthavong suggesting that nanotechnologies are being 'thrown' in with ongoing discussions about ethical responsibility in science merely "for good measure".

One area of tension remains around whether nanotechnologies raise bioethical concerns that extend the controversies within the GMO debates regarding the manipulation and modification of natural things. Understandably, bio-nanotechnologies were at the heart of this discussion (Changthavorn; Thajchayapong). In this sense, there was consensus that nanotechnology will include biotechnology debates, by default, given that nanotechnology complements biotechnology. In fact, Pothsiri saw a danger in terms of not making the link between biotechnology debates, such as those for GMOs, and nanotechnology, saying: "at the moment they[10] are debating this in isolation; they are not linking it with the nanotechnology, as yet...they are not realising that it is part of the application of nanotechnology...".

Bioethical discussions commenced with consideration for 'atomic manipulation' of organisms, outside the process of genetic manipulation. Cooper, a Research Fellow in risk at Macquarie University in Australia, suggested that cellular 'interventions' on the nanoscale are an "...unnatural attempt to reinvent what should not be tampered with". As discussed in my review of the literature, this issue came to prominence in Thailand when the Fast Neutron Centre in Chiang Mai announced they had fired a neutron into a rice cell, heralding what the ETC Group (2004e) called the "...brave new world of atomically manipulated organisms". In response, Yuthavong, from Thailand's BIOTEC, said that this research was not worth considering in terms of ethics because it was "...almost non-scientific...[and] really to demonstrate...the technical ability of the Chiang Mai people".

Scientifically, he responded that "...something as foreign as a foreign atom like that, inserted in a ruthless manner, like that, does not stand a chance, the cell will just say, 'o.k. I quit' and die". In this respect, he said he "...does not see any science for AMOs right now because when you modify organisms you do it through the gene".

Yuthavong's comments highlighted a fundamental belief shared by Ford: that there is a crucial distinction between nanotechnology and genetic engineering, given that the overarching goals of nanotechnologies do not include the production of GMOs. Yuthavong elaborated:

> The engineering of genes would take us into the same controversy as the GMOs...[but] nanotechnology will not really be concerned with creating new organisms as much as using the gene technology for diagnostic purposes or biosensors.

In this sense, and building on the earlier comment about nanotechnologies being more focussed on industrial products than biotechnology, Ford dismissed ethical debates relating to atomic manipulation by declaring:

> There is nothing mysterious about...atomic manipulation, in the case of nanotechnology...I do not think that there are things we need to worry about, ethically. We are not doing anything that crosses strange, ethical boundaries...writing 'I.B.M' in atoms is not an ethical issue.

However, despite his earlier assurances about AMOs, Yuthavong said that the ethical implications raised by nanotechnologies are "...less clear than [with] the case of genetic engineering". He added that cases such as a nanotechnology-based therapeutic mechanism infiltrating and creating an "...absolute coup of the cell mechanism" would have to be "...thought about really hard" because, no matter how unlikely, the magnitude of potential risk would mean its eventuality would raise "...a lot of ethical considerations".

Further bioethical issues included concerns that nanotechnologies "...may infringe human rights..." by accelerating and maximising current threats to individual liberties, such as those relating to the control of personal genetic information (Changthavorn). In explaining how this might be possible, Songsivilai also alluded to his underlying concern for the future: "...because it is so cheap to do diagnostics...instead of testing one or two things they test 100 things at the same time...Most of this information may not be needed for those people...".

Similarly, interviewees voiced concerns about privacy. Ratanakul, for example, spoke of nanotechnologies facilitating increased monitoring that could lead to "...the control of the lives of the general population by unscrupulous parties". Weckert agreed that nanotechnology-based sensing could create a sinister situation, explaining:

> It will be a lot easier to...control workforces simply because monitoring techniques will be able to be much more invasive than they are now...it will be very easy to have enormous databases to do a lot data mining very quickly...it is also going to make it a lot easier for governments, big businesses and so on, to control us in ways that we do not like.

The trialling of nanotechnologies in the South was another bioethical issue raised as a considerable concern. A number of Thai interviewees (Panitchpakdi; Ratanakul; Yuthavong) were concerned that the South would be 'guinea pigs' for the testing of nanotechnologies, mimicking situations that have occurred with antiretroviral drugs for the treatment of HIV/AIDS.[11] Selgelid outlined a similar concern for "...throwing some new technology at the problems of the developing world without taking the normal precautions that we would take before we would throw a new technology at the developed world".

For testing, interviewees saw relaxed ethical norms and susceptibility to monetary incentives in the South creating an ideal situation for Northern companies and researchers (Deutchmann; Ratanakul). This raised the larger concern about who will ultimately benefit most from the knowledge arising from Southern testing (Tegart).

However, interviewees (Deutchmann; Ratanakul; Selgelid) noted that issues such as careful monitoring of human studies, informed consent and standards of care, whilst important bioethical concerns, are not unique to nanotechnologies.

Following discussion about the impacts of research on Southern populations, interviewees considered the potential progression of nanotechnology-based diagnostics ahead of nanotechnology-based therapeutics potentially leading to the ability to detect a disease before being able to treat it. Bioethically, the issue received a mixed response, with Songsivilai saying it was a concern, whilst Ratanakul said he had never considered it. Deutchmann, an Australian healthcare practitioner who has spent a great deal of his life working in Southern countries, was not so concerned, saying that, in certain settings — particularly communicable disease settings where diagnosis does not provide a means of therapy, it remains important to diagnose, irrespective of the ability to treat, as this can mean that other aspects of the patient's life can be adjusted to maintain quality of life.

A few of the more mainstream, ethical debates in the North, such as spending on warfare or the potential for nanotechnologies to service bioterrorism, received little interviewee consideration. Although there was a query about how global military research might be "...limited to 'the good guys'" (Braach-Maksvytis), Ratanakul said that Thais are not very concerned about nanotechnology being steered towards military ends, or the subsequent possibility for misuse. Damrongchai was the only one to have made any strong comments, presenting military research into nanotechnologies as something that would lead to "...peace[ful] views of technology" with his only worry being that the military budget might be used to "... merely buy finished products".

Similarly, the issue of cybernetic humans was presented by two Australians (Cornell; Ford) as a long way off, whilst the debate did not appear to be on the Thai radar. Only Yuthavong commented, saying that that any negative developments in this area would have less to do with the science and more to do with a scientist who wants to "...become like Dr Frankenstein or someone who really want[s] to do something successfully, technically, but then forgot good old commonsense".

Combined, the ethical discussions above point to a lack of Southern engagement with the ethical implications of nanotechnologies as well as associated significant gaps in knowledge (Changthavorn; Charinpanitkul). In this sense, Panitchpakdi says it is difficult to judge ethical implications arising from nanotechnologies without clear examples of tangible applications. Subsequently, there is a severe void of literature on the ethical implications of nanotechnologies, making it difficult for social scientists and ethicists to engage with relevant debates (Charinpanitkul; Ratanakul; Thajchayapong). But, Yuthavong says that the lack of knowledge is part of a natural 'ethical lag' and that it is merely a matter of 'catching up'. In this respect he noted that Thai ethicists "...have not really got to that [nanotechnology] problem yet. Right now they are only at the problem of cloning and GMOs...they will still need months, or a few years to 'digest' the implications".

8.3.3 Legal

In terms of legal implications, there was concern that patenting on the nanoscale will provide "...too broad [a] protection" (Changthavorn). Furthermore, Changthavorn believed that nanotechnologies could accelerate the emerging problem of 'process patenting'.[12]

The issue of patenting signalled a return to ongoing debates experienced with biotechnologies regarding the patenting of natural matter and processes resulting from some form of 'novel' scientific modification. Although seen as a generic technological issue, given there will be cases for many new technologies where modifications to natural occurring matter or processes will be negligible and questionable (Chirachanchai), future problems were envisaged for the delineation between nanotechnologies' 'inventions' and discoveries' (Changthavorn). However, Changthavorn saw no immediate concerns, given Thai law permits patenting "...new applications of natural phenomena" as long as they have technical merit and human intervention. In this sense self-assembly, for an area such as drug discovery, is believed to clearly satisfy such criteria:

> ...assembling such structures — drug delivery vehicles — always includes the right steps, includes the right mixture and that is the technology and that will be easy to patent...[it] is easy to define what it is about and what the new steps are, what is included (Radt).

On the other hand, given its complementary nature in terms of building on and engaging with a range of scientific disciplines, Changthavorn believed working on the nanoscale makes it easier to add technical merit to applications, thereby increasing the ability to satisfy the 'novelty' criteria within a patent application. As an example, he says that working on the nanoscale:

> ...may enhance the position of patenting biotechnology products... fulfil[ling] the second requirement of patentability which is 'inventive step'...because it adds technical merit to biotechnology invention[s].

As with the area of ethics, there was concern about the lack of legal knowledge in the field and the lack of understanding amongst Thai lawyers (Changthavorn). Changthavorn was blunt in his appraisal of the Thai situation: "in terms of legal perspective[s], I think Thai lawyers have no idea about this technology". He explained that this was part of a broader problem, whereby extremely few lawyers, including IP specialists, have a scientific background.

The concern for a lack of legal knowledge and understanding extended to regulatory bodies such as the Thai Food and Drug Administration (FDA). Here Changthavorn spoke about the inability for the FDA to respond to a nanotechnology-based application:

> ...for the time being they are not very well adjusted...if you file [an] application now, you would have difficulty because they do not understand it...they do not know how to investigate the products.

However, he said that, if a nanotechnology application could be explained scientifically, then the FDA would accept it.

When combined with the surprisingly fast rate of technological development (Panitchpakdi; Pothsiri), there was a belief that these factors are indicative of a global regulatory lag in nanotechnology (Changthavorn). According to Weckert, "things just keep getting ahead of us...", with Yuthavong adding that this obviously means "... the [regulatory] gap will be even bigger".

8.3.4   Social

When it comes to the broader societal implications of nanotechnologies, interviewees largely focussed on the operational implications for scientific institutions and the associated implications for the workforce. To start, working with an interdisciplinary-based paradigm was seen as presenting difficulties in assessing national nanotechnology capabilities (Charinpanitkul) and coordinating research (Chirachanchai). In this respect, Chirachanchai believed that Thailand is struggling, given that the country "...did not start from gathering and thinking about 'what is the goal?' and separate the work out". Radt added that difficulties in coordinating interdisciplinary research could hinder R&D outcomes.

Difficulties in coordinating nanotechnology research also stem from the resistance of people in the individual sciences to the interdisciplinarity required by working on the nanoscale (Kanok-Nukulchai; Tanthapanichakoon). Interviewees said that such resistance would be a natural reflex, given the research culture in the individualised sciences supports a protective mindset and fear of "...being branded in the same basket" (Kanok-Nukulchai). Tanthapanitchakoon explained further: "the typical mindset is that civil and electrical engineers are not concerned with other issues beyond their field". Ford saw this as a challenge for those seeking to develop nanotechnology research, saying: "the novelty arises from trying to mount research that is truly cross-disciplinary in a university that still thinks along the lines of 'physics, chemistry and biology'...".

The development of nanotechnology R&D capabilities founded on interdisciplinary research was viewed as an area where certain Southern countries could face greater difficulties than their Northern counterparts (Radt; Tegart). Radt explained that the intensified struggle for funds and the departmental coveting of funding means "...it would be, perhaps, harder to break down these barriers". Compounding this problem could be conflict over access to necessary, yet limited infrastructure given "...there are a whole range of different tools you need to 'do' nanotechnology that are traditionally in very different parts of the university and having access to all of them becomes problematic" (Ford).

The issues of 'single-silo' mentalities and access to infrastructure were seen as permeating interdisciplinary nanotechnology education as well. Ford, drawing on his experience in co-establishing the world's first nanotechnology degree, said that any country developing educational programs for nanotechnology could face the negative reaction from within institutions: "that you are no longer teaching 'real' science". He continued, saying that this was the biggest hurdle to overcome but that "...[it] came, almost entirely, from established scientists who had been working in their particular discipline for 'x' years...". Building on this point, there were concerns about where those in training will go when they graduate from nanotechnology courses, considering the lack of graduate nanotechnology jobs in Thailand (Charinpanitkul; Kanok-Nukulchai).

The issue of graduate employment was the lone issue raised with respect to nanotechnology's potential impact on employment and local workers in the South, despite a belief that nanotechnology will require a highly skilled workforce (Radt). Conversely, and building on previous comments about the displacement of indigenous technologies, Cooper wondered whether the push to high-tech and subsequent patenting might turn traditional workers to "...kind of come back to the use of technologies and relearn them and also invent new ways [of] merging these traditional knowledges with new technologies?".

8.4 Conclusion

Nanotechnologies are seen as offering numerous technical advantages, but any associated benefits are set against numerous imponderables relating to risks and implications, as well as economic imperatives that can mean nanotechnologies are oriented away from Southern needs. The wide variety of responses reinforces my findings in the last two chapters: that nanotechnology can be conceptualised in very different ways, even when there is evidence that its distinguishing characteristics are commonly understood.

In terms of benefits, nanotechnology is seen as creating new opportunities based on the introduction of a novel paradigm for science. Whilst there is agreement about the diversity of potential nanotechnologies for the South, Thai interviewees generally presented the greatest potential lying with food and agricultural technologies, thin films, coatings and textiles, whilst Australian interviewees generally presented the greatest potential residing with water- and energy-related applications. However, when it came to healthcare, there was a joint belief that nanotechnologies offer new opportunities to empower communities through simple and user-friendly diagnostic and therapeutic applications.

On the other hand there is broad concern that stated opportunities are likely to prove more rhetorical than realistic. Whilst significant hype for nanotechnologies is said to obscure significant scientific barriers, the main concern is that opportunities for nanotechnologies to address Southern needs in a locally sensitive manner will be confounded by various systemic barriers. The concern is largely that the market will ensure Southern-developed nanotechnologies are orientated towards industrial applications for economic gain. Whilst this, in itself, supports the argument that nanotechnology R&D provides new opportunities in terms of market development, the early orientation of nanotechnology researchers in Thailand reinforces a tension between Northern idealism and Southern pragmatism with respect to Southern implications relating to nanotechnologies.

In looking at the social implications raised by nanotechnologies, interviewees identified a wide range of issues that, importantly, are said to cut across the various sectoral applications. As with the literature, responses regarding whether nanotechnologies create new risks to human and environmental health were clearly polarised, although there was agreement that the debates are constrained due to a lack of available knowledge and numerous imponderables. On the whole, nanotechnologies were viewed in a different ethical light to GMOs and were not seen as raising any new legal implications in near term, aside from the introduction of more broadly applicable patents — a point questioned by many in terms of its implications for control over technological development. Interestingly, interviewees did not raise any of the prominent concerns highlighted in the literature regarding the implications of nanotechnologies for the human-machine-nature interface. However, interviewees did speak at length about the impact of nanotechnology research on the way science is conducted, highlighting the tensions surrounding interdisciplinary demands.

It is also important to highlight a few of the prominent issues relating to appropriateness that were raised in my review of the literature yet were barely explored by Thai and Australian interviewees. In particular, beyond the familiar rhetoric of sustainable development, there was scant demonstration that reflexivity to the limits to growth is central to the thinking about the appropriateness of nanotechnologies. In a similar light, I was surprised by the lack of discussion around issues of local ownership, the ability for nanotechnologies to provide productive and fulfilling employment, and the potential for nanotechnologies to displace existing Southern industries and automate labour.

This chapter has advanced this book by giving a much clearer picture of the foreseen implications of nanotechnologies for the South. It has also raised a number of issues about how nanotechnology will be applied that I will next explore, along with other issues relating to Southern approaches to nanotechnology's governance.

Endnotes

1 See http://siamcement.com.

2 RNA (riboneucleic acid) is a biological molecule central to the process of protein synthesis.

3 Formally known as 'American trypanosomiasis', this tropical parasitic disease is potentially fatal and currently without a vaccine.

4 As I have already alluded, the highly regular demands for therapeutic intervention associated with both the standard and short-course treatments for tuberculosis are contributing factors in management challenges such as generally high drop-out rates amongst patients.

5 A technique in which a dispersion of colloidal particles are synthesised to form a ceramic material.

6 The transactional equivalent of approximately $12,375,000, as of September 2004.

7 Despite the then Prime Minister's key role in driving Thai nanotechnology up to that point, the Thai government's commitment to nanotechnology has continued just as strongly since 2006.

8 A group of viruses causing, amongst other things, warts and genital infection.

9 The kind of disruption referred to here differs from the traditional understandings as defined by Bower and Christensen (1995).

10 It was not made clear to whom Pothsiri was referring.

11 See Annas and Grodin (1998) regarding informed consent, placebo-controlled trials and inhibitive costs of effectively trialled drugs for Southern populations.

12 According to Sterckx (2004), "process patents offer protection for the way in which the final product is made and for the way in which the product is used to reach certain goals (e.g. the treatment of specific diseases)" (p. 60).

Chapter 9

Approaches to Technological Governance

In order to further expose who will likely hold the power and influence in nanotechnology's emergence, in this chapter I will complete my qualitative research by investigating interviewee perspectives on Southern approaches to the governance of nanotechnology. In particular, I will explore the extent to which the present and foreseen approaches to nanotechnology's governance in the South enable empowering, democratic processes.

9.1 How to Become 'Nano-Innovators'

Interviewees presented many strategies for the South to build nanotechnology research capabilities. These strategies can be grouped into three areas that loosely reflect stages of the capacity building process. I will first explore how it is believed the South can strategically plan for nano-innovation, looking at the advantages of a coordinated approach, the ability to learn from global efforts, methods to evaluate domestic capabilities as well as strategies to develop a nanotechnology vision, grounded in the selection of appropriate research areas. Having established how Southern countries might successfully plan for nano-innovation, I will then look at how to develop adequate and appropriate human resources to support such innovation. Here the focus will be on the various

Nanotechnology and Global Equality
Donald Maclurcan
Copyright © 2012 by Pan Stanford Publishing Pte. Ltd.
www.panstanford.com

ways in which education might be used to develop human resource capabilities, the proposed nature of nanotechnology curricula and how to best harness international support for human resource development. To conclude discussion on the innovation-building process, I will look at strategies and methods for developing supportive capacity for nano-innovation. In particular, I will look at whether or not nanotechnology demands reorganised infrastructure, the suggested methods for ensuring cross-disciplinary research, the specific means by which to engage the private sector, and the various kinds of strategic international partnerships and ways to facilitate international interaction.

9.1.1   Planning

A number of strategies were proposed by interviewees for Southern countries to improve planning for nano-innovation, exposing an underlying belief that nanotechnology demands planning that is solely focussed on the challenges of nano-innovation (Tanthapanichakoon; Thajchayapong). Yuthavong went further, suggesting that planning for nano-innovation requires a unique response and that it demands reorganising and reorientating efforts to harness the arising opportunities.

Globally, specific planning for nano-innovation has been most successful when it has been part of a formally coordinated approach (Warris). Both Thai and Australian interviewees (Berwick; Tanthapanichakoon; Thajchayapong) agreed that government bodies are the best vehicle to coordinate and drive initial engagement with nano-innovation. More specifically, interviewees pointed to the importance of establishing a coordinating, national body, such as NANOTEC (Charinpanitkul; Dutta; Tanthapanichakoon; Thajchayapong; Warris), to plan, implement and control a 'national nanotechnology program' (Warris).

In addition to the need for a coordinating body, Thajchayapong suggested, as a useful measure, the establishment of a supporting committee, such as the Thai National Nanotechnology Committee. This committee, he noted, is made up of subcommittees for the fields of electronics, materials, biotechnology and academia. Given these committees are chaired by a range of experts from various universities, NANOTEC and the private sector, he said their usefulness was seen particularly in terms of the ability to draw together individuals representing a diverse range of backgrounds. Although noting the need to incorporate more social scientists in the committee compositions, Thajchayapong saw the current level of diversity as a good means to ensure balanced planning for nano-innovation. Adding prestige, power and drive to the committee's usefulness, and potentially expediting the planning process, he said, was the agreement from the Thai Prime Minister to chair the committee.

The utility of a coordinated approach was first seen in terms of the ability to efficiently conduct formal evaluations of nanotechnology capabilities and areas of strategic engagement (Charinpanitkul; Dutta; Tegart), therein providing the foundations for the development of a nanotechnology "master plan" (Tanthapanichakoon). A coordinated approach was also seen as providing the initial, national momentum for nanotechnology research (Dutta), thereby reducing risks and increasing the likelihood of a government's commitment to nano-innovation (Panitchpakdi). In the Thai situation, it was believed that momentum will primarily stem from the ability of the national nanotechnology centre to conduct its own research (Charinpanitkul). But interviewees also spoke of the national centre's ability to administer research funds (Charinpanitkul; Tegart) and facilitate both cross-disciplinary research interactions (Tegart; Warris) and broader interactions between Thai nanotechnology's many 'stakeholders' (Panitchpakdi).

In terms of conducting formal evaluations of nanotechnology capabilities, interviewees were resolute that the process should commence with an assessment of global capabilities. Three desired outcomes were discussed as the driving rationale for a global evaluation. The first is to map the global status of scientific knowledge and investigation, identifying what research is being conducted and then to 'benchmark' this research, both internationally and domestically (Ratanakul; Tanthapanichakoon; Thajchayapong). This was seen as a key strategy for avoiding unnecessary duplication of research and the futile development of technologies already under patent (Dutta), whilst simultaneously establishing the realms in which Southern countries might leapfrog existing and emerging knowledge (Ratanakul; Tanthapanichakoon). The second desired outcome of a global evaluation is to map the distribution of relevant resources, in order to ascertain points for Southern leverage. As Dutta explained, this process could be particularly important for some, claiming: "least developed countries can enter the nanotechnology revolution by trying not to emulate what others have already done but, instead, try to look around for resources that are available and in place". The third desired outcome of a global evaluation is to map the strategies and plans adopted by the leaders in nano-innovation. By learning from both the initial mistakes of others (Dutta; Ratanakul; Tanthapanichakoon), as well as those strategies proving most effective (Berwick), Southern countries can gain ideas on how to save time, resources and money (Dutta). More specifically, interviewees spoke of the mapping process exposing both the broad nature of research strategies, such as Singapore's heavy focus on developing nanotechnology IP (Thajchayapong) or the E.U.'s focus on developing 'functional nanotechnology' (Dutta),[1] as well as specific matters of implementation, such as how a country invests in nanotechnology R&D or incorporates local industries (Thajchayapong). However, Thajchayapong noted that the mapping of strategies and plans utilised by nano-innovation leaders merely provides ideas for consideration, rather than a blueprint for what is to be copied or avoided.

Whilst a few interviewees (Ratanakul; Tanthapanichakoon) spoke of an obvious need to be informed by the current literature, others, such as Panitchpakdi, said that sending delegations to assess international nanotechnology capabilities was the most useful strategy for a country to assess what knowledge could be adapted to its situation. Dutta noted that Thailand is particularly interested in this approach, considering "Thai administrators of nanotechnology, or the people who are deciding on nanotechnology research, are quite open to ideas, and like to hear what others are doing". Subsequently, Thai researchers, government ministers and NANOTEC delegates have been part of a number of official visits to countries, including Singapore, Taiwan, Japan and the U.S., with plans for delegations to visit France and Australia (Tanthapanichakoon). Similarly, according to Dutta, Vietnamese delegates have been to the U.S. and Japan. Furthermore, in a point that highlights the ability for Southern countries to leverage developments from other Southern countries, a number of Vietnamese delegations have visited the Thai NSTDA (Tanthapanichakoon).

Beyond national efforts to map international nanotechnology capabilities, Dutta proposed the establishment of an independent, international evaluation committee. Such a committee, he envisaged, would be charged with assessing the global nanotechnology R&D situation, with an aim to highlight potential duplication of research and thereby "...sieve out nanotech[nology]-related work which is not necessary...". An international, nanotechnology evaluation committee, he continued, would need people with substantial nanotechnology experience who were also 'in touch' with recent, global trends in associated fields.

Following an international evaluation and the identification of points of leverage for Southern countries, interviewees, such as Tegart, saw an evaluation of domestic nanotechnology capabilities as the next phase in developing Southern innovative strategies and directions. This was seen as providing an important benchmark of strengths and weaknesses upon which academia and government could apportion appropriate support to nanotechnology (Warris), as well as a clearer understanding about how a country might best adopt or adapt nanotechnology and the feasibility of its engagement with various areas (Tegart).

In terms of how to evaluate domestic nanotechnology capabilities, interviewees (Dutta; Tegart; Warris) suggested that Southern countries could start with an audit of the country's resources and capabilities. According to Warris,[2] this audit should initially focus on resources and capabilities in academia. Interviewees suggested that the audit assess the situation in terms of both "...the people who are there and the facilities that are available..." (Dutta), as well as the level of existing expertise and areas of strength (Warris).

In 2002, at the behest of the Japanese Government, a research team at Chulalongkorn University evaluated Thailand's nano-technology capabilities, resources and knowledge. Charinpanitkul, lead researcher in the study, outlined the following four stages in the process:

- Members from the key Thai science agencies[3] met and brainstormed to develop basic nanotechnology information to inform the study;
- Representatives from academia, government and private sector were identified using an NSTDA database, combined with contacts gained from prior nanotechnology-related conferences and exhibitions;
- Representatives were provided with a questionnaire asking for their perspectives on nanotechnology, its promising areas and their research interests; and

- Results were collated and statistically analysed.

Although a number of limitations[4] were recognised for the study, Charinpanitkul said it provided a greater picture of "...the present situation of understanding, information and knowledge of nanotechnology in Thailand". This, he said, prompted crucial action, including the development of a database of 'nanotechnology people', or those in associated fields. The four-stage process was presented as a "good model" for evaluating nanotechnology capabilities in other countries, such as Laos, where he felt it might be useful sometime around 2010.

Given the baseline knowledge resulting from an evaluation of international and domestic nanotechnology capabilities, Thajchayapong suggested the development of a national nanotechnology strategy as the next stage of planning for Southern nano-innovation. By articulating a clear vision and specific plan for engagement (Chirachanchai; Kanok-Nukulchai), as well as a 'master plan' that considers a 'longer-term view' (Arya; Kanok-Nukulchai), a national strategy can establish how a country might best commence its own nanotechnology program and 'catch up' (Tanthapanichakoon). In this sense, a national strategy underpins the development of endogenous R&D; seen as the key means by which to reduce Southern technological dependency and address the problems of underdevelopment (Arya; Bryce; Charinpanitkul; Coyle; Lynskey).

The identification of strategic areas of R&D is viewed as the foundation on which a national nanotechnology strategy should be based (Arya; Changthavorn; Thajchayapong). Yet, considering no country can be "...world leaders in all areas..." (Warris), interviewees spoke of the occasional need to license, rather than always attempt to lead innovation (Charinpanitkul; Turney). Licensing nanotechnologies will allow countries to modify and adapt both knowledge and technologies to their own situation and needs (Berwick; Charinpanitkul; Turney). Berwick elucidated with an example:

> If there are issues of environmental remediation [or] waste management, look at what is the world's best technology...[and consider,] could that be applied to their own infrastructure, waterways?...[in this way a country can] get the best of both worlds: a local solution using a global technology.

In Thailand's case, the private sector was seen as the crucial vehicle to increase support for the licensing of nanotechnologies (Arya). However, although interviewees identified the need to embrace technological licensing in certain circumstances, Yuthavong said there was a generally recognised need for a balance between a country licensing and developing its own nanotechnologies.

Furthermore, interviewees strongly felt that, given the lack of Southern resources to engage in all applied areas and because it is such a wide field, it would be strategic for Southern countries to focus on niche areas (Arya; Changthavorn; Charinpanitkul; Chirachanchai; Dutta; Kanok-Nukulchai; Tegart; Warris). Dutta explained: "...there is a lot of scope to develop certain aspects of nanotechnology, but due to the limited number of scientists around here...[we need to] focus on the areas that we would like to develop".

The selection of niche areas will depend on a number of key factors that will differ in nature from country to country. The first is what the country is hoping to achieve from engaging in nanotechnology (Dutta; Warris). As an example of the difference between countries, Dutta noted that Singapore is focussing its research on bio-nanotechnology, whilst Japan has a more 'balanced' program, incorporating research into energy and materials in addition to bio-nanotechnology.

The second factor in a country's selection of niche areas is its ability to be competitive or excel in certain things (Warris). Berwick went so far as to say that competitive advantage, in terms of having something to offer the world, is a necessity for smaller Southern economies seeking a point of leverage for growth. Building on comments made in Chapter 7, interviewees (Berwick; Charinpanitkul; Kanok-Nukulchai; Tanthapanichakoon; Warris) first spoke of competitive advantage in terms of comparative advantage. In this respect, interviewees suggested that Southern countries select areas for nanotechnology R&D based on natural strengths, such as the potential cost-savings associated with the utilisation of natural resources:

> ...in Thailand, we have the natural resource[s]. If we produce some product...at [the] nanoscale...[such as a] nanotechnology drug delivery base, the products, compared to other countries, may be cheap if we use our natural abundance for that (Chirachanchai).

In this light, Thailand's selection of niche R&D areas was, and continues to be, strategically influenced by the nature of pre-existing resources and knowledge, as well as the strength of production capacity. Hence, the Thai focus is on food, agriculture and textiles (Kanok-Nukulchai; Tanthapanichakoon; Turney) as well as "...medicines and herbs, packaging and automobiles" (Turney). Highlighting the emergence of nanotechnology R&D from within areas of existing strength, Panitchpakdi spoke of nanotechnology-based diagnostic test kits, the development of which stemmed from previous Thai successes in diagnostics with microelectromechanical systems[5] and 'lab-on-a-chip'[6] devices.

However, interviewees such as Tanthapanichakoon noted the strategic danger in relying on natural resource abundance, outlining its potential to perpetuate the 'nut-cracker' effect, via which Thailand is currently "...being squeezed between the low...commodities, and the high-value products" (Thajchayapong). According to many, counteracting such a danger demands a move to a 'knowledge-based' economy (Tanthapanichakoon) in which knowledge creation is a tool for 'value-adding' to exports (Panitchpakdi; Sawanpanyalert; Tegart; Thajchayapong; Turney). In such a case, scientific or technical knowledge is used to upgrade products and therein industries, making both more internationally competitive (Thajchayapong). To explain how this process might work, Dutta used the example of rubber, a global commodity in which Thailand has a very strong position:

> Rubber should be something that needs to be urgently looked at, in terms of nanotechnology applications, because Thailand is manufacturing most of the rubber for the world. However, this rubber is sold at dirt-cheap prices...if we can functionalise the rubber at the root, by a process that will avoid future processing in more expensive plants, then we have a niche there. We can sell the rubber at a price.

However, whilst developing niche export markets was seen as Thailand's only option to "...survive...[and] be competitive" (Dutta), there was also a belief that, strategically, Thailand should "...focus on the domestic [market] first, because the world market arena is very tough, [and it] is quite difficult for Thailand to compete with other countries..." (Charinpanitkul).

Irrespective of the areas selected for nanotechnology R&D, or whether technologies are developed for an export or domestic market, interviewees (Damrongchai; Thajchayapong) spoke of the need for a larger 'road-mapping' process to outline paths for how a country's nanotechnology research might best progress through the various R&D phases. According to Tanthapanichakoon, in the early stages of Thai engagement with nanotechnology, such processes were informal:

> When we first try to set up our national nanotechnology centre we have had several brainstorming meetings and sessions with the experts and professors in Thailand; all over Thailand we invited them and even representatives from other government agencies. Of course we also gather[ed] information from our contacts and our colleagues.

However, in 2004 the Thai NSTDA conducted a formal technology road-mapping exercise that involved 117 participants from all over the country (Charinpanitkul), with sectoral representation largely from academia (Thajchayapong). Although interviewees did not discuss the outcomes from this process, Damrongchai alluded to the need for a narrowing of scope, saying that "...road-maps [are] best for focussing on certain technology or certain products because you will be more focussed on what will happen...what can be the possible road-blocks".

In addition to mapping out paths for nanotechnology R&D, the following, four 'key performance indicators' were presented as useful for any country wishing to evaluate its nanotechnology progress: the number of nanotechnology publications[7] and patents (Charinpanitkul; Warris); the number of researchers (Charinpanitkul); the level of government funding (Warris); and the level of private sector funding (Charinpanitkul).

9.1.2 Resourcing

Having considered planning, in terms of evaluating global and domestic nanotechnology capabilities and selecting strategic areas and directions for research, the focus moved to developing adequate and appropriate resources for nano-innovation. Interviewees spoke particularly strongly about the importance of building human resources, with Arya saying:

> ...if we want to succeed, we have to prepare the human resource[s]... we have to have a clear plan...how many researchers, students, etcetera [we need]...[we] cannot [take the] short-cut...[a] human resource component in the planning is very important.

When it came to discussing what number of researchers would be adequate for a Southern country to meaningfully engage in nanotechnology R&D, many spoke of the need to develop a 'critical mass', rather than single researchers (Berwick; Charinpanitkul), noting that Thailand had set itself ten years to achieve this goal (Arya; Thajchayapong). However, Kanok-Nukulchai argued that, in terms of nanotechnology's human resources, quality is more important that quantity.

This raises the important issue of the kind of human resources that need to be developed. As I have already alluded in Chapter 6, resourcing nano-innovation requires a range of skill sets, though these are rarely envisaged as being at the 'lower' end, in terms of their level of complexity (Kanok-Nukulchai). Tanthapanichakoon outlined what he saw as necessary for Thai human resource development:

> At the R&D level we have to rely more on the researchers or highly educated people, but when you want to go into mass production or commercialisation of prototype[s], then you need good technicians, good supporting staff and a large number of engineers.

In addition to the focus on researchers, technicians and engineers, Dutta spoke of a need to train nanotechnology business managers because "...if they do not understand nanotechnology, whatever the engineers might find, they will not invest in it".

In terms of how to develop such human resources, interviewees presented a range of suggestions, many of them part of contemporary Thai discussions given, "the National Science and Technology Development Agency has already drafted a program to develop this lack of scientific personnel" (Dutta). With the 'importing' of human resources only ever seen as supplementary measure (Thajchayapong), strategies focussed on building domestic capacity, therein highlighting the importance of nanotechnology education (Tegart). Ford explained the significance of education in an historical light: "education is the number one issue because, in much the same way that education has been the driver or the inhibitor behind biotechnology, the same is going to happen in nanotechnology".

Investigating education further, interviewees (Braach-Maksvytis; Dutta; Yuthavong) felt that nanotechnology demands a reoriented approach to science and subsequent restructuring of education (Chirachanchai; Kanok-Nukulchai). Dutta, for example, saw an "...

urgent need around the world to look at nanotechnology as a totally different science", with Tegart explaining things in greater detail:

> The conventional educational systems work very much in a discipline-oriented way and keep people in boxes. Now you do need basic science capabilities, but one of the things in the nanotechnology area is the ability to look outside the box and see where other things can come together...

Dutta took this point one step further, arguing that, before getting into nanotechnology, students should "...probably forget what is physics or what is chemistry...".

Education was seen as necessary across three formal settings. The first is schools, with a couple of interviewees (Charinpanitkul; Kanok-Nukulchai) speaking of the importance of 'starting young' and providing students with basic information that will prove useful should they go on to higher education. In this sense, it is proposed that education focusses on:

> ...the concepts that underlie what is important about nanotechnology or...what defines nanotechnology...some of the basic concepts like self-assembly...and its various manifestations in areas of science (Ford).

Whilst Ford saw high schools[8] as the most appropriate point to introduce nanotechnology education in the South, Charinpanitkul envisaged, at the primary school level, a broad introduction to science and technology, incorporating nanotechnology education. He added that this would be all the more crucial as an initial human resource development strategy for a country such as Laos, to ensure students "...get used to science and technology as soon as possible". In support, Turney pointed out that Taiwan has established "...a great k-12[9] educational program for nanotechnology".

The second setting in which formal education was seen as necessary is technical colleges. Although seen as most applicable for Southern countries with well established nanotechnology R&D programs, it was believed that technical colleges could provide significant human resources for industry. According to Ford, countries "...might want to educate people though some sort of TAFE system[10] that is very highly directed to[wards] industry...". He envisaged this form of education involving short, practical- and technical-based courses to reorientate existing scientists in relevant areas, and others with science and technology awareness, into nanotechnology. In this sense, utilising existing knowledge and retraining people through technical colleges was presented as a useful strategy to expediently develop human resources (Ford; Kanok-Nukulchai). Damrongchai saw this as particularly pertinent to the Thai situation, given the early 'shift-in' of researchers from materials science.

The third setting in which formal education was seen as necessary is in universities, given "...good universities have all the disciplinary areas that are required to offer nanotechnologies..." (Kanok-Nukulchai). Ford saw this as most critical for those countries seeking to develop a nanotechnology industry considering "... you need to turn out people who are going to take a much more involved and 'leadership' role in developing it". Kanok-Nukulchai described the Thai situation at the university level as inadequate, explaining that students seeking to learn about nanotechnology are offered courses that merely merge the basic subject areas, rather than being provided with new courses that are specifically focussed on nanotechnology. Some efforts are being made to address this situation, with Charinpanitkul noting that Chulalongkorn University is trying to establish a new nanotechnology course, and Tanthipanichakoon adding that there are: "...several universities considering new curriculum of courses in nanotechnology, and our NANOTEC centre will [be] trying to facilitate or trying to help them as much as we can".

Interviewees saw a new, university nanotechnology curriculum incorporating two key features. As with education at technical colleges, the first feature is practicality, with nanotechnology teaching needing to occur 'hand-in-hand' with industrial experience and ensure students gain skills specific to working on the nanoscale (Ford). Ford explained this as part of the shift to producing 'nanotechnologists', rather than specialists in the traditional disciplines:

> It is simply a different way of structuring the science that you teach...traditionally you can take a physics degree and you come out a specialist in physics. Now you take a nanotechnology degree and you are not a specialist in physics, you are a specialist in nanotechnology.

The second, key feature of a new nanotechnology curriculum is that it will have to be designed to account for students coming together from different disciplines to exchange views (Ford; Kanok-Nukulchai). In this sense, a nanotechnology curriculum will need to be more diversified than its biotechnology equivalent (Kanok-Nukulchai). Having said this, interviewees (Berwick; Ford; Kanok-Nukulchai; Warris) firmly believed it necessary to initially train students in the basic, scientific disciplines, building up 'the fundamentals' in chemistry, physics and biology (Ford; Warris), engineering (Kanok-Nukulchai) and mathematics (Ford).

The necessity of a strong grounding in a diverse range of traditional sciences explains why Ford saw nanotechnology as best taught at the undergraduate level. However, Kanok-Nukulchai said that if nanotechnology is to be taught at the undergraduate level, it will only work as a five-year degree, with "...a focus on microscopy and similar things in the final year". In this light, he saw nanotechnology as more appropriately taught at the postgraduate level, explaining how it would build on undergraduate knowledge:

> Today technology change[s] very fast. If you start to teach technology at the undergraduate level, sooner or later it will always be obsolete...knowledge does not change, technology change[s]. So, for anybody in hi-tech, they should have [a] very strong fundamental background: physics, chemistry, engineering and then at the graduate level it is like different layers. The bottom layer is knowledge, then the second layer is technology. You have the knowledge in different discipline[s] and you combine them and make them [in]to a technology.

With this in mind, Kanok-Nukulchai envisaged nanotechnologists maintaining a professional connection with at least one of the traditional disciplines.

Providing support for universities was seen as a key means by which Northern, or even other Southern countries could contribute to human resource development in a country like Thailand (Dutta; Kanok-Nukulchai; Panitchpakdi; Tanthapanichakoon). Interviewees spoke of inviting academics from countries such as Japan, Australia and the U.S. to work with Thai scientists (Kanok-Nukulchai), as well as flying teachers in from overseas to provide courses, and encouraging overseas students from Japan, Australia, Singapore and Vietnam to do courses with Thai students (Dutta). More specifically, Dutta felt that Thailand should focus on integrating international students into Thai research degrees, given their ability to promote research and provide an important platform for inter-country relations:

> Research projects are started by students, not by professors...the students will certainly make intercultural contacts, intra-country contacts and that could develop into further ties between the countries...

Given the importance of forming and developing programs that facilitate an exchange of students between the North and the South (Radt), many interviewees (Charinpanitkul; Dutta; Kanok-Nukulchai; Tanthapanichakoon; Thajchayapong) mentioned the Thai Government's sponsorship offer for Thai students to study nanotechnology. Noting it is part of a broader move to develop human resources in science and technology, Tanthapanichakoon outlined the scheme in greater detail: "the government has set up a five year program to send 1500 young students abroad to study science and technology, and at least two- or three-hundred of them will specialise in nanotechnology".

In response to the 'brain drain' fears that I raised in Chapter 7, there was a firm belief in Thailand's ability to retain students and researchers who travel abroad to further their education and professional development (Dutta; Kanok-Nukulchai; Yuthavong). This was seen as particularly true if Thailand can improve its PhD programs (Thajchayapong) and provide strong, postdoctoral experience and support for nanotechnology graduates (Dutta). On this issue, Dutta said he expected visible differences in the coming five years, given positive efforts by the Thai government to provide support for Thai nationals, such as a recently established mentoring program for material science and electronics researchers returning from abroad.

To help facilitate an easy flow of researchers between countries and "...open up the possibility for greater international collaboration", Dutta suggested a move towards an internationally standardised nanotechnology curriculum. One step in this direction, he commented whilst explaining his early work in this area, would be to produce online, open-access materials:

> ...[I have] been writing a manuscript on nanomaterials...I have developed a course on nanotechnology for our Masters students... [that is] up and running on my website...we are developing some multimedia courses along with Uppsala University,[11] part of which we will leave accessible to people, worldwide.

Adding to this, Ford said that universities could electronically share nanotechnology courses across countries, noting that, in Australia, "...there are a reasonable amount of resources that have been developed across a number of universities here that could, in principle, be used in other countries, at a university level...". Kanok-Nukulchai agreed, saying that partnering with or gaining educational licences from universities with established nanotechnology curricula could be an expedient means for building educational capacity, with a longer-term potential for Thailand to look at exporting any materials it develops.

9.1.3 Supporting

Having explored how to plan and resource nano-innovation, interviewees presented some strategies and methods for developing supportive capacity for nano-innovation. In line with the new approach nanotechnology demands for education, there was said to be a general need to reorganise infrastructure to help unify capabilities and draw together a 'critical mass' (Charinpanitkul; Thajchayapong).

Given the need for the knowledge base to serve industry problems, Dutta said academic interactions are "...possibly the most important" in the early stages of developing R&D capabilities. Interviewees therefore focussed on opportunities to draw together researchers for collaboration. Fundamentally, opportunities formally begin through the creation of clearly identified nanotechnology groups, such as in Vietnam where certain universities plan to "...start several research groups or nanotechnology centres..." (Tanthapanichakoon). Beyond this is the need for people to come together, cross-institutionally, with the Thai government having funded a "...network for different universities to work together..." (Kanok-Nukulchai). Yet, Dutta saw "active...[and] informal" interactions as the best way to nurture useful collaboration, and provided his vision for an environment in which such interactions could occur:

> At these early stages one should...start a 'nanotech coffee club', a 'nanotech beer club', meeting once a week or once a month. Have a meeting in which there are no formal talks...people just raise their hands...

Given nanotechnology's interdisciplinarity, Dutta said, it is all the more important that these meetings be less virtual and more, "...eye-to-eye communication".

Rather than just focussing on supportive infrastructure for academics, interviewees also explored what is needed to bring together the work of government, industry and academia. In this sense, the Thai NSTDA is reported to have a range of strategies, with 'clustering' — grouping professionals from different disciplinary areas and sectors — the most prominent (Berwick; Thajchayapong). As I discovered (see Table 9.1), clustering presents three main benefits for nano-innovation: centralised coordination, network and partnership building and improved efficiency.

Table 9.1   Three Key Benefits of Clustering for Nano-Innovation.

Centralised coordination	Creates a new level of coordination for dealing with nanotechnology's diversity (Kanok-Nukulchai; Thajchayapong)
Network and partnership building	Draws together academia, government and industry, bridging the lack of understanding amongst these groups (Tanthapanichakoon)
	Assists in the formation of partnerships around traditionally sector-specific issues. For example, BIOTEC could act as the mediator between NANOTEC and the Ministry of Public Health with respect to developing biosensors for detecting bird flu (Tanthapanichakoon)
	Fosters a situation where researchers and entrepreneurs "cross-fertilise" and 'feed off' each other's ideas, therein providing mutual drive for a diverse range of science and technology areas (Berwick)
	Facilitates networking that increases the potential for international partnerships (Berwick)
Improved efficiency	Increases R&D efficiency via a pooling of resources, such as nanotechnology-specific characterisation equipment, thereby improving value for investments in scientific equipment and facilities (Berwick; Dutta)
	Raises the potential for improved technology transfer from academia to industry and subsequent ability to take products 'to the market' (Tanthapanichakoon; Tegart)

With many of these points in mind, the Thai Ministry of Science and Technology has made a serious effort to use clustering (Dutta; Thajchayapong) as a "...tool of national development" (Tanthapanichakoon). Clustering is occurring at three different levels. At the level of technology-specific clusters, the Ministry of Science and Technology has established a 'nanotechnology cluster' as one of 12 such groupings (Tanthapanichakoon). At the level of field-specific clusters, NANOTEC has established three, collaborative research networks in nanoelectronics, bio-nanotechnology and nanomaterials (Thajchayapong).

Finally, as I ascertained (see Fig. 9.1), at the level of industry-specific clusters NANOTEC has established six strategic groupings, cutting across all of Thailand's major industries (Tanthapanichakoon).

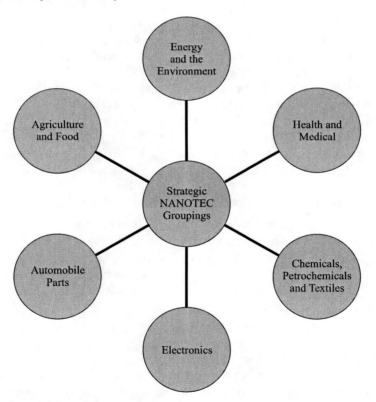

Figure 9.1 Strategic Nanotechnology Clusters Coordinated by the National Nanotechnology Centre of Thailand.

Damrongchai, Tanthapanichakoon and Thajchayapong all saw the cluster model extending to the village level, with Tanthapanichakoon adding that, in addition to the six clusters identified (Fig. 9.1), NANOTEC has also established a group to work on nanotechnology's integration into the 'One Tambon One Product' (OTOP) model[12] in which every village (tambon) improves or refines locally available resources, producing internationally accepted goods that help build local tourism and economic development.

Science parks were suggested as the best vehicle to facilitate multilevel clustering. Based on the principles of shared resources, cross-sectoral integration and concentrated knowledge, these parks enable direct partnering between technicians, researchers and industry, resulting in the efficient development of technologies (Kanok-Nukulchai; Thajchayapong). Some interviewees saw science parks as a particularly suitable venue for nanotechnology R&D, given complementary objectives around the principles outlined above. Furthermore, Thajchayapong saw the science park partly as an attempt to build a 'science community'.[13] In Thailand, nanotechnology has already been integrated into the science park model. In Bangkok's North, the 'Thailand Science Park' merges NANOTEC with BIOTEC, the National Metal and Materials Technology Centre (MTEC) and the National Electronics and Computer Technology Centre (NECTEC), whilst the Asian Institute of Technology, Sirindhorn International Institute of Technology and Thammasat University are in close proximity. As with the general science park model, the Thailand Science Park also integrates the private sector, providing a crucial environment in which local industries can drive forward:

> ...[in the] science park, we...work with local industry, [to] try to induce them to come to observe technology...[We] send expert[s] to them...whatever they want to upgrade their technology. So far, we have been working with...600 or 900 companies...to sector upgrade the...existing industries and induce the new one (Thajchayapong).

Providing a collaborative platform between ideas and commercialisation is particularly relevant for 'new' fields such as nanotechnology, given that industry requires a critical mass of nanotechnology researchers and facilities to conduct technical services such as nanofabrication (Tanthapanichakoon). Furthermore, the inclusion of NANOTEC laboratories in the Thailand Science Park provides new means for dealing with scientific problems beyond the capabilities of Thai universities, BIOTEC, MTEC and NECTEC (Thajchayapong). More generally, given its cross-sectoral nature, the NANOTEC centre can play a coordinating role between the universities, national laboratories and the private sector for certain R&D projects, with the potential to act more broadly as the nucleus, or "focal point" for science park activities (Thajchayapong).

Just as the Thailand Science Park draws its inspiration from the previously formed 'Software Park Thailand',[14] it also offers a foundation for future 'parks' that need to emerge "...throughout the whole country" (Thajchayapong). In this way, researchers and industry could be supported in more remote settings, such as Thailand's North-East or South, or around educational institutions such as Chiang Mai University, in the county's North (Thajchayapong).

As highlighted here, the private sector will be a crucial driver for nano-innovation (Arya). In light of "the unknown", it is suggested that governments will need to provide various assurances to ensure industry's engagement with nanotechnology R&D (Berwick). In addition to supportive infrastructure, such as that provided by science parks, governments will have to reduce business risks associated with nano-innovation. According to Tanthapanichakoon, one way to increase industry confidence is by taking the lead in nanotechnology investment. However, to maximise private sector engagement, Arya believed that the Thai government must demonstrate a broader "change in attitude" to signify it wants to "...'join the club'...of 'innovator', or 'producer'". To do this, it was said, requires a general increase in science spending; a move that would be both relevant and appropriate for Thailand, according to Yuthavong:

> We can afford to spend more...Thailand only spends 0.3 per cent of our gross national product to R&D...with our economic status we need to be spending one per cent like India or like Malaysia (Malaysia is spending a little bit less than one per cent).

In addition to increased R&D spending that includes targeted investments towards nanotechnology, industry will require venture capital "...in order to start up new things" (Tanthapanichakoon). In this sense, the Thai government is providing local industry with tax concessions and extremely low-interest loans of up to 30 million baht[15] for investments that are steered towards research into emerging areas such as nanotechnology:

> ...we now have 200 per cent tax exemption on any R&D project, that [is] approved by [the] NSTDA and we are talking about setting up... [a] research fund...talking to groups like [the] fashion, software or shrimp group...'if you put money into this [nanotechnology] R&D...[you] will get [an] exemption' (Thajchayapong).

Industry also requires information and guidance about prospective areas of nanotechnology R&D. In this respect, results from the planning phase, such as the identification of potential products, need to be shared with industry (Tanthapanichakoon). This will also allow companies to assess their ability to integrate nanotechnology by seeing "...how it actually relates to their day-to-day operations" (Turney).

Effective communication between government and industry is an important foundation on which assurances need to be made. Consequently, Thai interviewees highlighted the usefulness of a peak body, such as the Federation of Thai Industries, in acting as an intermediary for industry's nanotechnology-related communication with government (Tanthapanichakoon; Thajchayapong). In this light, Thajchayapong — Chair of the NANOTEC Board — added that engagement between NANOTEC and the Federation of Thai Industries needs to increase. From the government side, Panitchpakdi suggested that there is a particular need to incorporate representatives from the Ministry of Finance, particularly those who manage the government's budget.

Strategic international partnerships are suggested as another key means and area requiring Southern investment in order to support nano-innovation and develop Southern innovative capacity (Dutta; Panitchpakdi). As alluded earlier in this chapter, if countries can be convinced of the mutual advantage in "...'labouring-up' their neighbours" (Tanthapanichakoon), then international partnerships can support the development of Southern human resource capabilities (Arya; Tegart). But partnerships can also play an important role in increasing access to facilities (Tegart), offering resource-poor countries a means by which to develop "competitive advantage" (Turney). Tanthipanichakoon elaborated on this point within the Thai context:

> In order to be able to compete or catch up with other countries, or not fall behind, we need...to make strategic alliance[s] with our partners...Since we are a small country and we do not have all the resources, we should try to be part of some network or strategic alliance that is beneficial to all the members...

Moreover, according to Turney, LDC entry into nanotechnology R&D may be contingent on the formation of "...appropriate alliances with R&D providers and, indeed, corporations in the developed world...".

What, then, are the envisaged ways in which international partnerships could strengthen nano-innovation? Most interviewees saw partnerships in terms of "specific, bilateral arrangements" between countries from the North and the South (Turney). For these arrangements, some general guidelines were proffered for the respective input of Northern and Southern countries. According to Braach-Maksvytis, Northern countries could help establish "...the science and technology base in nanotechnology" by assisting with a range of broad capacity building measures. In return, Southern countries could help drive Northern innovation by undertaking various measures to support Northern R&D. Measures by both the North and the South fell within the three areas of human resources, R&D capacity building, and commercialisation (see Table 9.2).

However, partnerships to strengthen Southern innovation are not limited to arrangements between the North and the South. Rather, they extend to alliances within the South, with countries such as India playing the 'capacity building' role in partnerships with countries such as Thailand (Dutta). Moreover, partnerships are not restricted to bilateral arrangements, with their nature depending on circumstance (Tanthapanichakoon). In this sense, interviewees envisaged the South engaging in multilateral nanotechnology partnerships, such as collaborations to develop regional plans or strengthen regional capacity, through activities like regional road-mapping (Damrongchai).

Yet, referring to partnerships as collaborations between 'countries' disguises the reality that they can involve different players relevant to the partnership's aims. In addition to government organisations and agencies acting as a medium for Thailand to establish its initial international nanotechnology partnerships, interviewees also spoke of the potential for international partnerships between universities (Tanthapanichakoon) and international partnerships amongst industry, given "...alliances and joint ventures between companies...[are] an excellent way of getting technology into the marketplace" (Turney).

Table 9.2 Guidelines for Respective Contributions to North-South Partnerships.

Area	North assisting South	South assisting North	Example
Human Resources	Resourcing and providing a catalyst for Southern nano-innovation through the development of base knowledge and skills (Braach-Maksvytis; Tanthapanichakoon)	Supplementing Northern nano-innovation with human resources from existing areas of R&D strength (Damrongchai)	One agreement between Japan and Thailand[16] focusses on Thai human resource development in nanotechnology (Tanthapanichakoon)
R&D Capacity Building	Supporting Southern nano-innovation through the development of basic infrastructure (Damrongchai) and transfer of relevant technology (Tanthapanichakoon)	Providing access to specific, local Southern knowledge (Damrongchai; Songsivilai), technical infrastructure, practical experience and population samples to conduct clinical trials (Songsivilai)	Thailand has strong knowledge of Northern-prevalent diseases, such as HIV, as well as "...inside knowledge on how tests or services can be deliver[ed]..." (Songsivilai)
Commercialisation	Assisting with the commercialisation of Southern nano-innovation (Tanthapanichakoon) and support with product distribution (Damrongchai)	Acting as a manufacturing base for Northern nano-innovation, providing the appropriate infrastructure for making products (Songsivilai)	None provided

Interviewees focussed on facilitated interaction as a method for fostering international nanotechnology partnerships. Here it was expected that different sectors will be involved in these interactions, depending on the outcome sought. In terms of partnering regional governments and the private sector, interviewees returned to the example of the Asia Nano Forum. By drawing together 13 different countries, including LDCs such as Vietnam,[17] the forum is said to create a necessary and effective space for communication between governments and industries from various countries (Dutta; Turney), allowing mutual benefits to flow through the region (Turney). Furthermore, the forum promotes discussion on ways to raise government, public and private sector awareness about nanotechnology, and how to use this awareness to drive the formation of relevant education programs (Turney). These discussions reinforce another of the forum's objectives — to stimulate industry interest in graduates of nanotechnology courses (Dutta).

Of particular note with the Asia Nano Forum is the attendance and input of politicians, often ahead of bureaucrats, with Damrongchai reporting that speakers have included members of the Thai Cabinet, such as the Minister for Science. For some (Damrongchai; Turney), it is crucial that nanotechnology dialogue occurs at this 'government level'. Damrongchai elaborated, saying that regional or global political initiatives are critical in achieving certain outcomes such as:

> ...lower[ing] the obstacles of transferring technology, especially from the developed countries to developing countries; and that can only happen in the political area, because it needs strong leadership to push forward...

Providing a more consistent form of interaction, networks were seen as another important means for developing international partnerships. Dutta said that Thailand has been leading efforts to draw together fragmented research in the region, with the NSTDA "... trying to form [a] network of excellence between Japan, Singapore, Korea, Hong Kong, Australia, Taiwan and Vietnam...". In terms of forming networks within industry, Turney said that Southern countries in Asia may want to draw inspiration from existing structures in the U.S. and E.U., such as the NanoBusiness Alliance.[18] Such structures would provide important places for businesses to "...get together, within the region, in order to be able to exploit the specific expertise within individual companies..." (Turney).

A less-formal place for facilitated networking is at global trade shows, such as the former Asia Pacific Economic Co-operation (APEC) Technomart[19] and Japanese-initiated 'Bio Forum',[20] where displays and exhibitions open new avenues for industry collaboration (Damrongchai; Tegart).

In all of these interactions, Northern countries, as well as international bodies such as the E.U. and APEC, are seen as having a responsibility to actively engage Southern countries (Braach-Maksvytis; Damrongchai; Tegart). But the South must also actively seek opportunities. One suggested method for Southern countries to maximise these opportunities for Southern countries is to ensure that, where possible, publications are written and translated into English (Dutta). With this in mind, Dutta said that NANOTEC has already enforced this policy with most of the emails they generate.

9.2   Ensuring Appropriateness

As raised in the previous chapter, many interviewees hold significant concerns about the appropriateness of nanotechnologies to development settings. The main tensions surround hype in terms of application utility, orientation and end-user, as well as the potential for nanotechnologies to drive greater inequality. The economic motivation behind many of the strategies raised near the beginning of this chapter, further highlight these tensions. How, then, did interviewees suggest nanotechnology might be best governed to ensure the emergence of appropriate applications? In this section I will look at three strategies suggested in response to these challenges.

9.2.1   Keeping Open to a Balanced Approach

Many interviewees, even those such as Cooper who were most sceptical about the overall benefits offered by nanotechnologies, spoke of the need to be constructive in order to effect any sort of meaningful change in the way nanotechnology develops. As Cooper noted:

> There is a real danger of becoming anti-technology, which would be silly, you do not want to get into this 'let us defend our traditional values against technology', you want to politicise

all kinds of technologies and find ways they could be used in a collective, on-the-ground way.

Ratanakul was of a similar mindset, adding that Buddhist monks, with whom he had spoken, suggested that the best way to deal with nanotechnology, in light of imminent implications, is to "...learn to live with its powers, and try as much as possible to steer it away from selfish uses".

In this light, Arya proposed that a central tenet of governing nanotechnologies, in terms of appropriateness, is to ensure an integrated approach that considers a range of technologies. Ultimately, such an approach will provide citizens with alternatives, which Arya says is fundamental in the quest to ensure the appropriateness of nanotechnologies. In terms of engaging with new research, Selgelid said that it is about "...striking a balance between...[nanotechnology and] other promising technologies". Yuthavong agreed, saying that the hype around nanotechnologies must not drive disproportionally appropriate funding, and that nanotechnologies must fit within an overall science and technology framework. He noted: "we want a stable situation where we have a balance between all types of science and technology and nanotechnology would fit into one important corner but it is not the whole room".

Similarly, given the complexity of social challenges, nanotechnologies will need to find where they fit within relevant response frameworks (Braach-Maksvytis; Lynskey), considering their technical roles as "...part of the broader solution which involves biotechnology, [and] maybe information systems" (Braach-Maksvytis).

9.2.2 Working at the Intersection of Market and Community

Given it is important to define the roles nanotechnologies might play in development, some (Tanthapanichakoon; Turney) believed that there is important space to explore at the intersection of the market and social needs, with potential for certain technologies to be of social and export benefit.

With this in mind, interviewees suggested that Southern countries make formal efforts to ensure that nanotechnology is geared towards economic gain through R&D that targets social needs (Braach-Maksvytis; Thajchayapong; Turney). According to a couple of interviewees (Deutchmann; Lynskey) the best way to ensure nanotechnologies fulfil both social and market needs is to ensure local participation in technological creation. For some (Bryce; Deutchmann; Lynskey), this means going beyond consultation and ensuring the development of technologies is driven, owned and controlled by local end-users and/or producers. Rather than just a consultative process, Bryce said that this implies a "...ceding of control...to the people within the context who have that complementary knowledge..." that, for Arya, equates to a shift away from the 'consumer' paradigm. Arya placed these statements in perspective by saying that, whilst it is not possible for all citizens to participate in the development of nanotechnologies, it may be possible for members of the broader population to develop components for nanotechnology-based products or provide related services. He particularly spoke of the need to "...allow small entrepreneurs to play a certain role in this area" and the importance of engaging local artisans in the development of nanotechnologies:

> ...if we can develop some materials using nanotechnology, can these materials be used by artisans, or by...people who need not know how to produce the material but would have the capability of changing that material for human use?

On this note, Charinpanitkul reported that locals in Thailand's North East are already working at a grassroots level to integrate nanotechnologies into local products such as wine, with potential envisaged for medicines and herbal drugs (Charinpanitkul; Sawanpanyalert; Turney) as well as Thai silks and fabrics (Thajchayapong). Early efforts form part of the 'One Tambon One Product' (OTOP) model. Such a model was presented as a useful vehicle for integrating work on the nanoscale into local production (Charinpanitkul; Thajchayapong) and part of a bigger effort to integrate the work of the middle class and local agricultural communities (Damrongchai). As Damrongchai reported, the model is already being actively promoted by the Thai government to help develop export markets:

> What the government tends to do is strengthen the manufacturing commercial capability of villages throughout the country by encouraging them to come up with one product of themselves and sell it to the world...

In this sense, the OTOP scheme is geared towards developing "best selling product[s]" for niche export markets, as part of national economic development (Damrongchai). This fits with the views of others, such as Ratanakul, who raised indigenous and local technologies as distinct areas in which value-adding to nanotechnologies can create competitive niche exports. With these points in mind, Damrongchai said that a fundamental aspect of appropriateness, for nanotechnologies that build on natural resources or knowledge, is that local populations "...keep the traditional value, local wisdom, [and] maintain the good ways of life". Furthermore, Sawanpanyalert said that where technology is integrated into a traditional medicine, it will only be appropriate if is used in the traditional manner.

Given the OTOP model is already part of Thai government policy, Damrongchai suggested that, as was the case with biotechnology, a working group should be established to explore how to best integrate nanotechnology into the model.

9.2.3 Building on Equity-Enhancing Resources

The final approach suggested for ensuring the emergence of appropriate nanotechnologies in the South was to build on existing resources that already assist with a more equitable distribution of innovation outcomes.

To start, some interviewees (Coyle; Dutta; Tegart) focussed on the power of the Internet to disseminate nanotechnology-related information and avoid concentration of knowledge. In this sense, Coyle spoke of the increasing availability and accessibility of forums such as the Global Forum for Health Research's Health Information Forum,[21] suggesting that "...it would be excellent for nanotechnology to be discussed on some of these chat lines". Similarly, Dutta suggested that popular science sites with dedicated nanotechnology content, such as "...smalltimes.com...[and] nanotech.com, keep posting recent advances in nanotechnology" that may prove useful to professionals in a wide range of areas.

In addition to global tools such as the Internet, a couple of Thai interviewees (Panitchpakdi; Pothsiri) identified the integration of nanotechnologies into their country's '30 Baht[22] scheme' as a means to ensure health-related nanotechnologies reach Thai people at low costs. Under this scheme, all Thais are heavily subsidised for most care associated with each hospital visit or admission. Whilst more advanced treatments, such as kidney dialysis and antiretroviral drugs, are currently excluded from the scheme, Pothsiri was confident that if nanotechnologies could improve low-cost diagnosis and drug delivery, then the resulting applications "...would be appropriate technology to be applied in our 30 baht scheme". Alternatively, Coyle stated that, in cases where nanotechnologies are only available in centralised settings, such as in 'referral hospitals', "...people have to be referred there or somehow it has to be shunted out to the local hospital for that particular case".

Whilst in this chapter I have already explored various partnerships to support nano-innovation, hope for the development of needs-driven nanotechnologies was also seen through emerging and alternative forms of partnership. Here interviewees gave little attention to the possibilities presented by working with the large, multilateral institutions such as the World Bank. Rather, it was suggested that the strongest prospects for socially focussed nanotechnology partnerships lie with Southern countries themselves, government agencies (internationally), international philanthropic institutes and organisations, and NGOs.

Focussed on regional issues, Panitchpakdi believed Southern countries can partner together around mutual concerns to ensure they have the market indicators and incentives to drive innovation and sales. In the field of healthcare, for example, he claimed that "...[if] at least three or four countries put their resources together, they could produce [a] drug...". In this light, interviewees such as Damrongchai morally and practically supported the Southern development of generic drugs and granting of compulsory licences as a response to the 'access' problem. Alluding to compulsory licences to ensure that "...access is not disrupted", Pothsiri, a health policy official, said:

> ...if it comes to such a situation whereby we may have some new drugs which have been innovated by using nanotechnology...if there is a public health need then...they [other Southern countries] should try to access that new drug by using whatever option they have...

Pointing to an emerging partnership between Australia's Commonwealth Scientific and Industrial Research Organisation and Thailand's NSTDA, Turney saw government agencies from both the North and South partnering on meaningful research relating to human needs. Similarly, Braach-Maksvytis suggested the Global

Research Alliance[23] as exemplary of an untapped critical mass of researchers who are capable of steering nanotechnologies towards social needs in the South.

Focussed on global concerns, international philanthropic institutes and organisations were also said to be prime candidates for alternative forms of partnership. As a more formal way of increasing access to knowledge, Damrongchai felt that there needs to be:

> ...new research environment[s] for researchers to come and exchange their skills, knowledge, [and] their research in a very liberal way...where it does not matter which country you are from or which agency you work for...

Dutta imagined institutes of this nature providing groups, such as international health organisations, with avenues to propose problems and ask for a range of responses. In this light, the Bill and Melinda Gates-driven 'Grand Challenges for Global Health' initiative[24] was said to offer an exciting opportunity for nanotechnology to be explored in this manner, with Yuthavong adding: "...you need to persuade Bill Gates to say, 'not just biotech[nology] but perhaps, also, nanotech[nology]'".

Non-governmental organisations were presented as the final 'alternative' partner in the development of appropriate nanotechnologies. Following proof that a nanotechnology-based application is safe and not too expensive, Panitchpakdi believed that NGOs could play a critical role in increasing public access to the technology.[25] In furthering hopes for access, means for this partnership would include R&D through "non-commercial channel[s]" (Damrongchai), and alternative "modes of distribution" such as the "...collective sharing of technologies" (Cooper). According to Lynskey, the role for certain NGOs thus becomes one of providing "...a bridge between local community, local or national government and the outside world, in terms of helping bring new ways to do things into the country".

9.3   Responding to Risks

The potential for nanotechnologies to raise new or inadequately explored societal implications prompted discussions about how to appropriately respond to challenges. In this light, three issues emerged, with interviewees presenting diverse views on whether nanotechnology requires new strategies and frameworks for its implementation. I will first investigate whether the foreseen risks arising from nanotechnologies justify new regulation and precaution. I will then look at how best to develop knowledge about the societal implications of nanotechnologies and who to involve in the process. Lastly, I will assess the different strategies and methods for public engagement with nanotechnology.

9.3.1 Appropriate Regulation

According to Thajchayapong, the first issue that needs to be addressed with respect to the risks raised by nanotechnologies is whether they can be regulated under existing domestic or international legislation or whether new, specific regulatory measures are required. Regulation proved central to discussions on risk, largely because of the faith some placed in the ability for strong regulation to adequately and appropriately mitigate risks. Changthavorn's comments on the Thai situation exemplified this attitude: "...if we come up with the good guideline[s], good regulation, that should be okay". These issues of regulation proved central to discussions about responding to health, environmental, ethical and legal risks, particularly given that, as of 2004, Thailand did not have standards for nanometrology and consideration for regulatory measures were not seen as a priority (Changthavorn; Dutta; Thajchayapong).

With respect to human and environmental health risks, the 'R&D', 'production' and 'use' phases of nano-innovation were each considered to require different responses. For example, Changthavorn said that whilst he felt relaxed about the development of nanotechnologies in the present Thai context, he felt that the use of nanotechnologies should be "carefully under control". This highlighted a general lack of consideration for regulating nanotechnology in the R&D phase that contrasted with some consideration for regulating nanotechnology's production and more significant consideration for regulating nanotechnology's use.

Regulatory issues in the production phase focussed on safety standards in terms of handling nanomaterials. For Thajchayapong, nanotechnology reinforced the need for factories to have "...good quality control...during the production phase". Yet for Radt, this is simply a case of nanotechnology materials being "...treated as any other toxic substance...". Furthermore, he felt that existing regulation and guidelines were adequate given that, globally, there are "...well established procedures on how to work with toxic materials...".

Regulatory issues in the use phase focussed on the assessment of nanotechnologies and their controlled application. Whilst some (Chirachanchai; Pothsiri) felt that Southern engagement with nanotechnologies can be adequately governed by existing regulations, a greater number (Damrongchai; Deutchmann; Sriyabhaya; Weckert) felt that new regulations are required. Yet, for Deutchmann, the extent to which a country needs new nanotechnology regulations may be relative to the country's innovative status. Whilst nano-innovation in the U.S. and Australia may be governed by existing legislative and risk frameworks, he felt that countries with poorly developed regulatory frameworks will require new regulations. Others (Changthavorn; Sriyabhaya; Weckert) saw limitations in generalising about regulatory requirements, suggesting that regulation needs to be considered on a field-specific or particular-issue basis, given the wide-ranging nature of nanotechnologies. In this sense, interviewees saw a need for 'adjustments' to the regulation of food and drugs (Changthavorn), consideration for issues of "privacy and control" (Weckert), and international laws restricting "...the use of nano-devices in war" (Weckert). Also implied in the response of many was that industry should voluntarily self-regulate, with Weckert saying that scientists have to take more ownership over the implications and risks associated with emerging technologies.

The process for developing appropriate regulation is said to start with a review of existing legislation and assessment of the legislative compatibility of nanotechnologies (Changthavorn). As has been the case with GMOs in Thailand, Changthavorn said that regulators should then develop guidelines for specific areas, such as bio-nanotechnology, with these regulations addressing health and toxicological risks as part of the formal 'examination' process for products seeking to reach the market.

Irrespective of the need to regulate, responses to health and environmental risks need to be investigated and coordinated by national bodies such as the Thai FDA (Chirachanchai; Thajchayapong). According to Changthavorn, such measures, and indeed all Southern responses to risk, should constitute part of an overarching process of developing a "...clear national policy on nanotechnology".

With respect to ethical risks, there was a belief from the ethicists I interviewed that nanotechnologies would need different, much tighter "sorts of legislation" or "generally accepted rules" (Weckert), but that existing international frameworks, such as the UNESCO's Universal Declaration of Bioethics,[26] could guide the development of such regulations (Ratanakul).

The need for new regulations aside, Ratanakul firmly believed that the development of nanotechnologies enters a Thai context that is highly conducive to ethical evaluation and that two points must be harnessed to ensure appropriate responses are developed. Firstly, that the problems raised by biotechnologies, with respect to IP ownership and dominant orientation towards serving Northern markets, have left the Thai population "more watchful" of new technologies, ensuring that "...in the near future, there will be more discussion on this new technology [nanotechnology], to consider its ethical applications...". Secondly, Ratanakul noted that previous ethical debates about stem-cell research and gene therapy have ensured the establishment of a modern framework for discussions, whilst the case of cloning has proven the ability for academic, ethical and religious people to discuss things in a productive manner.

With respect to legal risks, the general feeling was that nanotechnology can be governed by existing legislation (Changthavorn; Chirachanchai; Dutta). Patent laws that had evolved from the introduction of biotechnology were seen as adequate (Chirachanchai), with Changthavorn, an intellectual property specialist with the Thai National Centre for Genetic Engineering and Biotechnology, explaining:

> For intellectual property law, especially patents, we do not have to change it [legislation] because it already provides [a] big enough space for this. If you look into the definition of 'invention'...this can accommodate these nanotechnology products or process[es].

Similarly, he believed that international regulations, such as those proscribed by the TRIPs agreement, already provide "...legal protection for all kinds of technology, so we do not have to change it". Here it is interesting to note how Changthavorn reframed the issue of responding to potential risks in terms of whether regulation will be adequate to facilitate nano-innovation.

The other aspect of legal responses to nanotechnologies relates to the issue of standards. In this sense, Dutta saw no immediate need for universal measures such as nanometrology or chemical provisions in international law, saying: "...nanotech is still between research and reality". Chirachanchai added that if new laws relating to standards are needed, they will not be for "...about 10 or 20 years...".

As a more general response to the health, ethical and legal risks posed by nano-innovation, a number of interviewees quite strongly ruled out a moratorium. The main reason was less a matter of whether a moratorium is warranted and more a matter of its "impossibility" in the current climate (Ratanakul; Weckert). According to Yuthavong, this is compounded by the practical problem of identifying the scope for a moratorium and how it would be regulated:

> ...I do not think it can be done in the sense that it is not like a moratorium on genetic engineering that is very precise, very clear, 'you should not work with putting genes into cells' and so on...But nanotech[nology]: everything is nanotech[nology]! How can you have a moratorium on nanotech[nology]? Does it mean tomorrow I cannot go into my lab and do my own thing? I have been doing it for...all my life...It is not possible, unless you say, 'moratorium on...[the] utilisation of nanoparticles, less than 10 nanometres', or something like that...but then it is too specific.

If there is a need to regulate in some areas but a broad moratorium is infeasible, what kind of precaution did interviewees suggest Southern countries take when engaging with nano-innovation? Generally, interviewees such as Selgelid and Yuthavong spoke of exercising "reasonable precaution", seen as all the more sensible in light of Southern difficulties to assess nanotechnology risks (Arya). But what does 'reasonable precaution' really mean? For Weckert, this meant focussing on the identification of where the greatest risks lie. According to Ratanakul, this would best involve an analysis of benefits versus risks at the community level:

> In the process of adopting nanotechnology in any nation, there are certain precautionary measures that need to be taken into account to prevent the abuse of its application...before introducing any new technology into any community you have to be very cautious, thinking of the benefits and the harm that it would make to the people.

Yet, others, such as Yuthavong, were keen to point out that a strict adherence to a literal interpretation of the precautionary principle is impractical: "...if precautionary means you should not do anything at all until you absolutely prove that it is absolutely safe, then it is too much precaution...". Similarly, Weckert pointed out that risks can also arise "...if certain developments do not happen", with Ratanakul adding his belief that certain risks are necessary and justifiable "...for the benefit of mankind". Hence, Selgelid spoke of finding a balance: "...you do not want to be too cautious, you do not want to be too careless, you want have a degree of caution that is reasonable and maybe moderate...".

9.3.2   Developing Knowledge about Implications

Building knowledge of the risks and benefits associated with nanotechnologies was presented as the best, and often necessary, way for countries to identify appropriate precautionary measures for engaging with nanotechnology (Braach-Maksvytis; Changthavorn; Ford; Ratanakul; Selgelid; Weckert). According to Changthavorn, Thailand will only require one or two years to build enough knowledge to ensure appropriate initial responses. However, it is important that the knowledge building process continue indefinitely, with interviewees calling for ongoing surveillance and monitoring of nanotechnologies (Coyle) to ensure "constant [regulatory] revision" for ethical, legal and social considerations (Deutchmann). In this respect, Ratanakul suggested that the Thai government would benefit from establishing a "...responsible advisory body of experts to monitor the use and the effects of nanotechnology".

Interviewees identified three methods for building knowledge of the societal implications of nanotechnologies. The first is to harness existing and emerging knowledge from international leaders in nano-innovation. Given the difficulty of regulating nanotechnologies in a void of endogenously produced products, some interviewees such as Weckert, spoke of Southern countries drawing regulatory inspiration and ideas from the E.U. and the U.S. Panitchpakdi felt it was particularly important to see what standards and quality benchmarks countries are setting for new nanotechnology products. According to Ratanakul, in the Thai case, drawing regulatory inspiration and ideas from abroad translates to a kind of "...'wait and see',...if it work[s] well in other societies, it may work here" attitude. However, it is important to note that the origin from which regulatory ideas and inspiration are drawn could have a significant influence on the implementation of nano-innovation in the South, considering the U.S. approach differs greatly to that of Europe, where countries are "...much more in favour of the precautionary principle" (Weckert).

The second method for building knowledge of the societal implications of nanotechnologies is to harness the collective wisdom of the international community through international discussions, meetings and institutionalised bodies (Braach-Maksvytis; Changthavorn; Yuthavong). In this respect, Pothsiri said there is early work occurring within the Commission on Intellectual Property Rights, Innovation and Public Health, the WIPO, the WTO and the WHO to develop understanding about the ethical, legal and social implications of nanotechnologies, and ensure that knowledge reaches the "public domain". Often this work is collaborative, Pothsiri said, citing the example of the Commission that will order the development of a WIPO study to collect and collate information about the societal implications of nanotechnologies, to be shared with the World Health Assembly.[27]

The third method for building knowledge of the societal implications of nanotechnologies is for Southern countries to actively engage in endogenously focussed research. For this to occur, Southern governments must be interested enough to play a driving role in the process (Ratanakul). To ensure nanotechnology is placed on a government's agenda, politicians require relevant, accessible literature, with Thajchayapong explaining: "if you want politicians to understand it you do not give them scientific book[s]...". He continued by saying that the Thai Prime Minister has therefore set as obligatory reading for Thai government ministers, two straightforward books from the U.S.: 'As the Future Catches You'[28] and 'The Next Big Thing is Really Small'.[29] But some Southern countries may already possess enough useful information to stimulate government interest, he said, noting that his motivation to explore nanotechnology came from a report written by the Bangkok-based and Thai-staffed APEC Centre for Technology Foresight.[30]

Once interest has been generated at the government level, the wider knowledge building process should commence with forums and conferences that draw key people together to discuss the societal implications of nanotechnologies (Berwick; Braach-Maksvytis; Turney). In this respect, Changthavorn believed that the South should follow the North's lead, given:

> ...the initial National Nanotechnology Initiative in the [United] States was conscious of this and they did, in fact, have a...big meeting in the United States on social and ethical issues. Those meetings have actually continued with the Europeans...

In drawing key people together, meetings must ensure cross-disciplinary interactions to stimulate both a widening and balancing of perspectives (Turney). In the case of scientists, it is about integrating views on societal implications with thoughts on innovation:

> ...if you have to organise a conference or a meeting today, we should not be talking about technology or its applications alone, we should mention something like legal implications, social implications or economic implications... (Changthavorn).

For ethicists, cross-disciplinary interactions ensure work that is more grounded in the realities of nano-innovation, with Weckert saying: "...what we do not need is people trying to just sit around and think about the ethics of it without actually spending a lot time talking to scientists and so on to see exactly what is happening".

Interviewees thus saw three professions having a key role in these discussions. The first are scientists and technologists (Coyle; Weckert; Yuthavong), with particular effort needed to ensure people come from a variety of scientific backgrounds (Berwick). Next are ethicists (Ratanakul; Weckert; Yuthavong), followed by a certain need for people with legal knowledge (Braach-Maksvytis; Weckert; Yuthavong). Lesser, but still important roles were seen for economists (Braach-Maksvytis; Lynskey; Thajchayapong), planners (Lynskey); social scientists (Braach-Maksvytis; Ratanakul), medical doctors (Coyle; Ratanakul), Buddhist monks (Ratanakul) and government representatives (Lynskey; Weckert).

To be most useful, these kind of meetings need to flow into different kinds of knowledge-building pursuits. These could include basic research to "understand the mechanism[s]" of nanotechnology-based drugs (Chirachanchai), long-term testing for nanotechnology-based drugs (Ratanakul), environmental and health impact assessments (Arya) and foresight studies (Damrongchai). The latter was seen as an area where Southern countries such as Thailand could really take the initiative (Damrongchai). Supporting this belief was the early work conducted by the Thai-based, APEC Centre for Technology Foresight that involved both scientists and social scientists and identified several good and bad scenarios for nanotechnology's future (Damrongchai).

9.3.3 The Public and Nanotechnology

Underlying any serious exploration of the societal implications of nanotechnologies is the importance of meaningful public engagement, incorporating issues of awareness and understanding as well as genuine dialogue and participation. These aspects, in turn, influence aspects of technology, such as the rate of diffusion and the overall trajectory of its development. What became quickly apparent in speaking with Thai interviewees is that there is currently a considerable lack of meaningful public engagement with nanotechnology in Thailand. At the base of poor social engagement are low levels of awareness about nanotechmology amongst Thai citizens (Ratanakul), with Yuthavong estimating the number of Thais who have heard about nanotechnology or know a little about it at "... less than one per cent". Yuthavong and Chirachanchai believed that only the wealthy, well educated, members of government or those in high-tech areas really have any awareness of nanotechnology's existence.

Evidence of poor social engagement also surfaced in terms of the poor and/or limited understandings associated with nanotechnology. The envisaged problem of people not understanding what nanotechnology actually is (Cooper) was confirmed by those who said that nanotechnology is suffering from lingering confusion with previously introduced terms such as 'genetically modified organisms' (Pothsiri; Ratanakul). Unlike awareness, the lack of clear understandings extends to those with high levels of exposure (Sawanpanyalert; Sriyabhaya; Yuthavong). Yuthavong elaborated: "...even those in...high-tech areas, or the educated people, or people in government...when you go and talk to them it is really partly true and partly not true — what they think...is nanotechnology...". He explained that the government situation is one driven by hype and the expectations of the Prime Minister, without any correlating development in levels of understanding amongst politicians. Even more concerning was Thajchayapong's claim that this lack of understanding extends to government boards that are making national decisions about nanotechnology. Thai high-school graduates were also seen to be lacking knowledge and understanding about nanotechnology (Charinpanitkul), but this is not surprising, given a general lack of public understanding about science, post high-school (Yuthavong). Panitchpakdi, speaking from his experiences in communities he has visited, claimed that understandings are most poor amongst local communities.

As I outlined in the previous chapter, there is simply not a basic level of awareness and understanding in Thailand around nanotechnologies foreseen ethical and legal implications, with ethicists, lawyers and regulators just "not aware yet" (Chirachanchai). Subsequently, ethical and legal matters are barely ever raised (Changthavorn; Charinpanitkul). Arya saw these outcomes meaning that there will be insubstantial debates about already marginal issues in the North, such as the foreseen environmental implications of nanotechnologies.

Combined, the lack of awareness and understanding equates to a lack of public discourse regarding nanotechnology (Yuthavong), which Turney claims is "...not even on the radar" in the South. Furthermore, there was scepticism that community organisations and NGOs will be consulted about the orientation of nanotechnology research and roll-out of applications (Panitchpakdi) and that Southern citizens will "...not really participate in the debate..." (Arya). A lack of purposeful public engagement by government was viewed by Ratanakul as the status quo: "...when the government introduce[s] technology, the government never ask[s] the people whether they would accept it or not". Yuthavong explained this by saying that "...in countries like Thailand, it is easier to manipulate public opinions".

Rather, many interviewees envisaged that government, scientists and industry would utilise the existence of a knowledge void by dictating the terms by which nanotechnology is discussed and therein understood. Yuthavong said that there is a general reluctance for scientists to engage with the public, whom they view as ignorant and potentially creating unwanted delays for their own research. Additionally, Yuthavong spoke of challenges for such engagement to occur successfully, considering "...few scientists are good communicators or feel communication is worthwhile" and that the transfer of concepts through the media breaks down because explanations are literally lost in translation, given "...everything that comes from the researchers about nanotechnology is in English" (Panitchpakdi).

In this light, there was a real fear that the nanotechnology discourse would essentially be a discussion between industry and government (Weckert), or that scientists would dominate proceedings at forums (Damrongchai). Speaking from experiences in discussing nanotechnology in South East Asia, Braach-Myksvitis said scientific views are "hijacking" discussions, leading to a lack of "genuine dialogue...[creating] almost a one-way monologue of 'let us get the information out to the public so they can understand what is happening'". Arya described how this one-way monologue develops:

> ...sometimes, it is a kind of 'road-show', you come to explain that 'this is good', you have a position, you want to defend nanotechnology or you are already partial: 'what is new is good', so there is no change or space for debate.

As a result, both Arya and Panitchpakdi said that Thai nanotechnology is mainly experiencing 'positive talk'.

In recent times, a global lack of meaningful social engagement around emerging technology has been cited as a key reason for a backlash to emerging technologies (Ford; Yuthavong). A lack of public awareness, understanding and inadequate citizen engagement with nanotechnology in the South raises the issue of whether nanotechnologies will be accepted or if there is a potential backlash brewing. Despite some early indicators of unrest, most saw a smooth transition to nanotechnologies, with issues of public acceptance hardly raised (Damrongchai; Thajchayapong). However, there were notable differences in how interviewees envisaged nanotechnology entering the mainstream, with different justifications for why nanotechnology would be accepted.

Some justified the expected acceptance of nanotechnologies by speaking of the incremental, somewhat inconspicuous nature of their introduction. Despite the speed of technological developments, nanotechnologies were seen as emerging quite slowly (Tegart; Weckert), in a large number of places and through a large range of application (Tegart). But the main comment was that this will be a subtle revolution (Tegart; Weckert), in that nanotechnologies will not necessarily be recognisable (Coyle; Panitchpakdi; Tegart). This would seem to hold true for Thailand where "most of the Thai[s] do not know about it, even if they are using the technology!" (Ratanakul). Ratanakul elaborated:

...our student[s], [a] few days ago...I asked them about this technology, 'do you know about this technology?' They say they know nothing about it but when I said this technology is...'what, what, what...' they said, 'oh, I know it, because I'm using it'.

Interviewees spoke of utilitarian rationale for nanotechnology's subtle introduction, whereby the public would focus on the product outputs, rather than the embedded componentry. Cornell said he saw it as unlikely that Southern populations would even be "...aware of something as running under the heading of nanotechnology as compared to the fact that 'this particular source of electricity seems to be cheap and convenient'". Berwick agreed, claiming:

> ...at the end of the day the customer probably is not even aware that nanotech was involved in the development of the products, they just know the product works better, looks better, is cheaper, lighter, what have you, and for them, that is the important thing...

As an example, Thajchayapong admitted that whilst the thought of isolated nanoparticles could raise public concerns, his experiences suggested that the embedded nature of work on the nanoscale will be used to maintain public ignorance: "when I talk to our Taiwanese friend it is said that: 'of course people are afraid of...small science and nanopowders but the finished product...it is already integrated into the product'...". Others justified the expected acceptance of nanotechnologies by referring to a belief that associated hype will mean the revolution is conspicuous and outwardly focussed on public education about the benefits.

Most felt the Thai public would welcome nanotechnologies, with Damrongchai speaking of blind faith in technology and its producers as a general trait of the masses:

> ...the middle class are ready to accept any technology: computer; Internet; biotechnology and nanotechnology [if] adapted to their lifestyle...people accept things that are usable no matter what technology lies beneath.

Interviewees provided a couple of reasons for these attitudes. Pothsiri spoke of citizens "...recognizing the importance of this advance[d] technology in helping them to develop faster". Whilst Tanthipanichakoon said "the tendency for Thai people or many people in developing countries is that they want to have a good image or good access to all the new things". In this light, it was believed to be just a matter of "...some time for the general public to get accustomed to the concept of the word 'nanotechnology'" (Damrongchai).

The sum of arguments relating to public engagement with nanotechnology led many to believe that public acceptance for nanotechnologies will not be as bad as with GMOs or cloning (Thajchayapong). Songsivilai even went so far as to assert that "the Thai public are not against medical nanotechnologies...[as] compared to agricultural biotechnologies...".

However, others argued that, whilst it is too early to judge, the lack of meaningful social engagement suggests debates will be "... much the same as the GM ones" (Weckert), leading to the possibility of a GMO-styled backlash (Changthavorn; Damrongchai; Turney; Weckert). This possible backlash was justified by four factors, with the first revolving around issues of participation. At its most basic level, interviewees returned to matters of awareness and understanding which they saw compounded by paternalistic 'one-way' instruction about nanotechnology (Selgelid). Based on the 'participation' argument, a lack of understanding and missing information was seen as a crucial factor in catalysing a potential backlash (Changthavorn; Ford; Ratanakul). In this way, some were concerned that nanotechnology could suffer similar confusion about its nature as has occurred with GMOs, thereby reducing any possibility for meaningful social engagement (Ford; Yuthavong). Furthermore, the argument was made that the subtle nature of nanotechnologies could actually prove the field's undoing. Weckert, for example, saw future problems given people will be using and potentially ingesting nanotechnology without their knowledge of it. Similarly, Songsivilai believed that the pervasive nature of nanotechnology-based testing could generate anxiety.

The second justification for a possible backlash involves inequities around orientation and distribution. Here there was a belief that various inequities can influence the level of public backlash:

> If nanotech[nology] only brings benefits to the rich, then more than half of the population who is not rich they will think: 'it is a good technology but what do we have sacrifice to get this technology? Do we get anything from this technology?' If they do not see it and use something like testing, certain type[s] of experiment, [or] certain kind of applications and business, they will bring out the safety issues and so on (Damrongchai).

The third justification for a possible backlash involves issues of risk. Interviewees such as Cooper spoke of the potential for an "instinctive resistance" to nanotechnologies, as witnessed with biotechnologies. Moreover, and particularly in terms of safety and matters of risk, a number of interviewees (Changthavorn; Weckert) envisaged nanotechnologies being placed in the same 'basket' as GMOs, irrespective of whether nanotechnology raises distinct issues (Ford). According to Weckert, this creates an environment in which the public might focus on the association between nanotechnology and food, setting a platform for "enormous resistance".

Building on these points, interviewees such as Lynskey envisaged a great amount of caution to something new from the point of the associated risks. Contrary to arguments I explored earlier in this section, Ratanakul spoke of the Thai population having a basic mistrust of science. Perhaps resulting from this, there is said to be a concern from the greater community about issues of safety relating to nanotechnologies and their creation (Tanthapanichakoon), with people "...starting to be a little bit nervous about the liabilities from nanotechnology..." (Yuthavong). Furthermore, Changthavorn claimed that there are big ethical concerns in Thailand that nanotechnology could be used for cryogenics, thereby propelling a fear that "humans 'play god'".

In a similar vein, Yuthavong believed that the new word 'nanosafety' has mimicked the word 'bio-safety'. He added that things will be more difficult once the word 'nanosafety' enters the public lexicon, creating the potential for a similar phenomenon to what has happened to GMOs:

> ...thirty years ago, when genetic engineering started, the word 'GMO' was a beautiful word...[then] public relations went badly wrong until GMO is now a dirty word...So, be careful; 'nanotech[nology]' can become a dirty word.

The final justification for a possible backlash relates to the issue of public expectation. According to some (Lynskey; Tanthapanichakoon; Yuthavong), the combination of setting high hopes in the face of significant and often long-fought challenges raises the 'old' problem of unrealistic public expectations. Placing this argument in the Thai nanotechnology context, Sawanpanyalert said that:

> ...because of the high popularity, the high expectations, when people [are] talking about nanotechnology (even though they

do not understand what it is)...they will expect too much, unrealistically.

An "unstable situation" is seen as a long-term repercussion of people being 'oversold' (Yuthavong). Yuthavong outlined his concerns, noting the similarities to past technological hype and the danger of the public realising the long-term nature of investing in nanotechnology:

> ...it could be like the 'dot com' disappointment because [with] nanotechnology...the main benefits will come...[in] ten years, fifteen years...so, in three years people will start to ask, 'aye, you took a lot of our money, where's the product?'...That is really the big threat...because there is so much hype, when the pendulum swings back it could be bad.

Adding to these concerns is a belief that "...people will be sceptical" about Thailand's ability in the global nanotechnology market (Panitchpakdi) and that the Thai nanotechnology 'master plan' is too ambitious (Arya; Ratanakul).

As a result of these challenges, Weckert envisaged a "...fairly uninformed and fairly emotional...enormous debate, like with GM". In such circumstances, Panitchpakdi felt that, in terms of the response to social concerns, many members of the Southern public would look to see what is happening in other countries. Others (Damrongchai; Yuthavong) felt NGOs would lead the debate and any potential backlash. In this light, Yuthavong was worried that nanotechnologies could offer new 'fodder' for NGOs to target MNCs, with the potential for companies such as L'Oreal to "take a hit":

> ...if it becomes a real concern and if there are evidences of liabilities...it could be very bad and the NGOs would come in, like Greenpeace. Greenpeace like[s] to 'bash' GMOs but they know, as well as we know, that you cannot keep on the same theme for ever, so they are looking for new heads to bash!...with NGOs participation, you could really come in for a bad time for nanotechnology, down the road...there could be some groups which say, 'O.k. Let us make Thailand a showcase...let us show that nanotech[nology] is bad'. So, I am really worried about these anti-technology groups, anti-science groups.

But Panitchpakdi countered this argument with his belief that Southern NGOs will only work against nanotechnologies, such as pharmaceuticals, if they are not proven safe or affordable.

The ramifications of a backlash on the ability for a country such as Thailand to develop nanotechnology capabilities are not to be underestimated (Changthavorn). As Warris noted, with genetic engineering: "...[the] public reception has...led to a lot of research programs being stopped because there are concerns about its effects...". His thoughts represented those of many others when he spoke of a genuine concern that "if nanotechnology just goes forging off there without thinking about how people are going to react to it, then it could be in trouble as well".

Up to this point, knowledge building has been presented as an exclusive process for professionals. Some interviewees (Berwick; Coyle; Ratanakul) felt that certain professions must develop basic understandings about nanotechnology before the wider public can be engaged in the knowledge building process. Once this knowledge has been cultivated it was believed that specific expert interactions, on topics such as nanometrology and measurement standards, must continue outside the public arena, given the limits to public knowledge and interest:

> ...this is something for the nanotechnologists to hash out among themselves and the public cannot be, really, all that interested. They can say, 'O.k. let us have one product, one standard, a universal standard'... but then it is for the technologists to decide among themselves (Yuthavong).

Yet most (for example Braach-Maksvytis; Ford; Panitchpakdi; Thajchayapong; Weckert) felt that engagement must move beyond specialists to include the general public as early as possible. This tension about whom to involve and at what stage exemplifies the wide differences of opinion as to the public's role in nanotechnology's development and the best strategies for public engagement. Approaches presented by interviewees broadly fell into two models that I have classified as the 'instructive' and the 'participatory'.

The instructive model is grounded in a 'top-down' philosophy whereby a government and the science community use controlled methods to create public awareness about nanotechnology. This process of manufacturing consent involves "...making people understand what you are doing" (Thajchayapong), without needing to ensure they understand everything about the technology (Damrongchai). Somewhat paradoxically, the public is viewed as both inept, technological recipients and a group important enough to persuade, considering that ongoing innovation "...need[s] the whole nation to go along with you" (Thajchayapong).

In the instructive model, early explanation and ongoing communication from the science community to the public is seen as a way to 'defuse' fear or misunderstanding about things such as molecular manufacturing (Thajchayapong), whilst reducing damaging over-expectations about technological potential (Sawanpanyalert). Here there are a range of methods proposed for public engagement with nanotechnology, including the promotion of discussion within managed constructs. In this sense, there is talk of information being "put out there" (Weckert), through mediums such as the television and newspaper (Arya; Yuthavong) or conference seminars that are open to students and the public (Charinpanitkul).

The Thai government would appear to have adopted the instructive model, carefully controlling the flow of information about nanotechnology to the public as part of the 'public awareness' policy stipulated in the Thai Government's National Strategic Plan for Nanotechnology (Thajchayapong). Given current government efforts to increase nanotechnology media coverage (Charinpanitkul), the Thai Prime Minister takes an active role in disseminating information, "...talking to the general public about the importance of nanotechnology..." (Damrongchai). Furthermore, the NSTDA regularly feeds the media information (Panitchpakdi), whilst articles directed towards the public are generally written by Thai scientists (Damrongchai). As Damrongchai explained, this is a chance for leading scientists to take a firm position on the direction of nanotechnology to avoid problems with public understanding.

Given the need for Southern governments to reassure the public that nanotechnology is safe and proven (Berwick; Panitchpakdi), a couple of interviewees (Chirachanchai; Damrongchai) highlighted the importance of exposing successful research and visible applications to the public. As Chirachanchai explained, in the Thai case:

> The first priority [is]...to prove that nanotechnology function[s] in our country...If you presented people with a visible application such as a 'nanosensor'...then people [would] start to believe or to have the image of nanotech[nology].

According to Damrongchai, any further unrest can be mitigated by largely ensuring that applications are not just visible but provide widespread benefits.

The antithesis of the instructive model, the participatory model involves bottom-up, stakeholder-focussed initiatives, framed in terms of a broad, two-way dialogue, consultation or democratic debate between the public and other stakeholders (Arya; Braach-Maksvytis; Ford; Selgelid; Turney). Here the public is viewed as having a moral right to engagement (Weckert), given "the public needs to participate in all things that have to do with their own destiny, their own well being, their own safety" (Yuthavong). The public is also seen as having a critical role in knowledge building — able to guide R&D (Arya) by "...informing the scientists and the direction that they take" (Braach-Maksvytis).

Arya felt that an increased "...possibility of sharing the gain[s] of the new technology" will accompany inclusive discussions that frame "ordinary people" as nanotechnology beneficiaries and stakeholders. Furthermore, in contrast to the instructive model, dialogue and participation are viewed as the means of increasing the chances of nanotechnology being accepted and supported (Arya; Ford; Ratanakul). As Tegart noted, the lesson from the GMO debate is not to foist technology on people without their knowledge of it:

> ...the lesson is that you have to involve [the] public in the exercise...the backlash that has occurred against GMOs, some of it is very emotive, some of it is certainly based on the fact that things have been forced on people against their will or without them knowing...GMO foods on the shelves without people being alerted to them...people object to that.

In Thailand's past, public discussions about the introduction of technology have broken down because of a lack of basic knowledge required for meaningful engagement. Ratanakul provided the example of biotechnology debates about cloning that proved fruitless because "...some people criticise[d] the cloning technology without knowing about it...". With the public's right to knowledge being a crucial aspect of the participatory model, meaningful public participation in the development of nanotechnologies must therefore be grounded in good understandings about the technologies themselves (Ratanakul; Tegart).

The public will best understand and engage with nanotechnology when the information provided is most comprehensive (Cooper) and its distribution most equal (Panitchpakdi). Four aspects of knowledge were recognised as important to developing comprehensive public understandings. Firstly, the public must develop an understanding of what nanotechnology is and how it works (Arya; Kanok-Nukulchai; Ratanakul; Selgelid).[31] Secondly, the public must develop an understanding of how nanotechnologies can and will be applied (Cooper; Tegart; Pothsiri). Thirdly, the public must be openly and duly informed of the potential benefits and threats of nanotechnologies, any relevant research findings that emerge, new products that reach the market as well as ongoing updates on technological performance (Arya; Lynskey; Panitchpakdi; Ratanakul). Finally, the public must be presented with the "various policy options" that are available (Selgelid).

In addition to comprehensive information, Cooper said that the public must have the opportunity to develop "...correct perception[s] regarding this technology". An assessment of advantages relating to the adoption of nanotechnologies, for example, will also logically include an indication of the disadvantages of non-engagement (Coyle). Whilst information will ideally be "...correct and reliable, not something misleading or distorted" (Cooper), the public needs to receive 'both sides of the story' in a rational, balanced manner (Cooper; Tegart). For Yuthavong, a "neutral platform" for assessment is particularly important so that scientists are not unfairly "...perceived as Dr Frankenstein". The stark differences I have highlighted in the past three chapters, in terms of how nanotechnology is perceived, reinforce this need, as does Thailand's recent history with GMOs:

> ...in the case of GMO[s] we used to have Nobel prize winners come and say, 'hey GMOs are great' and then the NGOs would come in and say, 'they are terrible, they are very dangerous'; so people are confused (Yuthavong).

In the participatory approach, the development of public knowledge about nanotechnology involves engaging people in face-to-face exchanges as part of public 'outreach' (Ford). This can involve methods such as the government, NANOTEC or the universities establishing discussions within schools and universities (Arya; Yuthavong), public 'nanogroups' that link in with the NSTDA's 'Public Understanding in Science, Technology and Innovation' program (Yuthavong), or a space within the media open to public commentary (Arya; Sriyabhaya). Critically, these forms of outreach are open to participant input and steer clear of didacticism, with some, such as

Yuthavong, rejecting the use of the word 'education' and the concept of teaching nanotechnology content. Kanok-Nukulchai questioned whether a formal way of teaching the public is even possible given "nanotechnology is not just something very casual. You need really good knowledge in the basic field[s]". Whatever form it takes, Dutta believed that nanotechnology outreach in Thailand will require scientists letting go of their resistance to actively engaging with the public. He explained: "...the scientists think that 'we know too much, we cannot go and talk to the people'...that is the wrong perception. We will never get nanotechnology down [to] the people [that way]".

In terms of whom outreach should target, Panitchpakdi felt that, in order to effectively stimulate debate, the focus should be on "...the people who can afford it, like the middle class". Yet Turney suggested that, to maximise public participation and the potential for nanotechnology to be widespread, the greater challenge is to engage with the most marginalised:

> ...when you are hungry and your children and your parents are dying of AIDS and you have got tuberculosis and malaria you have other things on your mind, but we need to find some way to try to get that level of debate [happening] that is [already] occurring within the middle classes and the decision makers in those countries.

Others (Arya; Sriyabhaya) agreed, saying that Thai outreach should commence with the general population, at the village level. As I have already mentioned in this chapter, there is a belief that this kind of outreach would seem a natural fit with models such as OTOP. Yet, ultimately, there was a strong belief that engagement must cross the spectrum, from the grassroots level through to very high, international settings (Arya; Braach-Maksvytis). If outreach can result in widespread engagement with nanotechnology discussions, Ratanakul suggested the implementation of formal mechanisms to monitor public satisfaction throughout various phases of technological development.

Compared to the instructive model, the participatory model advocates a more democratic engagement with technological governance. Historically, however, the instructive model has been most widely adopted, given that public discussions in the early stages of biotechnology, the nuclear debate, stem cell research and other areas started after the technology had arrived (Braach-Maksvytis). Interviewees saw great possibility for this to change, with the milieu in the South suggested as all the more conducive to productive debate given three main factors. Firstly, Braach-Myksvitis pointed to recent precedents, such as the broader dialogue surrounding access to essential drugs for HIV/AIDS patients, saying that these have opened up a "...tremendous opportunity with nanotechnology to...talk with this very wide section of the community". Secondly, Changthavorn spoke of increased public awareness and interest in technology-related issues arising from crises such as the case of Jasmine Rice and IP in Thailand.[32] Cooper added: "...maybe the anti-GM movement has prepared the ground for creating a more level playing field for this kind of technology to be distributed". Thirdly, Charinpanitkul suggested that "very useful" public engagement frameworks have emerged from experiences with recent technologies, such as with GMOs in Thailand. Combined, interviewees (Braach-Maksvytis; Weckert; Yuthavong) saw these features providing a unique opportunity to learn from the past and ensure that, in the case of nanotechnology, the public is informed and engaged in meaningful ways.

A number of other factors present opportunities for nanotechnology's early decision-makers to "...think about a different model" (Ford). According to Ford, public knowledge about nanotechnology is currently greater than at comparable stages for previously introduced technology. Reasons suggested for this phenomenon include the establishment of Northern programs to develop basic levels of awareness around nanotechnology, and the emergence of regular 'social' forums discussing the implications of working on the nanoscale (Braach-Maksvytis; Changthavorn; Ford). This position of relative knowledge is said to mean that a country's nanotechnology strategy can "...develop from the bottom upwards, more from a public base, not necessarily a research base" (Ford). Furthermore, whilst there may be a relatively large amount of knowledge about nanotechnology that is circulating globally, the lack of public awareness and understanding in Thailand, for example, is believed to provide a 'clean' platform for debate (Braach-Maksvytis; Weckert; Yuthavong). Weckert said that such a platform is an opportunity to particularly ensure the public receives balanced information about the scope and potential implications of nanotechnologies, summarising: "...[in] areas of the world, [where] people have not come across these more sensational aspects or supposedly sensational aspects, you could probably have a much more sensible and informed debate".

Irrespective of whether the instructive or participatory model is pursued, the general feeling was that now is the appropriate time for proactive strategies around public engagement with nanotechnology, given it "...does not really dominate the world, at this stage, in the way that, say, IT [information technology] does..." (Weckert). Early engagement with the critical issues was said to provide societies with the flexibility to "...choose whether nanotechnology has a positive or negative disruptive influence" (Turney). In this respect, Braach-Maksvytis suggested that nanotechnology offered an opportunity to "...be a bit smarter...[and] create a different way of moving forward...a whole different scenario we could go down".

9.4 Conclusion

Southern approaches to nanotechnology's governance are found to largely focus on supporting innovation through regulation and managing risk at the expense of meaningful public engagement.

According to many of the interviewees, strategic planning should occur as part of a coordinated approach through a central body, with an ability to learn from global efforts using a 'pick and choose' philosophy. Domestic capabilities should be audited and countries must select appropriate research areas, with Southern countries needing to focus on niche product development. In this light, there is a preoccupation with infrastructural and managerial approaches to nanotechnology's challenges. Developing adequate and appropriate human resources to support such innovation, it is believed, must ultimately focus on endogenous training, commencing at schools and engaging all tertiary levels. Nanotechnology curricula must be broad in the initial stages and then specialised, whilst international support for human resource development can best be harnessed through aid and student exchange. Furthermore, nanotechnology demands reorganised infrastructure. Cross-disciplinary research can be stimulated through the science park model, whilst engaging the private sector requires financial incentives such as lead investment from government, venture capital and tax breaks. Strategic international partnerships can be focussed on human resources, R&D capacity building and commercialisation, with contributions to each of these areas coming from both South and North.

Strategies suggested in response to the challenge of ensuring appropriateness in the development of nanotechnologies include keeping an open mind and balanced approach to engagement, merging market and social needs and building on equity-enhancing resources.

Rather than speak broadly of nanotechnology regulation, regulatory responses must be considered for each of the specific phases of nano-innovation. Here there is seen to be adequate existing regulation in the 'R&D' and 'production' phases, whilst most believed new laws need to be explored with respect to the 'use' phase. A moratorium, however, is seen as infeasible, although 'reasonable precaution' must be exercised. Thus, knowledge about nanotechnology's societal implications becomes paramount. Here the belief is that this knowledge will be best developed by harnessing existing and emerging knowledge from international leaders in nano-innovation, utilising the collective wisdom of the international community through international discussions, meetings and institutionalised bodies. Active Southern engagement in endogenously focussed research is also seen as naturally leading to greater knowledge regarding nanotechnology's societal implications. Scientists, ethicists and lawyers are highlighted as key players in the exploration of nanotechnology's implications for the South. Corresponding with this outstanding consideration for the involvement of professionals, there was considerable concern for a lack of meaningful public engagement with nanotechnology, particularly in terms of awareness and understanding. Whilst this was not novel, and most anticipated a relatively smooth acceptance of nanotechnology, others felt that nanotechnology's subtle introduction, amongst other things, could lead to a GMO-style backlash.

Reflecting on the perspectives presented, it would appear that countries can adopt either an instructive or participatory model for engaging the public with nanotechnology. Whilst there appear exciting new possibilities for a participatory approach to materialise in the wake of recent experiences with emerging technology, the signs suggest an instructive approach is already being adopted in Thailand.

Having completed the qualitative research phase of my study, in the coming Chapter I can now draw some overall conclusions about the extent to which nanotechnology offers hope for a more equitable world.

Endnotes

1   Exemplified by technology such as diagnostic nanoparticles that are functionalised to look for specific diseases or conditions (Dutta).

2   Warris held experience in assessing nanotechnology capabilities on a national level, having conducted a 2002 assessment of Australian capabilities and performance (see Warris, 2004).

3   NSTDA, BIOTEC, the National Metal and Materials Technology Centre and the National Electronics and Computer Technology Centre.

4   For example, the study was finished in two months, yet Charinpanitkul said that a comprehensive survey would have required "...at least four months or a half year".

5   Electrical systems made up of components between 1 and 100 micrometres, fabricated on a silicon substrate.

6   A device integrating multiple laboratory functions on a single, semiconductor chip.

7   In high impact journals (Warris).

8   More specifically year 10 (Ford), as defined by the Commonwealth system (approximately 15 – 16 year olds).

9   Kindergarten to year 12, as defined by the Commonwealth system (approximately 4 – 17 year olds).

10   An Australian Government training system providing "...courses for vocational education and training, apprenticeships and traineeships, support for workplace training and commercial courses as needed by business and industry" (Department of Education, 2008).

11   A Swedish university. See: http://uu.se.

12   Full information about the OTOP concept can be seen at: http://thaiotop-city.com/background.asp.

13   A group of science professionals and their families that live and interact largely within a confined space but are provisioned with most of the amenities representative of broader community living, such as banks and shopping centres.

14   A government agency under the NSTDA that seeks to stimulate the development of the Thai software industry. Its facilities house 50 different software related companies and its administration coordinates a number of virtual information technology clusters.

15 The transactional equivalent of approximately $825,000, as of September, 2004.

16 With whom there are very strong nanotechnology ties (Dutta).

17 Vietnam joined the Asia Nano Forum in 2002 (Tanthapanichakoon).

18 The NanoBusiness Alliance is a not-for-profit association dedicated to promoting the commercialisation of nanotechnology and helping companies bring products to the market. See: http://nanobusiness.org.

19 A now defunct conference, previously aimed at boosting technological and business contacts between APEC members.

20 The Bio forum is "...a comprehensive bio event gathering all bio-related people, products, technologies and information from all across Asia" (Reed Exhibitions Japan, 2008).

21 See: http://dgroups.org/groups/hif-net.

22 The transactional equivalent of approximately $0.82, as of September 2004.

23 An alliance focussed on addressing the U.N. MDGs by drawing on the capacity of approximately 50,000 scientists from around the world.

24 The Grand Challenges in Global Health initiative is a partnership dedicated to supporting scientific and technical research to solve critical health problems in the developing world. See: http://gcgh.org.

25 As occurred with antiretroviral drugs in Thailand (Panitchpakdi).

26 See: http://unesco.org/new/en/social-and-human-sciences/themes/bioethics/bioethics-and-human-rights.

27 The World Health Assembly is the decision-making body of the WHO, comprising of representatives from 193 member states.

28 See Enriquez (2005).

29 See Uldrich and Newberry (2003).

30 See: http://164.115.5.161/apec/.

31 Working towards ensuring that nanotechnologies can be understood and operated by local, indigenous staff (Coyle).

32 Variants of jasmine rice have been patented by U.S. companies, leading to widespread public outrage in Thailand (Kerr, Hobbs and Yampoin, 1991).

Chapter 10

Conclusions

In recent years, nanotechnology has emerged as an area of hot debate with respect to development and global inequity. Key perspectives are significantly polarised between those who claim nanotechnology is part of the 'development solution' and others who claim it is part of the 'development problem'. At the heart of my research, therefore, has been an exploration of these claims in order to judge the extent to which nanotechnology offers hope for a more equitable world.

In this chapter I will bring this book to a close by reflecting on the key themes of my research and associated findings and then use my new knowledge to map some logical pathways forward.

10.1   Key Themes and Findings

In this book I have mapped an historical and contemporary narrative of development, technology and inequity. By reviewing the continual fault lines for debates about technology and inequity, three broad themes emerged as a useful interpretive framework for my assessment of nanotechnology. In turn, these themes pointed to what I call the 'legitimate requirements of equitable development' that then formed the basis for my research sub-questions. My assessment, therefore, specifically looked at reflexivity around my emerging themes in order to assess evidence of the legitimate requirements of development being factored in to nanotechnology's development.

Nanotechnology and Global Equality
Donald Maclurcan
Copyright © 2012 by Pan Stanford Publishing Pte. Ltd.
www.panstanford.com

My first theme, *innovative capacity*, led to my research sub-question: 'will nano-innovation and innovative capacity be globally and locally decentralised and autonomous?' My second theme, *technological appropriateness*, led to my research sub-question: 'do nanotechnologies offer appropriate technologies for the South?' My third theme, *approaches to technological governance*, led to my research sub-question: 'do the present and foreseen approaches to nanotechnology's governance in the South enable empowering, democratic processes?'

These three themes shaped my analysis of the literature and my interviews. The associated questions became key organising tools, answered in the literature on nanotechnology, development and inequity and then revisited in my quantitative and qualitative research.

When applying these three themes to my review of the literature about nanotechnology, development and inequity, I was able to synthesise previously disparate debates and build an even more comprehensive context in which I could assess relevant claims. Whilst this review confirmed the appropriateness of my three key themes and associated questions for my assessment of nanotechnology, I discovered the critical importance of another theme: *understandings*. This, then, became foundational to my new four-theme framework and led to my final research sub-question: 'is nanotechnology understood in ways that allow common discussion about its implications for global inequity?'

My key findings cover these four themes and associated research questions as well as my central research question, more broadly.

10.1.1   Understandings

My review of the literature showed obvious confusion about how nanotechnology is understood, yet a dearth of research into this phenomenon and its possible implications. As a result, there is a distinct lack of clarity as to whether nanotechnology is understood in ways that allow common discussion about its implications for global inequity.

In my research I found that, whilst there is agreement about nanotechnology's common characteristics, simultaneously there are substantially different ways in which it is conceptualised, placing limitations on common discussions about its implications for global inequity.

I firstly discovered that, across Thai and Australian key informants, there is agreement about six common characteristics for nanotechnology. These characteristics surround issues of size, 'control' on the nanoscale, exploitation of unique nanoscale properties, nanotechnology's practical nature, its use of 'old science' in a new way and its ability to naturally draw together disciplines into a new field. These commonly identified characteristics preface the Thai focus on near-term nanotechnology, making the assumption that Southern debates will reflect Northern hype around molecular manufacturing appear misplaced.

However, agreement about common characteristics does not translate into agreement about how nanotechnology is conceptualised. Although some interviewees acknowledged the wide range for interpretation, most referred to nanotechnologies of a very different nature in their representations of the field as a whole, particularly in terms of the associated scope and complexity of applications. These differences were more distinct between those who are 'nano-engaged' and those who are 'nano-disengaged', than by country, suggesting that, with greater information, comes a more circumspect perspective on nanotechnology's diversity of application.

As reinforced throughout my research, differences in conceptualisation have big implications for debates about global inequity. The result of differences in the way interviewees conceptualised nanotechnology shall be discussed soon.

Do the different forms of conceptualisation deny common discussions? The more people generalise in matters such as trade norms for nanotechnology patent regimes, technology transfer liability, labelling, international agreements and national regulations, the more likely important differences will be lost in meaning. Clarifying scope or utilising specific case studies would, therefore, appear to be important to discussions, negotiations and debates about specific issues relating to nanotechnology.

10.1.2 Innovative Capacity

My review of the literature showed signs of budding nano-innovation and associated capacity in parts of the South but fear for further concentration of capacity and influence — loosely termed 'the nanodivide'. My review also showed that, given an unclear picture of which countries are engaging with nanotechnology R&D, the nature of such engagement and the implications of this engagement for the South, there was a need to map the global 'state of play', accompanied by perspectives 'on the ground'.

In my research I found that, whilst there is evidence of widespread engagement and feasible, largely autonomous R&D entry points for some Southern countries, innovative capacity is increasingly centralised and disengaged from 'the local'.

My search engine research provided greater clarity about the nature of global engagement with nanotechnology R&D by revealing a more internationalized scene than previously reported. Comparing favourably to the equivalent stage of biotechnology's emergence, numerous claims materialised about Southern nanotechnology R&D having already having existed for decades. China convincingly leads Southern engagement, with its second-place in global, health-related nanotechnology patenting demonstrating the country's interest and strength in the area as well as the serious 'threat' it poses to U.S. and E.U. dominance in this field.

Supporting claims in the literature, most saw nanotechnology presenting new potential for Southern innovation, with a strong belief in the existence of feasible R&D entry points, even for LDCs. Greater autonomy over innovation is seen as stemming from an ability to leapfrog and harness the benefits of globalisation that did not exist in the early stages of biotechnology's emergence. Perhaps, as Cooper says, early widespread engagement "totally changes the picture".

Yet there is a large variation of opinion surrounding issues of technological capacity, such as the expected entry costs and infrastructural requirements of nanotechnology R&D. This highlights different conceptualisations for nanotechnology, in terms of scope and complexity, as well as the potential for strikingly different types of engagement with nano-innovation.

Understandably then, there was a great deal of disagreement in terms of where the strongest barriers to Southern nano-innovation lie. Interestingly, though, views were almost universally split between Australian and Thai interviewees who, respectively, saw the largest barriers as exogenous and endogenous.

Overall, however, both my quantitative and qualitative results support the position that, despite its potential, nanotechnology will continue to concentrate control over innovation, increasing Southern technological dependency and exploitation through the international division of labour. The concentration of innovation, nonetheless, may be in slightly different hands. Early private sector engagement from MNCs is more pronounced than with biotechnology. Furthermore, the evidence suggests that emerging gaps in capabilities are as much South-South as North-South.

When it comes to 'local' nano-innovation, there was a noticeable void of information and evidence to assess. In this sense, debates were almost exclusively framed and focussed on the national-international sphere at the expense of consideration for local or community engagement with nano-innovation. In practice, innovative capacity would appear increasingly disengaged from 'the local'.

10.1.3   Technological Appropriateness

My review of the literature showed strong, yet often speculative disagreement between those who saw nanotechnology providing exciting new applications for development settings and those who questioned the ability for the rhetoric to reconcile with the realities. The need for further research was clear, given that, beyond identifying potential benefits, consideration for Southern perspectives remained largely absent, particularly in terms of investigating the foreseen implications of nanotechnologies.

In terms of appropriateness, nanotechnology is seen as offering numerous technical advantages, but any associated benefits are set against numerous imponderables relating to nanotechnology's risks and implications, as well as economic imperatives that can mean nanotechnology is oriented away from Southern needs.

Nanotechnologies appear to offer simple, efficient and user-friendly applications across a range of areas, with my investigation of healthcare presenting a number of useful applications, such as easily managed diagnostics for remote locations. Furthermore, easily used and potentially inexpensive nanotechnologies seem representative of a new, mainstream understanding of what should be considered appropriate technology.

In this light, I sense the concept of appropriate technologies is being co-opted, forgetting some of the fundamental and holistic aspects Schumacher first presented. As Shiva (in Jamison, 2009) notes:

> The way in which nanotechnologies are being presented is a betrayal to the science on which they are based. At a time when science allows us to see the world in a deeper way, the quantums are brushed aside by the uses of the science (p. 135).

In particular, there seems scant concern for the appropriate means or ownership of the production of nanotechnologies. The OTOP program was the sole effort presented with respect to reconciling the modern with the traditional, the social with the market and the local with the national. Equally so, however, there seems little interest in critiquing nanotechnology from either labour or feminist perspectives, with a surprising lack of consideration, particularly amongst interviewees, for the challenges the South faces in terms of commodity substitution resulting from nanotechnologies.

Rather, the focus amongst proponents is similar to that witnessed with biotechnologies. In this sense, there is strong evidence of 'bandwagon science', fronted by the promotion of both the technical and development benefits to be gained by the adoption of nanotechnologies. The corresponding evidence and logic points to disappointment arising from these promises. Given I have concluded that nano-innovation is rarely localised, it is not surprising that neither is its orientation. This is particularly so, given the early evidence of Southern engagement being driven by the market's growth imperative, resulting in a focus on value-adding to existing export industries.

Thus, unless a market exists, socially focussed nanotechnologies are viewed as relevant only for 'development', missing the point that human survival and a more equitable world requires a global shift — particularly in the North — to truly appropriate technologies. Such a point will be missed as long as energy efficiency claims made with respect to nanotechnologies are used to hide confounded arguments about ecological modernisation.

On the rare occasions when nanotechnology is genuinely explored in terms of its potential to aid Southern development, the dominant approach is to return to a failed mode of transplanted techno-fixes, with little consideration for existing alternatives or social contexts. In this sense, aspects such as the gendered insensitivity associated with the introduction of new technologies, continue to go unnoticed.

In terms of the societal implications of nanotechnologies, my research shows a considerable tension remains around the novelty of risks to human and environmental health, compounded by a lack of data and numerous imponderables. Industry spin, however, would seem to be a key feature in these debates, given that the very characteristics said to make nanotechnologies distinct and the embodiment of such great potential are simultaneously denied as making them the riskiest in terms of human and environmental health, trade and commodity substitution.

Interestingly, nanotechnologies were viewed in a different ethical light to GMOs and were not seen as raising any new legal implications in the near term, aside from the introduction of more broadly applicable patents — a point questioned by many in terms of its implications for control over technological development.

10.1.4 Approaches to Governance

My review of the literature highlighted that surprisingly little research has looked at whether the present and foreseen approaches to nanotechnology's governance in the South enable empowering, democratic processes.

My research suggests that present and foreseen Southern approaches to nanotechnology's governance are focussed on supporting innovation at the expense of public participation and flexible trajectories.

More specifically, the focus of Southern engagement with nano-innovation is on State-driven initiatives that are motivated by the prospect of commercial gain. Just as the World Bank (1997) claims for development more generally, the State certainly has become the 'facilitator' for nano-innovation in many Southern countries. Here, immense efforts, across many countries, are being placed into building human resource capabilities through nanotechnology education, supporting the commercialisation of research through strategies such as 'clustering' and 'national systems of innovation', and seeking mutually beneficial partnerships across borders. Supporting arguments made in the literature, my research shows that 'heads of State' are playing a critical role in the genesis and early development of Southern engagement with nanotechnology R&D.

In light of the focus on applications for the market, a number of proposals are made to ensure nanotechnology is governed in a way that guarantees it serves human needs. Some interviewees say that nanotechnology should be assessed by its ability to merge market and social drivers. Others add that nanotechnology can build on existing endogenous and exogenous equity-generating resources and capabilities, such as the Internet, that offer avenues for more widespread R&D and policy input as well as technological distribution.

Of equal concern to the issues surrounding nanotechnology's orientation are Southern responses to the governance of risk. Highly polarised views emerged about whether nanotechnology is inherently regulated, with whom the onus should rest in relation to safeguarding against scientific risks, and the likelihood of a backlash — versus public acceptance — albeit conscious or unconscious. Thus, debate emerged at the instrumentalist/contextualist tension between slowing down the science versus speeding up the ethics, although there was broad agreement on the infeasibility of a blanket moratorium on the development of nanotechnologies. In this light, there would seem little evidence to suggest nano-innovation will be purposefully slowed down in the South, and there appears no evidence of steps for specific, prohibitory nanotechnology regulation in Thailand. Rather, the debate about risks would seem to be constantly manipulated to ensure desired outcomes, as demonstrated by the thinking of Chanthavorn who, in response to the envisaged challenges raised by patenting nanotechnologies, said that one of the greatest fears was actually how Thailand can ensure commercialisation of nanotechnologies.

There is a great deal of rhetoric and even aspirations about public engagement with nanotechnology in the South. However, as has often been the case with biotechnology, beneath the veneer of somewhat participatory engagement would appear to lie an instructive, disempowering process, set upon rigid, predetermined trajectories. Furthermore, my research suggests that, when it comes to participation in relevant debates, there is a particular failure by Southern governments to engage typically marginalised populations, such as women, people with disability, farmers, peasants and indigenous populations. My search engine data suggests that this inequity of input extends to the arena of international governance.

Evidence of an undemocratic approach to nanotechnology's governance in a country such as Thailand is all the more disappointing given the lessons arising from poor public engagement with biotechnology. In this sense, some interviewees expected new knowledge and practices around participatory governance and public engagement with the ethical, legal and social implications of emerging technologies. However, there are few signs that such wisdom will be proactively utilised when it comes to engaging with nanotechnology in the South.

10.1.5   Hope for a More Equitable World?

Combined, my findings suggest that nanotechnology presently offers little hope for a more equitable world based on the failed demonstration of a reflexive response to the legitimate requirements of equitable development. Overall, an increasing concentration of capacity and influence, simplistic hype that obfuscates key criteria of appropriateness, and a largely 'managed' process of public engagement with predetermined desirable outcomes, suggest that nanotechnology is likely to maintain and possibly amplify the inequities stemming from existing forms of technological innovation, such as biotechnology. Furthermore, debates surrounding nanotechnology and development remain so polarised that mainstream reflexive engagement seems unlikely.

Following Scrinis and Lyons (2007), my research shows that, characterised by cross-cutting control, a 'nanocorporate' era is emerging and therein reinforcing the mainstream economic paradigm and its associated inequities. The most obvious of these inequities relates to the how the possibilities compare with the probabilities when it comes to nanotechnology and 'the poor'. Here the hype around nanotechnology offering faster, stronger, cheaper, more efficient 'solutions' for development embodies the practices of mainstream alternative development, drawing on modernisation theory and enlightenment values. In this way, the term 'nanotechnology' is largely understood from the view of technological determinism — focussed on the artefacts and scientific benefits behind the meaning — without realisation that nanotechnology's development must be interpreted as a social process if its implications for Southern development are to be comprehensively understood. Understandably then, the mainstream approach to nanotechnology and the South equates the challenges of development to deficiencies with respect to technical solutions, whilst the South is still viewed as 'backward' and unable to independently address endogenous concerns. The process of addressing social needs is therefore underpinned by Northern-generated liberal-market ideology and practice, with programs such as One Tambon One Product showing that attempts to merge social and market needs will be primarily driven by the desired fulfilment of market outcomes.

Even if the distribution of nano-innovation and innovative capabilities can be more widespread amongst Southern countries, whether this will lead to a greater focus on social needs remains dubious. In this light, one of the most prominent characteristics of nanotechnology's inappropriateness is that the current uses of nanotechnology are unsustainable, perpetuating the 'treadmill of production' (see Schnaiberg, 1980). There is, therefore, a need to rethink nanotechnology's role in responding to the limits to growth.

Overall, nanotechnology presents some unique characteristics for consideration, such as the potentially low-costs associated with self-assembly manufacturing, the danger of broad-brush patenting and the early nature of engagement by multinationals in R&D. Whilst these may result in different considerations regarding various cross-cutting issues, issues of control, appropriateness and governance would seem to remain central to emerging debates. Thus, perhaps it is not so much nanotechnology itself that I am placing under critique, but what nanotechnology represents as one of the most recent and broadly encompassing iterations of inequitable science policy and practice.

10.2 Limitations and Further Avenues for Research

Any broad, cross-cultural study of an emerging technology will have inherent limitations. In terms of my research focus, it was necessary for me to look broadly at nanotechnology, rather than specifically at one of its applications, given its nascent state of development at the time of this study. Furthermore, there is said to be a need for more holistic research into nanotechnology's implications, integrating systems thinking in order to understand how technologies "...

spread out and create intended and unintended effects throughout the system" (Carroll, 2001, p. 192). Yet some, such as Rip (2006), claim that nanotechnology's broad nature makes it difficult to assess. This view is supported by my findings which demonstrate that nanotechnology can be conceptualised in many different ways, subsequently producing vastly different repercussions with respect to the technology's foreseen implications. In this light, Schummer (2007) argues that there is no quick study or investigation which can answer the question of nanotechnology's global implications:

> Given the diversity of nanotechnologies and the many different factors through which they can impact developing countries, it is obvious that no simple answer can be provided. Indeed we need hundreds of case studies that integrate all the available scientific, engineering, economical, political, legal, sociological, cultural, and ethical knowledge... (p. 295).

Building on these arguments, exploratory research infrequently yields definitive answers (Neuman 1997). Combined, these points suggest the benefit in future studies being context- and application-specific, thereby providing ever more tangible assessments with respect to attitudes and foreseen implications. This should work, given the field is now developed enough for useful, transnational case studies that might focus on specific sectors, if not specific technologies.

Given the limited amount and types of research that had been conducted into nanotechnology's implications for global inequity as of 2004, I also felt it important to provide a broad analysis that would necessarily include some generalisations in order to make sense of things. However, my extrapolations have their limits, particularly in light of my position as a researcher from the North and given Thailand has a higher Human Development Index ranking than most other Southern countries. Thus, I envisage more endogenous, detailed case studies as further avenues for research, utilising locally appropriate methods to improve the relevance and representation of studies by properly grounding debates in local struggles.

Regarding my own research methods, in order to provide a manageable and affordable mapping of Southern nano-innovation and innovative capacity I had to limit my quantitative research to assessing free search engine data. However, as shown by others (Marinova and McAleer, 2003; Huang et al., 2004; Zhou and Leydesdorff, 2006), many more avenues exist for exploring national engagement in greater detail. Zhou and Leydesdorff (2006), for example, look at:

> ...total publications, world share of publications, total citation rates, percentage of world share of citations, as well as the top one percent of most highly cited papers in order to measure scientific output (p. 3).

Yet, few studies of this more detailed nature seem to have been conducted since 2006. To provide greater clarity with respect to the strength of national engagement as well as longitudinal data important for measuring trends, further scientometric studies would therefore appear justified.

Similarly, whilst I needed to focus on a broad analysis with my qualitative study, there is potential, as well as a need, for other methods to be used equally successfully. Discourse analysis, for example, is one obvious option given indications of biases in language surrounding nanotechnology[1] and the history of this form of analysis being used for an area such as biotechnology (see, for example, Henderson and Kay Weaver, 2007; Leitch and Davenport, 2007).

Regarding my interviewee sample, I needed to assume the relevance of key informant views including those outside the immediate debates and those that may have been influential in the early shaping of nanotechnology's trajectory. Given the nascent point in nanotechnology's development and the documented gender biases in technology-related employment, including technology policy (Markert, 1996; Fountain, 2000), it was unsurprising that the large majority of my interviewees were male. Similarly, to provide a useful context for responses I needed to select countries for my study that were already engaging with nano-innovation and in which face-to-face interviews would be feasible. However, using a largely male, key informant population maps only certain perspectives; there is evidence to suggest, for example, that men associate less risk and more benefits than do women with claims made in the news about science and technology (see Hornig, 1992). In this sense, my research may be viewed as perpetuating the lack of diverse voices in debates surrounding nanotechnology and development. Similarly, selecting two countries already optimistically engaging with nanotechnology R&D may mean uncritical bias was built into my research design

(Throne-Holst and Stø, 2008). Hence, different studies, in terms of sample diversity and location, would appear appropriate and useful as part of engendering a more participatory discourse. More specifically, my experience — particularly in reviewing Grimshaw *et al.*'s (2006) work — suggests the usefulness of more local, subsistence, indigenous, Marxist, feminist and peasant perspectives and critiques. Such considerations would seem even more justified in light of the 2010 'People's Agreement[2]' at the World People's Conference on Climate Change and the Rights of Mother Earth,[3] which spoke of nanotechnology offering "...false solutions...that only exacerbate the current crisis". These proposals are enhanced given positions amongst the broader public may now be more formed than when I first commenced my research.

Following on from my broad approach, to constrict interviewee responses as little as possible, I tended towards expansive, rather than specific questions. However, at times this led to uncreative and non-reflexive responses from interviewees, perhaps also explained by nanotechnology's early stage of development. This rigidity may have been further influenced by my own dichotomous analysis of the literature on nanotechnology and development. In light of this limitation, future studies into nanotechnology's global implications could gain from exploring areas in which polarised thinking is being bypassed and, as Liao (2009) suggests, actively seeking approaches that combine instrumentalist and contextualist thinking. Since the interviews, I have been exposed to further, exciting avenues at the intersection of cutting-edge thinking and emerging practice[4] that appear to offer interesting angles for exploration. Such approaches could offer pragmatic attempts to create a truly reflexive dialectic.

10.3   Implications and Recommendations

I see my research holding serious, real-world implications, given the centrality of technology to debates about climate change (see United Nations Environmental Program, 1999; World Bank, 2006), the limits to growth (see Polimeni *et al.*, 2008) and 'making poverty history' (see Juma, 2005). Having highlighted an unhealthy connection with economic growth as the fundamental barrier to nanotechnology offering hope for a more equitable world, I therefore believe there must be consideration for whether nano-innovation could exist without growth. However, my experience and findings suggest that to adequately consider nano-innovation without economic growth, and the subsequent repercussions for equitable development, would require a more appropriate concept than reflexive development, given this falls back to developmentalism which, as Pieterse (1998) notes, is easily co-opted. My conclusions thus point towards the need for what I term *reflexive pluralisation*: a process of increasing the sociocultural, technological and resource autonomy of communities whilst effectively considering the importance of external change and critique (alternatively, I describe this as working towards *integrated autonomy*). Reflexive pluralisation would differ from reflexive development by problematising the underlying notions of modernity as well as development, acknowledging the ability for globalisation to lock locales into inequities and accepting 'thinking globally' as a natural by-product of increasing local autonomy. Reflexive pluralisation would bypass the unhelpful dichotomy of modernity versus tradition,[5] incorporating reflexivity to biophysical limits and moving away from ethnocentric language of power to a more equitable futures that are respectful of difference but cognisant of our shared humanity. Reflexive pluralisation would also shy away from universalism or the notion of 'solutions', respecting the interplay between multiple realities that make the impacts of change dynamic and contingent upon interactions within wider ecosystems.

Given that technology can both reflect and influence broader social phenomena, such as global inequity, I make the pragmatic recommendation that nanotechnology be considered, particularly by its critics, as a potential vehicle for reflexive pluralisation. Here it will be particularly interesting to keep an eye on explorations of nanotechnology in relation to: global economic futures; 'pro-poor innovation'; open source innovation[6]; village innovation; traditional knowledge systems; and industrial ecology.

Just as one of the most exciting aspects of nanotechnology is exploring what happens when researchers 'collide' at the edge of their respective experience, I believe certain approaches at the edge of mainstream thinking offer hope for nanotechnology to embody and champion a process of 'reflexive pluralisation'. In such circumstances, and as part of a broader reclamation of science for a more equitable world, nanotechnology would reveal paths for innovation that are autonomous yet responsive to external change and opportunities for mutually beneficial cooperation — delinked from national economic growth yet meaningful to people's lives. The field could then, as Schumacher proposed in 1973, blossom 'a new orientation of science and technology towards the organic, the gentle, the non-violent, the elegant and the beautiful'.

Endnotes

1 My interviewees favoured using the word 'technology' in its singular form, mostly associating its meaning with an understanding of technology that is representative of technological determinism. Use of various verbs, such as 'could', 'should' and 'would', also seem to provide indications of interviewee biases with respect to technology as well as Southern development.

2 See: http://pwccc.wordpress.com/2010/04/24/peoples-agreement/.

3 An event in Cochabamba, Bolivia that drew together over 30,000 people from more than 100 countries in response to the perceived failure of the 15th United Nations Conference of Parties in Copenhagen, 2009.

4 In fields as diverse as technology, social innovation and knowledge management it has been proposed that innovation occurs at 'the boundaries' (see for example Leonard-Barton, 1995; Gryskiewicz, 1999).

5 As Schumacher (1973) notes: ...it is not a question of choosing between 'modern growth' and 'traditional stagnation'. It is a question of finding the right path of development, the Middle Way between materialist heedlessness and traditionalist immobility, in short, of finding 'Right Livelihood' (p. 51).

6 An example of open source nanotechnology can be seen at: http://opensourcenano.net/, where anyone can view the simple instructions for inexpensively creating magnetite nanocrystals that can act as water filters, particularly for arsenic.

Appendix A

Justification of Interviewee Nationalities for My Qualitative Study

In addition to the practicalities associated with geographical distance, my research budget and strong connections between members of the University of Technology, Sydney's research faculty and Thai nationals, there were four main reasons I selected Thai and Australian key informants for my qualitative study. The first was a significant divide between the two countries on certain measures of 'development'. Thailand is a 'middle-income country', ranked 74th out of 175 countries on the Human Development Index (UNDP, 2003). Australia, on the other hand, is a 'high-income country' ranked 4th on the Human Development Index (ibid.). Thailand's greater population continues to face many significant challenges of a largely different nature to those in Australia. As of 2004, 21% of the Thai population earned less than $2 a day (World Bank Independent Evaluation Group, 2007), whilst financial inequity has increased over the past 40 years, particularly between urban and rural areas (Bhumiratana, 1991; United Nations Department of Economic and Social Affairs, 2001; UNDP, 2007b). Stark inequities are also evident in the distributed burden of the HIV/AIDS epidemic and general access to health services (UNDP, 2007b). Furthermore, various Thai populations still suffer from very high levels of child malnutrition and maternal mortality (ibid.). Significant divides also exist with respect to R&D expenditures as a percentage of GDP. Between the period: 1996–2000, Thailand committed 0.1% of GDP and Australia 1.5% (UNDP, 2004). Similarly, for every million people, Thailand has 74 scientists and engineers whilst Australia has 3,353 (ibid.).

My second reason for selecting Thai and Australian key informants for my qualitative study was that both countries share a common engagement with emerging technology, including nanotechnology. In terms of its engagement with emerging technology, Thailand has been classified as a 'dynamic adopter' (see UNDP, 2001), with a rapid increase having been registered in its high-technology exports from 1980 (1%) to 1999 (30%) (ibid.). In this light, Thailand has strong hopes for biotechnology R&D, as well as good levels of supportive infrastructure (Sahai, 1999). Australia, too, has a strong engagement with emerging technology. Since the 1980s, Australia has transformed itself from a 'classical' imitator economy into a 'second-tier' innovator economy, leading to "...a nascent capability for innovation throughout the life and agricultural sciences" (Gans and Stern, 2003, p. iv).

When it comes to nano-innovation, in an early study of Southern capabilities, Thailand was identified as a 'middle ground' country (Court et al., 2004). This analysis is supported by early evidence of Thai nanotechnology R&D (see Panyakeow and Aungkavattana, 2002; Liu, 2003; Thajchayapong and Tanthapanichakoon, 2003; Unisearch, 2004; Tanthapanichakoon, 2005). Such engagement includes the establishment of a national centre (Lin-Liu, 2003) with a proposed budget of $25 million for the period 2004–2008 and 300 personnel (Liu, 2003) as well as the development of a national nanotechnology strategy (Sutharoj, 2005). Furthermore, Chulalongkorn University, in Bangkok, is said to have launched South East Asia's first, international Bachelor's degree in nano-engineering (Bunnag, 2005) whilst, according to Liu (2003), Thailand's nanotechnology capabilities, in 2003, included "...14 laboratories in 6 universities and 5 laboratories in 2 government agencies" (p. 1). The commercial applications emerging from Thai research are said to include antibacterial surgery gowns (Thai Press Reports, 2006), plastic films for wrapping farm produce and hypo-allergenic cosmetics (Thai News Agency, 2007). Australia, too, has been firmly engaged in international nanotechnology developments, having developed the world's first 'nanomachine' in 1997 (Cornell, Braach-Maksvytis, King, Osman, Raguse, Wieczorek and Pace, 1997). According to a 2007 report, Australia has "...more than 75 nanotechnology research organisations and around 80 nanotechnology companies" (Invest Australia, 2007, p. 3).

My third reason for selecting Thai and Australian key informants for my qualitative study is that both countries have a common history of endogenous critiques relating to emerging technology.

From the perspective of ELSI, Thailand has a history of public and government protest relating to biotechnology, ranging from issues of morality (see Changthavorn, 2003) and environmental concerns (see Kachonpadungkitti and Macer, 2004), through to issues of intellectual property such as 'biopiracy' (see Kerr et al., 1991; Meléndez-Ortiz and Sánchez, 2005) and compulsory licensing (see Knowledge Ecology International, 2008). As Subramanian (2004) notes, Thailand is one of only three countries to authorise "...the production of patented drugs by their own firms to reduce the prices of AIDS drugs and help address their own public health challenges" (p. 24). Emerging technology has also been under considerable scrutiny in Australia, with the introduction of agricultural biotechnology having faced a number of regulatory challenges (Finkel, 2008). Furthermore, Australian biotechnology is seen by some local researchers, such as Salleh (2006), as perpetuating a destructive, neo-liberal ideology. Similarly, the diffusion of the Internet has been a topic for endogenous critique (see, for example, Willis and Tranter, 2006), with particular consideration shown for inequities across various Australian demographics.

Finally, my fourth reason for selecting Thai and Australian key informants for my qualitative study was that there are existing or anticipated critiques of nanotechnology in both countries. Nano-innovation is already said to face significant challenges in Thailand (Tanthapanichakoon, 2005; Sandhu, 2008). Protests around nanotechnology's ethical implications have already been brought to the fore in Thailand by controversy surrounding 'atomically modified organisms' (see ETC Group, 2004e). Australia has also faced challenges to engagement with nanotechnology, especially in areas such as health and safety (see Priestly, Harford and Sim, 2007). Compounding these challenges have been low levels of public understanding and knowledge about nanotechnology (see Bowman and Hodge, 2007a; MARS cited in Paull and Lyons, 2008).

Appendix B

Health-Related Patent Classifications

Criterion	Classification
The patent showed intent to improve or maintain human health	Included
The patent title and/or abstract incorporated the term 'DNA'	Included (considered health-related)
The patent incorporated reference to: Water purification relating to houses or sewage; Antibacterial clothing; or Antibacterial or antimicrobial surfaces relating to the inside of buildings, paints, water piping, toilets or kitchen items, such as fridges	Included (considered within the 'consumer health' category)
The patent incorporated reference to 'implants'	Included within the 'therapeutic' category, upon confirmation of its health-related nature
The patent incorporated reference to: Monitoring Sensors Immunoassay DNA sequencing	Included within the 'diagnostic' category, upon confirmation of its health-related nature
The patent focussed on the following biological materials or functions: Analytes Dendrimers Filtration mechanisms	Excluded, unless it had a specified relationship to health or was held by a medical or pharmaceutically related entity.

Appendix C

Patent Rules

Unique Patents

Criterion	Classification
Both a patent application and an assigned patent have been registered for the same patent	Include only the assigned patent
Two or more patents have identical titles and abstracts, except for one letter or hyphen	Consider the patents as the same and exclude the most dated patent
Two or more patents have different wording in the title, but the same abstract and owner	Consider the patents as the same and exclude the most dated patent
Two or more patents have the same title but a totally different abstract	Include both separately
Two or more patents have the same abstract but two different owners	Exclude the most dated patent
The patent is registered under joint nationalities	Record a result for each country referenced
The patent is registered under a certain entity's name that has since been acquired by a different entity, in an alternative country	Maintain the recorded result pertaining to the original nationality of the patent

Type of Ownership

Criteria	Classification
The patent appears, at first, to be registered to a private individual but, subsequently, a work address is found in the patent	Consider the patent as registered to a private company
The patent abstract and description is ambiguous with respect to its stated owner	Source the owner from the following documents, in this order: PCT patent EPO patent USPTO patent Any remaining documents

Link to a Health-Condition

Criteria	Classification
One or more patents mention one or more health-conditions in the same 'class'. For example: hepatitis, hepatitis A, hepatitis C	Record a result for the overarching 'class of patent'. For example: hepatitis, cancer or cardio-vascular disease (the latter including stroke, myocardial infarction, thrombosis, artherosclerosis and aneurisms)
The patent refers to bacteria. For example: staphylococcus aurelius	Excluded, given the reference is too ambiguous to be linked to any specific disease
The patent refers to general terms such as 'inflammation' and 'dermatological'	Excluded, although the term 'tumour' is considered as linked to cancer and 'insulin' is considered linked to diabetes

Appendix D

Top 10 Nanotechnologies for the Developing World (Singer et al., 2005)

Ranking (score)	Applications of Nanotechnology	Examples	Comparison with the MDGs
1 (766)	Energy, storage, production, and conversion	Novel hydrogen storage systems based on carbon nanotubes and other lightweight nanomaterials	VII
		Photovoltaic cells and organic light-emitting devices based on quantum dots	
		Carbon nanotubes in composite film coatings for solar cells	
		Nanocatalysts for hydrogen generation	
		Hybrid protein–polymer biomimetic membranes	
2 (706)	Agricultural productivity enhancement	Nanoporous zeolites for slow-release and efficient dosage of water and fertilizers for plants, and of nutrients and drugs for livestock	I, IV, V, VII
		Nanocapsules for herbicide delivery	
		Nanosensors for soil quality and for plant health monitoring	
		Nanomagnets for removal of soil contaminants	
3 (682)	Water treatment and remediation	Nanomembranes for water purification, desalination, and detoxification	I, IV, VII
		Nanosensors for the detection of contaminants and pathogens	
		Nanoporous zeolites, nanoporous polymers, and attapulgite clays for water purification	
		Magnetic nanoparticles for water treatment and remediation	
		TiO_2 nanoparticles for the catalytic degradation of water pollutants	

Ranking (score)	Applications of Nanotechnology	Examples	Comparison with the MDGs
4 (606)	Disease diagnosis and screening	Nanoliter systems (Lab-on-a-chip)	IV, V, VI
		Nanosensor arrays based on carbon nanotubes	
		Quantum dots for disease diagnosis	
		Magnetic nanoparticles and nanosensors	
		Antibody–dendrimer conjugates for diagnosis of HIV-1 and cancer	
		Nanowire and nanobelt nanosensors for disease diagnosis	
		Nanoparticles as medical image enhancers	
5 (558)	Drug delivery systems	Nanocapsules, liposomes, dendrimers, buckyballs, nanobiomagnets, and attapulgite clays for slow and sustained drug release systems	IV, V, VI
6 (472)	Food processing and storage	Nanocomposites for plastic film coatings used in food packaging	I, IV, V
		Antimicrobial nanoemulsions for applications in decontamination of food equipment, packaging, or food	
		Nanotechnology-based antigen detecting biosensors for identification of pathogen contamination	
7 (410)	Air pollution and remediation	TiO$_2$ nanoparticle-based photocatalytic degradation of air pollutants in self-cleaning systems	IV, V, VII
		Nanocatalysts for more efficient, cheaper, and better-controlled catalytic converters	
		Nanosensors for detection of toxic materials and leaks	
		Gas separation nanodevices	

Ranking (score)	Applications of Nanotechnology	Examples	Comparison with the MDGs
8 (366)	Construction	Nanomolecular structures to make asphalt and concrete more robust to water seepage	VII
		Heat-resistant nanomaterials to block ultraviolet and infrared radiation	
		Nanomaterials for cheaper and durable housing, surfaces, coatings, glues, concrete, and heat and light exclusion	
		Self-cleaning surfaces (e.g., windows, mirrors, toilets) with bioactive coatings	
9 (321)	Health monitoring	Nanotubes and nanoparticles for glucose, CO_2, and cholesterol sensors and for *in situ* monitoring of homeostasis	IV, V, VI
10 (258)	Vector and pest detection and control	Nanosensors for pest detection	IV, V, VI
		Nanoparticles for new pesticides, insecticides, and insect repellents	

Appendix E

World Bank List of Economies* (April 2004)

Afghanistan	Albania	Algeria	American Samoa
Andorra	Angola	Antigua and Barbuda	Argentina
Armenia	Aruba	Australia	Austria
Azerbaijan	Bahamas, The	Bahrain	Bangladesh
Barbados	Belarus	Belgium	Belize
Benin	Bermuda	Bhutan	Bolivia
Bosnia and Herzegovina	Botswana	Brazil	Brunei
Bulgaria	Burkina Faso	Burundi	Cambodia
Cameroon	Canada	Cape Verde	Cayman Islands
Central African Republic	Chad	Channel Islands	Chile
China	Colombia	Comoros	Congo, Dem. Rep.
Congo, Rep.	Costa Rica	Côte d'Ivoire	Croatia
Cuba	Cyprus	Czech Republic	Denmark
Djibouti	Dominica	Dominican Republic	Ecuador
Egypt, Arab Rep.	El Salvador	Equatorial Guinea	Eritrea
Estonia	Ethiopia	Faeroe Islands	Fiji
Finland	France	French Polynesia	Gabon
Gambia, The	Georgia	Germany	Ghana
Greece	Greenland	Grenada	Guam
Guatemala	Guinea	Guinea-Bissau	Guyana

Haiti	Honduras	Hong Kong, China	Hungary
Iceland	India	Indonesia	Iran, Islamic Rep.
Iraq	Ireland	Isle of Man	Israel
Italy	Jamaica	Japan	Jordan
Kazakhstan	Kenya	Kiribati	Korea, Dem. Rep.
Korea, Rep.	Kuwait	Kyrgyz Republic	Lao PDR
Latvia	Lebanon	Lesotho	Liberia
Libya	Liechtenstein	Lithuania	Luxembourg
Macao, China	Macedonia, FYR	Madagascar	Malawi
Malaysia	Maldives	Mali	Malta
Marshall Islands	Mauritania	Mauritius	Mayotte
Mexico	Micronesia, Fed. Sts.	Moldova	Monaco
Mongolia	Morocco	Mozambique	Myanmar
Namibia	Nepal	Netherlands	Netherlands Antilles
New Caledonia	New Zealand	Nicaragua	Niger
Nigeria	Northern Mariana Islands	Norway	Oman
Pakistan	Palau	Panama	Papua New Guinea
Paraguay	Peru	Philippines	Poland
Portugal	Puerto Rico	Qatar	Romania
Russian Federation	Rwanda	Samoa	San Marino
São Tomé and Principe	Saudi Arabia	Senegal	Serbia
Seychelles	Sierra Leone	Singapore	Slovak Republic
Slovenia	Solomon Islands	Somalia	South Africa
Spain	Sri Lanka	St. Kitts and Nevis	St. Lucia
St. Vincent and the Grenadines	Sudan	Suriname	Swaziland

Sweden	Switzerland	Syrian Arab Republic	Tajikistan
Tanzania	Thailand	Timor-Leste	Togo
Tonga	Trinidad and Tobago	Tunisia	Turkey
Turkmenistan	Uganda	Ukraine	United Arab Emirates
United Kingdom	United States	Uruguay	Uzbekistan
Vanuatu	Venezuela, RB	Vietnam	Virgin Islands (U.S.)
West Bank and Gaza	Yemen, Rep.	Zambia	Zimbabwe

*This table classifies all the World Bank member economies and all the other economies with populations of more than 30,000. Available at: http://worldbank.org/data/aboutdata/errata03/class.pdf.

Appendix F

Classification of Countries: Development Assistance Committee List of Aid Recipients, 2003 (adapted from OECD, 2003)

Part I: Developing Countries and Territories (Official Development Assistance) | | | | | **Part II: Countries and Territories in Transition (Official Aid)** | |

Least Developed Countries (LDCs)	Other Low-Income Countries (Other LICs) (per capita GNI < $745 in 2001)	Lower Middle-Income Countries (LMICs) (per capita GNI $746-$2975 in 2001)	Upper Middle-Income Countries (UMICs) (per capita GNI $2976-$9205 in 2001)	High-Income Countries (HICs) (per capita GNI > $9206 in 2001)	Central and Eastern European Countries and New Independent States of the former Soviet Union (CEECs/NIS)	More Advanced Developing Countries and Territories
Afghanistan	*Armenia	*Albania	Botswana	Bahrain	*Belarus	□ Aruba
Angola	*Azerbaijan	Algeria	Brazil		*Bulgaria	Bahamas
Bangladesh	Cameroon	Belize	Chile		*Czech Republic	□ Bermuda
Benin	Congo, Rep.	Bolivia	Cook Islands		*Estonia	Brunei
Bhutan	Côte d'Ivoire	Bosnia and Herzegovina	Costa Rica		*Hungary	□ Cayman Islands
Burkina Faso	*Georgia	China	Croatia		*Latvia	Chinese Taipei
Burundi	Ghana	Colombia	Dominica		*Lithuania	Cyprus
Cambodia	India	Cuba	Gabon		*Poland	
Cape Verde		Palestinian Administered Areas	Grenada			
		Paraguay				
		Peru				
		Philippines				
		Serbia and Montenegro				
		South Africa				

Part I: Developing Countries and Territories (Official Development Assistance)

Central African Republic	Indonesia	Dominican Republic	Sri Lanka	Lebanon
Chad	Kenya	Ecuador	St Vincent and Grenadines	Malaysia
Comoros	Korea, Democratic Republic	Egypt	Suriname	Mauritius
Congo, Dem.Rep.	*Kyrgyz	El Salvador	Swaziland	□ Mayotte
Djibouti	*Moldova	Fiji	Syria	Nauru
Equatorial Guinea	Mongolia	Guatemala	Thailand	Panama
Eritrea	Nicaragua	Guyana	□ Tokelau	□ St Helena
Ethiopia	Nigeria	Honduras	Tonga	St Lucia
Gambia	Pakistan	Iran	Tunisia	Venezuela
Guinea	Papua New Guinea	Iraq	Turkey	———————
Guinea-Bissau	*Tajikistan	Jamaica	*Turkmenistan	Threshold for World Bank Loan Eligibility ($5185 in 2001)
Haiti	*Uzbekistan	Jordan	□ Wallis and Futuna	———————
Kiribati	Viet Nam	*Kazakhstan		□ Anguilla
Laos	Zimbabwe	Macedonia (former Yugoslav Republic)		Antigua and Barbuda
Lesotho		Marshall Islands		Argentina
Liberia		Micronesia, Federated		Barbados
Madagascar				
Malawi				
Maldives				
Mali				
Mauritania				

Part II: Countries and Territories in Transition (Official Aid)

*Romania	□ Falkland Islands
*Russia	□ French Polynesia
*Slovak Republic	□ Gibraltar
*Ukraine	□ Hong Kong, China
	Israel
	Korea
	Kuwait
	Libya
	□ Macao
	Malta
	□ Netherlands Antilles
	□ New Caledonia
	Qatar
	Singapore
	Slovenia
	United Arab

Part I: Developing Countries and Territories (Official Development Assistance)			Part II: Countries and Territories in Transition (Official Aid)
Mozambique	States	Mexico	Emirates
Myanmar	Morocco	□ Montserrat	□ Virgin Islands (UK)
Nepal	Namibia	Oman	
Niger	Niue	Palau Islands	
Rwanda		Saudi Arabia	
Samoa		Seychelles	
Sao Tome and		St Kitts and	
Principe		Nevis	
Senegal		Trinidad and	
Sierra Leone		Tobago	
Solomon Islands		□ Turks and	
Somalia		Caicos	
Sudan		Islands	
Tanzania		Uruguay	
Timor-Leste			
Togo			
Tuvalu			
Uganda			
Vanuatu			
Yemen			
Zambia			

*Central and Eastern European countries and New Independent States of the former Soviet Union; □ Territory.

Appendix G

Thai Key Informant Details

Name/Title	Position, Affiliation	Sector	Nano-engaged
Gothom Arya PhD	Chair, Appropriate Technology Association, Thailand	N	N
Tanit Changthavorn PhD	Intellectual Property Specialist, National Centre for Genetic Engineering and Biotechnology, National Science and Technology Development Agency	G	Y
Tawatchai Charinpanitkul PhD	Associate Dean for Research Affairs, Faculty of Engineering, Chulalongkorn University	A	Y
Suwabun Chirachanchai PhD	Associate Professor, Petroleum and Petrochemical College, Chulalongkorn University	A	Y
Nares Damrongchai PhD	Policy Researcher, Asia Pacific Economic Cooperation Centre for Technology Foresight	G	Y
Joydeep Dutta PhD	Associate Professor, Microelectronics, Asian Institute of Technology	A	Y
Worsak Kanok-Nukulchai PhD	Professor, Structural Engineering, Asian Institute of Technology	A	Y
Promboon Panitchpakdi PhD	Director, Raks Thai Foundation	N	N
Pakdee Pothisiri PhD	Senior Deputy Permanent Secretary, Ministry of Public Health	G	N

Name/Title	Position, Affiliation	Sector	Nano-engaged
Pinit Ratanakul PhD	Executive Director, College of Religious Studies, Mahidol University	A	N
Pathom Sawanpanyalert PhD	National Professional Officer (Health Systems Development), World Health Organisation, Thailand	N	N
Sirirurg Songsivilai MD PhD	Chairman and Co-Founder, Innova Biotechnology Co. Ltd	P	Y
Nadda Sriyabhaya MD	President, Stop-Tuberculosis Association, Thailand	N	N
Wiwut Tanthapanichakoon PhD	Director, National Nanotechnology Centre, National Science and Technology Development Agency	G	Y
Pairash Thajchayapong PhD	Advisor to the Prime Minister on Science and Technology	G	Y
Yongyuth Yuthavong PhD	Senior Researcher, National Centre for Genetic Engineering and Biotechnology, National Science and Technology Development Agency	G	Y

A = Academic; G = Government; N = NGO; P = Private;
MD = Medical Doctor; PhD = Doctor of Philosophy

Appendix H

Australian Key Informant Details

Name	Position, Affiliation	Sector	Nano-engaged
Leigh Berwick	Investment Manager (Nanotechnology), Invest Australia	G	Y
Vijoleta Braach-Maksvytis PhD	General Manager, Global Aid, Commonwealth Scientific Industrial Research Organisation	G	Y
Paul Bryce PhD	Director, APACE-VFEG (Appropriate Technology for Community and Environment Inc — Village First Electrification Group)	N	N
Melinda Cooper PhD	Research Fellow, Department of Sociology, Macquarie University	A	N
Bruce Cornell PhD	Senior Vice President and Chief Scientist, AMBRI Pty Ltd	P	Y
Patricia Coyle MD	Medical Doctor, Department of Anaesthesia, Royal Prince Alfred Hospital	G	N
Peter Deutschmann MD	Director, Australian International Health Institute, University of Melbourne	N	N
Mike Ford PhD	Associate Director, Institute for Nanoscale Technology, University of Technology, Sydney	A	Y
Mike Lynskey	Chief Executive Officer, The Fred Hollows Foundation	N	N

Name	Position, Affiliation	Sector	Nano-engaged
Benno Radt PhD	Research Fellow, Department of Chemical and Biomolecular Engineering, University of Melbourne	A	Y
Michael Selgelid PhD	Sesqui Lecturer in Bioethics, Faculty of Medicine, University of Sydney	A	N
Greg Tegart PhD	Executive Advisor, Asia Pacific Economic Cooperation Centre for Technology Foresight	G	Y
Chris Warris	Researcher, Australian Academy of Science	N	Y
Terry Turney PhD	Director, Nanotechnology Centre, Commonwealth Scientific Industrial Development Organisation	G	Y
John Weckert PhD	Professor, Centre for Applied Philosophy and Public Ethics, Charles Sturt University Wagga	A	Y

A = Academic; G = Government; N = NGO; P = Private;
MD = Medical Doctor; PhD = Doctor of Philosophy

Appendix I

Key Informant Biographies (2004)

Dr Gothom Arya is Chair of the Appropriate Technology Association, Thailand (a non-profit organisation with a mandate to carry out research and development, and promote novel, appropriate technology for the betterment of rural society). Dr Arya received his Doctorate in Engineering from the Université de Paris in 1969. Dr Arya has worked as: Head of the Department of Electrical Engineering, Chulalongkorn University, from 1983–1987; Director of the Institute of Technology for Rural Development, Chulalongkorn University, from 1994–1997; and Registrar of the Asian Institute of Technology. Dr Arya is also a Council Member of the Pugwash Conference on Sciences and World Affairs and Chairman of the Asian Cultural Forum on Development Foundation.

Mr Leigh Berwick is Investment Manager with Invest Australia (Australia's national inward investment agency), specialising in nanotechnology. Mr Berwick received an undergraduate degree in Asian studies. Mr Berwick's work involves promoting foreign direct investment into Australia to support sustainable industry growth and development.

Dr Vijoleta Braach-Maksvytis is Director of Global Development at Australia's Commonwealth Scientific and Industrial Research Organisation. Dr Braach-Maksvytis received a Doctorate in Biophysics from the University of Sydney in 1992. Dr Braach-Maksvytis' work has involved convening the cross-CSIRO NanoScience Network and the first national 'Nanotechnology in Australian Industry Workshop', held in 2001, which brought together key players in industry, science and government. Dr Braach-Maksvytis also convened Australia's National Nanotechnology Network to harness the combined capability in science, industry, government, investment and social sectors as a model for driving the early uptake of emerging technologies in Australia. Dr Braach-Maksvytis holds over 25 patents in the field of nanotechnology, including the world's first example of a working nanodevice (see Cornell et al., 1997).

Dr Paul Bryce is an Adjunct Professor at the Institute for Sustainable Futures at the University of Technology, Sydney, and technical manager for Appropriate Technology for Community and Environment Inc. (a non-governmental development assistance agency that manages and implements renewable energy projects within communities in the South). Dr Bryce received a Doctorate in Sustainable Futures. Dr Bryce's work involves consultation, project management and community development facilitation in countries such as the Solomon Islands, Papua New Guinea, Indonesia, Cambodia, Vietnam and Lao People's Democratic Republic.

Dr Tanit Changthavorn is an Intellectual Property Specialist with the Thai National Centre for Genetic Engineering and Biotechnology. Dr Changthavorn received a Doctorate in intellectual property law from the University of London in 1998. Dr Changthavorn's work involves bioethical issues of intellectual property rights, with a research interest in nanotechnology.

Dr Tawatchai Charinpanitkul is Associate Dean for Research with the Department of Chemical Engineering at Chulalongkorn University. Dr Charinpanitkul received a Doctorate in Chemical Engineering from the University of Tokyo in 1992. Dr Charinpanitkul's work involves the synthesis and application of various nanoparticles. In 2003, Dr Charinpanitkul conducted a national assessment of Thailand's nanotechnology capabilities (see Unisearch, 2004).

Dr Suwabun Chirachanchai is an Associate Professor with The Petroleum and Petrochemical College at Chulalongkorn University. Dr Chirachanchai received a Doctorate in Applied Fine Chemistry from Osaka University in 1995. Dr Chirachanchai's work involves molecular and nanoscale polymer synthesis.

Dr Melinda Cooper is Postdoctoral Fellow in the Sociology Department of the Division of Society, Culture, Media and Philosophy at Macquarie University. Dr Cooper obtained her doctoral degree from the University of Paris VIII in 2001. Dr Cooper's work involves questions of growth, crisis, and limits at the intersection of economic and political theory and the life sciences.

Dr Bruce Cornell is Chief Scientist and Senior Vice President of AMBRI Ltd (an Australian Company pioneering the integration of biotechnology, nanotechnology and electronics with a major focus on the human medical diagnostics market). Dr Cornell obtained his Doctorate in Physics from Monash University in 1974. Dr Cornell's work has focussed on the development of novel diagnostic technologies based on the nanoscale function of biological membranes.

Dr Patricia Coyle is an Honorary Educational Fellow with the Department of Anaesthetics at the Royal Prince Alfred Hospital, Sydney. Dr Coyle received her qualification in medicine from the University of Sydney in 1960. Dr Coyle's work involves medical education and matters relating to appropriate healthcare technology for the South and remote areas more generally, having worked in Uganda, Pakistan, on the Thai-Cambodian border with International Committee of the Red Cross war surgery teams and most recently in East Timor with Medecins Sans Frontieres.

Dr Nares Damrongchai is a Senior Policy Researcher with the Asia Pacific Economic Cooperation Centre for Technology Foresight at Thailand's National Science and Technology Development Agency. Dr Damrongchai received his Doctorate in Engineering from Tokyo Institute of Technology in 1995. Dr Damrongchai's work involves policy studies and technology foresight projects in biotechnology and nanotechnology.

Dr Peter Deutschmann is Executive Director of the Australian International Health Institute and Associate Professor with the School of Population Health at the University of Melbourne. Originally a surgeon by training, Dr Deutschmann received his qualifications in medicine at the University of Melbourne in 1972. Dr Deutschmann's work has involved primary healthcare in international settings for over two decades.

Dr Joydeep Dutta is Associate Professor of Microelectronics at the Asian Institute of Technology. Dr Dutta received his Doctorate in Applied Physics from Calcutta University in 1990. Dr Dutta's work involves nanomaterials and their applications in electronics and biology. Dr Dutta has also been involved in the development of university nanotechnology courses.

Dr Mike Ford is Associate Professor of Nanotechnology, at the University of Technology, Sydney and Associate Director of the Institute for Nanoscale Technology. Dr Ford received his Doctorate in Physics from Southampton University in 1989. Dr Ford's work involves the study of fundamental electronic properties of materials and nanoscale systems and synthesising scanning tunnelling microscope images using quantum chemical methods as well as the development of nanotechnology educational initiatives. Dr Ford co-founded the world's first nanotechnology undergraduate degree at Flinders University, Adelaide, in 2000.

Dr Worsak Kanok-Nukulchai is the Dean of the School of Engineering at the Asian Institute for Technology. Dr Kanok-Nukulchai received a Doctorate in Structural Engineering and Structural Mechanics from the University of California (Berkeley), in 1978. Dr Kanok-Nukulchai's work involves engineering education with a specialisation in nanomechanics.

Mr Mike Lynskey is the Chief Executive Officer of The Fred Hollows Foundation (a non-governmental organisation seeking to eradicate avoidable blindness in Southern countries and improve the health outcomes of Indigenous Australians). Mr Lynskey received postgraduate degrees in History, Politics and Librarianship. Mr Lynskey's work involves development and Indigenous issues, with a focus on healthcare services delivery and capacity building.

Mr Promboon Panitchpakdi is the Executive Director of CARE Thailand/Raks Thai Foundation (a non-governmental organisation working on projects relating to: the environment and natural resources; health and HIV/AIDS; occupation and development; education; and emergency relief). Dr Panitchpakdi's work involves providing emergency relief in the form of food, clothing and medical assistance and projects in agroforestry and conservation, children's education, small enterprise development, and HIV/AIDS-prevention education and assistance to affected families.

Dr Pakdee Pothisiri is the Senior Deputy Permanent Secretary of Health in the Government of Thailand and President of The Pharmacy Council of Thailand. Dr Pothsiri received a Doctorate in Physical Chemistry from the University of Wisconsin and a Doctorate in Public Health Administration from Mahidol University. Dr Pothisiri's work involves public health research. Dr Pothsiri is a member of the WHO Commission on Intellectual Property Rights, Innovation and Public Health.

Dr Benno Radt is a Research Fellow with the Centre for Nanoscience and Nanotechnology at the University of Melbourne. Dr Radt received a Doctorate in Physics, from the Medical University of Luebeck, Germany, in 2003. Dr Radt's work involves the design of nanoparticulate drug delivery systems.

Dr Pinit Ratanakul is Professor of Philosophy and Director of the College of Religious Studies at Mahidol University, Bangkok. Dr Ratanakul received his doctorate in Philosophy at Yale University, c.1970. Dr Ratanakul's work focusses on bioethics from Theravada Buddhist perspectives.

Dr Pathom Sawanpanyalert is the National Professional Officer (Health Systems Development) with the World Health Organisation, Thailand. Dr Sawanpanyalert received a Doctorate in Public Health from Mahidol University in 1995. Dr Sawanpanyalert's work involves public health research particularly in HIV/AIDS prevention.

Dr Michael Selgelid is the Sesqui Lecturer in Bioethics with the Unit for History and Philosophy of Science and the Centre for Values, Ethics and the Law in Medicine at the University of Sydney. Dr Selgelid received a Doctorate in Philosophy from the University of California, San Diego in 2001. Dr Selgelid's work involves: the ethical aspects of eugenics; prenatal diagnosis and selective abortion; genetic enhancement; quality of life assessment; the healthcare situation in the South; social, political and economic causes and consequences of AIDS and other infectious diseases; the history of infectious disease; public health policy; bioterrorism; intellectual property rights in biological materials; drug resistance; and research involving human subjects.

Dr Sirirurg Songsivilai is co-Founder of Innova Biotechnology (a company that specialises in the development and manufacturing of rapid diagnostics for tropical infectious diseases), and Senior Expert at the National Center for Genetic Engineering and Biotechnology. Dr Songsivilai obtained his Medical Degree from Mahidol University in 1986 and Doctorate in Medical Science from Cambridge University in 1990. Dr Songsivilai's work involves molecular biology and genomics of infectious diseases, especially viral hepatitis and melioidosis, focussing on the understanding clinical characteristics from the genomics variations.

Dr Nadda Sriyabhaya is President of the Anti-Tuberculosis Association of Thailand. Dr Sriyabhaya is a qualified Medical Doctor. Dr Sriyabhaya's work involves community health, epidemiology, immunisation and vaccine development, focussed on tuberculosis.

Dr Wiwut Tanthapanichakoon is Founding Director of the Thai National Nanotechnology Center and Professor in Particle

Technology within the Faculty of Engineering at Chulalongkorn University. Dr Thanthapanichakoon received a Doctorate in Chemical Engineering from the University of Texas in 1978. Dr Tanthapanichakoon's work involves is the policy formulation and promotion of nanotechnology in Thailand.

Dr Greg Tegart is Founding Director and Executive Advisor to the APEC Center for Technology Foresight and Distinguished Visiting Fellow at the National Europe Centre at the Australian National University. Dr Tegart received a Doctorate in Metallurgy from the University of Sheffield c.1958. Dr Tegart's work involves conducting Foresight studies across the APEC region in areas such as nanotechnology and DNA diagnostics for human health, as well as consultation in several Australian and South East Asian countries on Foresight and strategic intelligence.

Dr Pairash Thajchayapong is Permanent Secretary of the Thai Ministry of Science and Technology. Dr Thajchayapong received a Doctorate in Electronics and Computer Engineering from the University of Cambridge in 1974. Dr Thajchayapong's work involves science policy and international cooperation in science and technology, with a focus on information technology.

Dr Terry Turney is the Director of the Commonwealth Scientific and Industrial Research Organisation Nanotechnology Centre. Dr Turney received a Doctorate in Chemistry from the Australian National University. Dr Turney's work involves directing emerging science policy and research directions and identifying and developing new market and commercial opportunities for innovative scientific research in advanced manufacturing. Dr Turney is co-Chairman of the Asian Nanotechnology Forum, which includes representatives from 13 economies in the Asian region.

Mr Chris Warris is a Research Officer with the Australian Academy of Science. Mr Warris received degrees in Science and Engineering from the University of Western Australia c.1999. Mr Warris' work involves science and technology policy and, in 2003, he conducted a nanotechnology benchmarking project for Australia (see Warris, 2004).

Dr John Weckert is Professorial Fellow at the Centre for Applied Philosophy and Public Ethics and Professor of Information Technology in the School of Information Studies, Charles Sturt University. Dr Weckert received a Doctorate in Philosophy from the

University of Melbourne. Dr Weckert's work involves information technology, computer ethics and nanotechnology.

Dr Yongyuth Yuthavong is a Senior Researcher at the National Center for Genetic Engineering and Biotechnology and National Science. Dr Yuthavong received a Doctorate in Organic Chemistry from Oxford University in 1969. Dr Yuthavong's work involves the development of antimalarial drugs, drug target interactions, drug resistance, molecular biology of malaria parasites, and the broad issues surrounding science and technology.

Appendix J

Responses from Thai Nanotechnology Practitioners*

Number: 24*

Average Time in the Field: 6 years (range: 1–14 years)

Gender: Male (10); Female (14)

Respondents Highest Level of Qualification: PhDs (91%); postgraduate study, non-PhD (9%)

Sector: academic (87.5%); government (12.5%)

Fields in which Respondents Trained: organic chemistry (1); polymer science and engineering (6); pharmaceutical science (especially drug delivery) (3); micro- and nanoelectronics (1); physics (including condensed matter physics) (2); materials science and engineering (2); chemical engineering (1); medical science (1); chemistry (1); ceramics (1); mechanical engineering (1); not stated (4)

Current Department or Group: physics (1.5); nanotechnology (0.5); petroleum and petrochemicals (2); pharmaceutical (3); polymer science (1); chemical engineering (2); materials engineering (1); electrical engineering (1); chemistry (3); chemical engineering (2); ceramic engineering (3); environmental engineering (1); agricultural and food engineering (1); industrial production (1); not stated (2)

Current Work: nanomaterials (including sensors and ultrathin films) (4); synthesis and applications of metal oxides (1); drug delivery and pharmaceutical science (2.5); microcellulose in nanoscale crystal and cellulose sponge (1); nanoelectronics (3); negative electrode materials for lithium-ion batteries (1); surface modification of polymers and polymeric thin films and polymeric biomaterials (1); membrane synthesis, membrane processes and proton exchange and zeolite membranes for fuel cells (1.5); human papillomavirus screening (1); zeolite catalysis and adsorption (1); manetoresistance and magnetoimpedence (1); ceramic powder processing and application (1); nanofiltration and adsorption of toxic substances (1); rubber technology and biodegradable polymer (1); coating of tooling or mould to prolong life (1); not stated (1)

Research Orientation: basic (4%); applied (37.5%); both (58.5%)

Collaborations with: academia (57%); government (17.5%); private sector (25.5%)

* Including 13 individuals claiming to be working in nanotechnology-related areas.

Bibliography

Acharya, T., Rab, M.A., Singer, P.A. and Daar, A.S. (2005), "Harnessing Genomics to Improve Health in the Eastern Mediterranean Region — An Executive Course in Genomics Policy", Health Research Policy and Systems, vol. 3, no. 1, p. 1.

Adams, T.K. (1998), "Radical Destabilizing Effects of New Technologies", Parameters: U.S. Army War College, vol. 28, no. 3, pp. 99–111.

Al'Afghani, M.M. (2006), "Developing Countries Must be Ready for Nanotechnology", The Jakarta Post, Available: http://old. thejakartapost.com/yesterdaydetail.asp?fileid=20060227.E02.

Alcott, B. (2005), "Jevons' Paradox", Ecological Economics, vol. 54, no. 1, pp. 9–21.

Alcott, B. (2010), "Impact Caps: Why Population, Affluence and Technology Strategies Should be Abandoned", Journal of Cleaner Production, vol. 18, no. 6, pp. 552–560.

Allier, J.M. (2009), "Socially Sustainable Economic De-Growth", Development and Change, vol. 40, no. 6, pp. 1099–1119.

Almeida, C. (2005), "Brazil and Argentina Launch Joint Nanotech Centre", SciDevNet, Available: www.scidev.net/en/news/brazil-and-argentina-launch-joint-nanotech-centre.html.

Almeida, C. (2006), "Brazil Launches Lab for Agricultural Nanotechnology", SciDevNet, Available: www.scidev.net/en/news/brazil-launches-lab-for-agricultural-nanotechnolog.html.

Altmann, J. (2004), "Military Uses of Nanotechnology: Perspectives and Concerns", Security Dialogue, vol. 35, no. 1, pp. 61–79.

Altmann, J. and Gubrud, M.A. (2002), "Risks from Military Uses of Nanotechnology: The Need for Technology Assessment and Preventive Control", in M. Roco and R. Tomellini (eds), Nanotechnology: Revolutionary Opportunities and Societal Implications, European Commission — NSF Report, Luxembourg, pp. 144–148.

Altwaijiri, A.O. (2000), "Thematic Meeting II.2 Science for Development: Introduction", in A.M. Cetto, S. Schneegans and H. Moore (eds), World Conference on Science: Science for the Twenty-First Century. A New Commitment (proceedings), Banson for the UNESCO, London, p. 302.

Alvares, C. (1997), "Science", in W. Sachs (ed.), The Development Dictionary, 6th Edn, Zed Books, London, pp. 219–232.

Anand, S. and Sen, A. (1995), Gender Inequality in Human Development: Theories and Measurement, UNDP, New York, 10 pp.

Andreasson, S. (2005), "Accumulation and Growth to What End? Reassessing the Modern Faith in Progress in the 'Age of Development'", Capitalism Nature Socialism, vol. 16, no. 4, pp. 57–76.

Annas, G.J. and Grodin, M.A. (1998), "Human Rights and Maternal-Fetal HIV Transmission Prevention Trials in Africa", American Journal of Public Health, vol. 88, no. 4, pp. 560–563.

Anton, P.S., Silberglitt, R. and Schneider, J. (eds) (2001), The Global Technology Revolution: Bio/Nano/Materials Trends and Their Synergies with Information Technology by 2015, RAND, Santa Monica, 69 pp.

Arirang News (2003), "Scholars from Two Koreas Meet on Nanotech", English.Chosun.com, Available: http://english.chosun.com/w21data/html/news/200608/200608210030.html.

Arvanitakis, J. (2007), The Cultural Commons of Hope: The Attempt to Commodify the Final Frontier of the Human Experience, Vdm Verlag, Saarbrücken, 276 pp.

Asgar, M.A. (2003), "Our First Step Towards Nanotechnology", The Independent, Available: http://independent-bangladesh.com/news/dec/28/28122005ft.htm.

Asian Technology Information Program (2006), Nanotechnology Infrastructure in China, Asian Technology Information Program, Tokyo, 11 pp.

Aubert, J.-E. (2005), Promoting Innovation in Developing Countries: A Conceptual Framework (World Bank Policy Research Working Paper 3554), World Bank Institute, Washington, 38 pp.

AzoNano (2006), "Laboratory for Biological Effects of Nanomaterials and Nanosafety Established in China", AzoNano, Available: www.azonano.com/News.asp?NewsID=2507.

Bachmann, G. (2000), "Nanotechnology: The Need for Interdisciplinary Cooperation", in J. Lievonen, J.C. Císcar and M. Rader (eds), IPTS-ESTO Techno-Economic Analysis Report 1999–2000, Joint Research Centre, European Commission, Seville, pp. 73–81.

Badran, A. (2000), "Building Capacity and Creativity in Science for Sustainable Development in the South", in A.M. Cetto, S. Schneegans and H. Moore (eds), World Conference on Science: Science for the Twenty-First Century. A New Commitment (proceedings), Banson for the UNESCO, London, pp. 310–312.

Bai, C. (2001), "Progress of Nanoscience and Nanotechnology in China", Journal of Nanoparticle Research, vol. 3, no. 4, pp. 251–256.

Bai, C. (2005), "Ascent of Nanoscience in China", Science, vol. 309, no. 5741, pp. 61–63.

Bai, C. (2008), "Nano Rising", Nature, vol. 456, no. 1, pp. 36–37.

Baran, P.A. (1952), "On the Political Economy of Backwardness", Manchester School of Economics and Social Studies, vol. 20, no. January, pp. 66–84.

Barker, D.K. and Feiner, S. (2004), Liberating Economics: Feminist Perspectives on Families, Work, and Globalization, The University of Michigan Press, Ann Arbor, 193 pp.

Barker, T., Lesnick, M., Mealy, T., Raimond, R., Walker, S., Rejeski, D. and Timberlake, L. (2005), Nanotechnology and the Poor: Opportunities and Risks: Closing the Gaps within and between Sectors of Society, Meridian Institute, Washington, 23 pp.

Barretto, P.M.C. and Rogov, A. (2000), "International Atomic Energy Agency Technology Transfer Facts & Trends. Pillars for Development", IAEA Bulletin, vol. 42, no. 1, pp. 8–16.

Bawa, R. (2007), "Patents and Nanomedicine", Nanomedicine, vol. 2, no. 3, pp. 351–374.

Baya-Laffite, N. and Joly, P.-B. (2008), "Nanotechnology and Society: Where Do We Stand in the Ladder of Citizen Participation?" CIPAST Newsletter, vol. March, p. 35.

Beck, U. (1992), Risk Society: Towards a New Modernity, Sage, London, 260 pp.

Beck, U. (1996), "Risk Society and the Provident State", in S. Lash, B. Szerszynski and B. Wynne (eds), Risk, Environment and Modernity: Towards a New Ecology, Sage, London, pp. 27–43.

Beck, U. (1997), The Reinvention of Politics: Rethinking Modernity in the Global Social Order, Polity Press, Cambridge, U.K., 216 pp.

Beck, U. (1999), World Risk Society, Polity Press, Cambridge, U.K., 184 pp.

Beck, U., Giddens, A. and Lash, S. (1994), Reflexive Modernization: Politics, Tradition and Aesthetics in the Modern Social Order, Polity Press, Cambridge, U.K., 225 pp.

Becker, M. (2001), "At Nanoscale, the Laws of Humans May Not Apply", Small Times, Available: www.smalltimes.com/document_display. cfm?document_id=1798.

Bell, M.M. (2004), An Invitation to Environmental Sociology, 2nd Edn, Pine Forge Press, Thousand Oaks, 325 pp.

Bello, M. (2007), "Nanotechnology and Low-Income Nations", Nanofrontiers, no. 2, p. 10.

Benetar, S.R., Daar, A.S. and Singer, P.A. (2005), "Global Health Challenges: The Need for an Expanded Discourse on Bioethics", PLoS Medicine, vol. 2, no. 7, p. e143.

Bennholdt-Thomsen, V., Faraclas, N. and von Werlhof, C. (eds) (2001), There Is an Alternative: Subsistence and World-Wide Resistance to Corporate Globalization, Zed Books, New York, 240 pp.

Berg, S. (1998), "Snowball Sampling", in S. Kotz and N. L. Johnson (eds), Encyclopedia of Statistical Sciences (Vol. 8), Wiley, New York, pp. 528–532.

Bernard, H.R. (2000), Social Research Methods: Qualitative and Quantitative Approaches, Sage, London, 659 pp.

Berry, W. (2002), "Hope", in A. Kimbrell (ed), The Fatal Harvest Reader, Island Press, Washington, pp. 317–322.

Bhumiratana, S. (1991), "Contribution of Biotechnologies to Sustainable Rural Development in Developing Countries: A Case Study in Thailand", in V. Costarini (ed), Biotechnologies in Perspective: Socio-Economic Implications for Developing Countries, UNESCO, Paris, pp. 155–166.

Biel, R. (2000), The New Imperialism: Crisis and Contradictions in North/ South Relations, Zed Books, London, 355 pp.

Birdsall, N., Rodrik, D. and Subramanian, A. (2005), "How to Help Poor Countries", Foreign Affairs, vol. 84, no. 4, pp. 136–152.

Blowers, A. (1997), "Environmental Policy: Ecological Modernisation or the Risk Society?" Urban Studies, vol. 34, nos. 5–6, pp. 845–871.

Bower, J.L. and Christensen, C.M. (1995), "Disruptive Technologies: Catching the Wave", Harvard Business Review, vol. 73, no. 1, pp. 43–53.

Bowman, D. (2007), "Patently Obvious: Intellectual Property Rights and Nanotechnology", Technology in Society, vol. 29, no. 3, pp. 307–315.

Bowman, D. (2008), "Governing Nanotechnologies: Weaving New Regulatory Webs or Patching up the Old?" Nanoethics, vol. 2, no. 2, pp. 179–181.

Bowman, D. and Hodge, G. (2006), "Nanotechnology: Mapping the Wild Regulatory Frontier", Futures, vol. 38, no. 9, pp. 1060–1073.

Bowman, D. and Hodge, G. (2007a), "Nanotechnology and Public Interest Dialogue: Some International Observations", Bulletin of Science, Technology & Society, vol. 27, no. 2, pp. 118–132.

Bowman, D. and Hodge, G. (2007b), "A Small Matter of Regulation: An International Review of Nanotechnology Regulation", Columbia Science and Technology Law Review, vol. 8, no. 1, pp. 1–32.

Brenner, M., Brown, J. and Canter, D. (eds) (1985), The Research Interview: Uses and Approaches, Academic Press, New York, 276 pp.

Britten, N. (1995), "Qualitative Research: Qualitative Interviews in Medical Research", British Medical Journal, vol. 311, no. 6999, pp. 251–253.

Brower, V. (2002), "Second Class Medicine for Developing Nations?" BioMedNet News, Available: http://www.vaccinationnews.com/DailyNews/December2002/SecondClass6.htm.

Brown, M.M. (2001), "Foreword", in UNDP (ed), Human Development Report: Making New Technologies Work for Human Development, Oxford University Press, New York, pp. iii-iv.

Brugger, B. and Hannan, K. (1983), Modernisation and Revolution, Routledge, London, 60 pp.

Brundtland, G.H. (1987), Our Common Future: Report of the World Commission on Environment and Development, Oxford University Press, Oxford, 383 pp.

Brundtland, G.H. (2002), Seminar on 'Globalisation and Health: The Way Forward'. Available: http://www.who.int/director-general/speeches/2002/english/20020228_globalizationoslo.html.

Bruns, B. (2004), "Applying Nanotechnology to the Challenges of Global Poverty: Strategies for Accessible Abundance", paper presented at 1st Conference on Advanced Nanotechnology: Research, Applications and Policy, Washington, January 15, 2005, Available: http://www.foresight.org/Conferences/AdvNano2004/Abstracts/Bruns/BrunsPaper.pdf.

Brus, L.E. (1984), "Electron–Electron and Electron-Hole Interactions in Small Semiconductor Crystallites: The Size Dependence of the Lowest Excited Electronic State", Journal of Chemical Physics, vol. 80, no. 9, pp. 4403–4409.

Bunnag, S. (2005), "Nano-Engineering Course to Open", Bangkok Post Online, Available: http://www.net-lanna.info/thaiscience/News/ThaiScienceNews/64000085.htm.

Burgi, B.R. and Pradeep, T. (2006), "Societal Implications of Nanoscience and Nanotechnology in Developing Countries", Current Science, vol. 90, no. 5, pp. 645–648.

Bush, V. (1945), Science: The Endless Frontier — A Report to the President on a Program for Postwar Scientific Research (Reprint, 1960), National Science Foundation, Washington, 220 pp.

Business Day Media (2007), "FG Approves Nanotechnology Plan", Business Day Online, Available: www.businessdayonline.com/?c=44&a=15429.

Carroll, J.S. (2001), "Social Science Research Methods for Assessing Societal Implications of Nanotechnology", in M.C. Roco and W.S. Bainbridge (eds), Societal Implications of Nanoscience and Nanotechnology, Springer, New York, pp. 188–192.

Carson, R. (1965), Silent Springs, Penguin Books, Harmondsworth, 317 pp.

Cascio, J. (2004), "Nanotechnology and the Developing World", WorldChanging, Available: www.worldchanging.com/archives/002105.html.

Cascio, J. (2005), "Nanotechnology and the South-South Divide", Available: www.worldchanging.com/archives/003045.html.

Castells, M. (1999), Information Technology, Globalization and Social Development, United Nations Research Institute for Social Development, Geneva, 15 pp.

Cavanagh, J. and Mander, J. (eds) (2004), Alternatives to Economic Globalization: A Better World Is Possible, 2nd Edn, Berrett-Koehler Publishers, San Francisco, 408 pp.

Cecchini, S. (2003), "Tapping ICT to Reduce Poverty in Rural India", IEEE Technology and Society Magazine, vol. 22, no. 2, pp. 20–27.

Cetto, A.M., Schneegans, S. and Moore, H. (eds) (2000), World Conference on Science: Science for the Twenty-First Century. A New Commitment, Banson for the UNESCO, London, 547 pp.

Changsorn, P. (2004), "Firms See Lower Costs, More Profit in Nanotech", The Nation (Thailand), November 22, p. unknown.

Changthavorn, T. (2003), "Bioethics of IPRs: What Does a Thai Buddhist Think?" paper presented at Roundtable Discussion on Bioethical Issues of IPRs, Cambridge, U.K., March 28–29, 2003, Available: http://www.shef.ac.uk/ipgenethics/roundtable/abstracts/TChangthavorn.doc.

Chenery, H. and Syrquin, M. (1975), Patterns of Development, 1950–1970, Oxford University Press for the World Bank, London, 234 pp.

Chennai Interactive Business Services (2007), "Rs 100 Cr Nanopark in Bangalore", Chennai Online, Available: http://chennaionline.com/colnews/newsitem.asp?NEWSID={604163E1-FFF9-46A3-B4C6-78760B13B957}&CATEGORYNAME=TECH.

Chinese Academy of Sciences (2007), "China Kicks Off Study on Bio-Safety of Nanomaterials", News Release, Available: http://english.cas.ac.cn/eng2003/news/detailnewsb.asp?infono=26361.

Choi, H., Kaplan, S., Mody, C.C.M. and Roberts, J.A. (2008), Setting an Agenda for the Social Studies of Nanotechnology: A Summary of the Joint Wharton-Chemical Heritage Foundation Symposium on the Social Studies of Nanotechnology, Chemical Heritage Foundation and Mack Center for Technological Innovation, Philadelphia, 20 pp.

Choi, K. (2003), "Ethical Issues of Nanotechnology Development in the Asia-Pacific Region", in UNESCO (ed), Regional Meeting on Ethics of Science and Technology (proceedings), UNESCO Regional Unit for Social and Human Sciences in Asia and Pacific, Bangkok, pp. 327–376, Available: http://www2.unescobkk.org/elib/publications/ethic_in_asia_pacific/327_376NANOTECH.PDF.

Chow, E. (ed) (2002), Transforming Gender and Development in East Asia, Routledge, New York, 268 pp.

Chrispeels, M.J. (2000), "Biotechnology and the Poor", Plant Physiology, vol. 124, no. 1, pp. 3–6.

Chung, D. (2003), "Nanoparticles Have Health Benefits Too", NewScientist, vol. 179, no. 2410, p. 16.

Cientifica (2007), Halfway to the Trillion Dollar Market? A Critical Review of the Diffusion of Nanotechnologies, Cientifica, London, 10 pp.

Clâemenðcon, R.G. (1990), Perceptions and Interests: Developing Countries and the International Economic System: A Comparative Analysis of Statements Given to the Plenary Meetings of UNCTAD in 1976, 1983, and 1987, Peter Lang, Bern, 306 pp.

Clark, N. (1985), The Political Economy of Science and Technology, Basil Blackwell, New York, 257 pp.

Cleaver, H. (1997), "Socialism", in W. Sachs (ed.), The Development Dictionary, 6th Edn, Zed Books, London, pp. 233–249.

Coalition Against Biopiracy (2004), "Captain Hook Awards 2004", Captain Hook Awards for Biopiracy, Available: http://www.captainhookawards.org/winners/2004_pirates.

Cobb, C., Halstead, T. and Rowe, J., T (1995), The Genuine Progress Indicator: Summary of Data and Methodology, Redefining Progress, San Francisco, 50 pp.

Coleman, W. (1976), "Providence, Capitalism, and Environmental Degradation: English Apologetics in an Era of Economic Revolution", Journal of the History of Ideas, vol. 37, no. 1, pp. 27–44.

Commoner, B. (1971), The Closing Circle: Nature, Man and Technology, Alfred Knopf, New York, 326 pp.

Compañó, R. and Hullman, A. (2002), "Forecasting the Development of Nanotechnology with the Help of Science and Technology Indicators", Nanotechnology, vol. 13, no. 3, pp. 243–247.

Corbett, J., Irwin, G. and Vines, D. (1999), "From Asian Miracle to Asian Crisis: Why Vulnerability, Why Collapse?" in D. Gruen and L. Gower (eds), Capital Flows and the International Financial System, Reserve Bank of Australia, Sydney, pp. 190–213.

CORDIS (2003), "Nanotechnology: Opportunity or Threat?" CORDIS News, Available: http://cordis.europa.eu/fetch?CALLER=EN_NEWS&ACTIO N=D&SESSION=&RCN=20401.

CORDIS (2004), Nanotechnology, Available: http://www.cordis.lu/ nanotechnology/src/sitemap.htm.

CORDIS (2006), "New Project Aims to Draw Women into Nanotechnology", CORDIS News, Available: http://cordis.europa.eu/fetch?CALLER=EN_ NEWS&ACTION=D&SESSION=&RCN=25308.

Cornell, B.A., Braach-Maksvytis, V.L.B., King, L.G., Osman, P.D.J., Raguse, B., Wieczorek, L. and Pace, R.J. (1997), "A Biosensor That Uses Ion-Channel Switches", Nature, vol. 387, no. 6633, pp. 580–583.

Corporate Bureau (2004), "Nanotech Promises Hopes for R&D Cost Cut for Drug Firms", The Financial Express, Available: http://www. financialexpress.com/news/Nanotech-Promises-Hopes-For-R&D-Cost-Cut-For-Drug-Firms/108880/.

Correa, C.M. (1998), "The South-South Dimension in Partnering: Strategic Alliances in the Biotechnology Sector", in UNCTAD (ed), Atas XI Bulletin: New Approaches to Science and Technology Co-Operation and Capacity Building, U.N., New York, pp. 289–299.

Coupe, G. (2004), "Shape of Things to Come", Manufacturing Engineer, vol. 83, no. 5, pp. 30–33.

Court, E., Daar, A.S., Martin, E., Acharya, T. and Singer, P.A. (2004), "Will Prince Charles Et Al Diminish the Opportunities of Developing Countries in Nanotechnology?" Nanotechweb.org, Available: http://nanotechweb. org/cws/article/indepth/18909.

Court, E.B., Daar, A.S., Persad, D.L., Salamanca-Buentello, F. and Singer, P.A. (2005), "Tiny Technologies for the Global Good ", Materials Today, vol. 8, no. 5, Supplement 1, pp. 14–15.

Cresswell, J.W. and Plano Clark, V.L. (2007), Designing and Conducting Mixed Methods Research, Sage, Thousand Oaks, 275 pp.

Crow, L. (1996), "Including All of Our Lives: Renewing the Social Model of Disability", in C. Barnes and G. Mercer (eds), Exploring the Divide, The Disability Press, Leeds, pp. 55–72.

Crow, M.M. and Sarewitz, D. (2001), "Nanotechnology and Societal Transformation", in M.C. Roco and W.S. Bainbridge (eds), Societal Implications of Nanoscience and Nanotechnology, Springer, New York, pp. 55–67.

Daar, A.S., Martin, D.K., Nast, S., Smith, A.C., Singer, P.A. and Thorsteinsdóttir, H. (2002), Top Ten Biotechnologies for Improving Health in Developing Countries, University of Toronto Joint Centre for Bioethics, Toronto, 132 pp.

Daar, A.S., Thorsteinsdóttir, H., Martin, D.K., Smith, A.C., Nast, S. and Singer, P.A. (2002), "Top Ten Biotechnologies for Improving Health in Developing Countries", Nature Genetics, vol. 32, no. 2, pp. 229–232.

Daly, H.E. (1973), Toward a Steady-State Economy, W. H. Freeman & Co, San Francisco, 332 pp.

Daly, H.E. (1977), Steady-State Economics: The Economics of Biophysical Equilibrium and Moral Growth, W. H. Freeman, San Francisco, 185 pp.

Daly, H.E. (2008), A Steady-State Economy: A Failed Growth Economy and a Steady-State Economy are Not the Same Thing; They are the Very Different Alternatives We Face, Sustainable Development Commission, London, 10 pp.

Daly, H.E. and Cobb, J. (1994), For the Common Good: Redirecting the Economy Towards Community, the Environment and a Sustainable Future, Beacon Press, Boston, 534 pp.

DaSilva, E.J. (2002), "GMOs and Development", Electronic Journal of Biotechnology, vol. 14, no. 2, Available: www.ejbiotechnology.info/content/issues/01/index.html.

Davies, G. (2004), Economia, ABC Books, Sydney, 496 pp.

Dayrit, F.M. and Enriquez, E.P. (2001), "Nanotechnology Issues for Developing Economies (Philippines)", in G. Tegart (ed), Nanotechnology the Technology for the 21st Century: Volume II the Full Report, The APEC Centre for Technology Foresight, Bangkok, pp. 115–120.

de Almeida, A.O. (2003), Responses to Questionnaire on Nanotechnology: Brazil, for the Royal Society and Royal Academy of Engineering Report on Nanoscience and Nanotechnologies: Opportunities and Uncertainties, Available: http://www.nanotec.org.uk/evidence/brazil.htm.

de Villiers, M.M., Aramwit, P. and Kwon, G.S. (eds) (2008), Nanotechnology in Drug Delivery, Springer, Berlin, 662 pp.

Department of Education, The State of Queensland (2008), What Is TAFE? Available: www.sqit.tafe.qld.gov.au/international_students/australia/what_is_tafe.html.

Department of Science and Technology, Republic of South Africa (2006), The National Nanotechnology Strategy, Department of Science and Technology, Republic of South Africa, Pretoria, 20 pp.

Dolmo, B.-C. (2001), "Examining Global Access to Essential Pharmaceuticals in the Face of Patent Protection Rights: The South African Example", Buffalo Human Rights Law Review, vol. 7, no. 2001, pp. 137–163.

Domar, E.D. (1946), "Capital Expansion, Rate of Growth, and Employment", Econometrica, vol. 14, no. 2, pp. 137–147.

Dos Santos, T. (1970), "The Structure of Dependence", The American Economic Review, vol. 60, no. 2, pp. 231–236.

Drexler, K.E. (1981), "Molecular Engineering: An Approach to the Development of General Capabilities for Molecular Manipulation", Proceedings of the National Academy of Science, vol. 78, no. 9, pp. 5275–5278.

Drexler, K.E. (1986), Engines of Creation: The Coming Era of Nanotechnology, Doubleday, New York, 320 pp.

Drexler, K.E. (1992), Nanosystems, Wiley and Sons, New York, 576 pp.

Drexler, K.E. (2004), "Nanotechnology: From Feynman to Funding", Bulletin of Science, Technology & Society, vol. 24, no. 1, pp. 21–27.

Dudley, J. P. (1995), "Bioregional Parochialism and Global Activism", Conservation Biology, vol. 9, no. 5, pp.1332–1334.

Dwivedi, K.K. (2004), Questionnaire Response for the International Dialogue on Responsible Research and Development of Nanotechnology, Available: http://www.nanodialogues.org/international.php.

Easterley, W. (2006), The White Man's Burden: Why the West's Efforts to Aid the Rest Have Done So Much Ill and So Little Good, Penguin Press, New York, 436 pp.

Eckaus, R. (1977), Appropriate Technologies for Developing Countries, National Academy of Sciences, Washington, 140 pp.

Economy Bureau (2006), "Tech-Savvy Kalam Advocates Nano Use, Favours Transgenic Route for Farm Research", The Financial Express, Available: http://www.financialexpress.com/news/Techsavvy-Kalam-advocates-nano-use-favours-transgenic-route-for-farm-research/183493/.

Ehrlich, P. (1986), The Population Bomb, Ballantine Books, New York, 201 pp.

Eigler, D.M. and Schweizer, E.K. (1990), "Positioning Single Atoms with a Scanning Tunneling Microscope", Nature, vol. 344, no. 6266, pp. 524–526.

Einsiedel, E.F. and Goldenberg, L. (2004), "Dwarfing the Social? Nanotechnology Lessons from the Biotechnology Front", Bulletin of Science Technology Society, vol. 24, no. 1, pp. 28–33.

El Naschie, M.S. (2006), "Nanotechnology for the Developing World", Chaos, Solitons and Fractals, vol. 30, no. 4, pp. 769–773.

Ellwood, W. (2001), The No-Nonsense Guide to Globalization, Verso, London, 143 pp.

Engdahl, F.W. (2007), "'Doomsday Seed Vault' in the Arctic: Bill Gates, Rockefeller and the GMO Giants Know Something We Don't", GlobalResearch.ca, Available: http://www.globalresearch.ca/index.php?context=va&aid=7529.

Enriquez, J. (2005), As the Future Catches You: How Genomics and Other Forces are Changing Your Life, Work, Health and Wealth, Crown Publishing Group, New York, 272 pp.

Escobar, A. (1991), "Anthropology and the Development Encounter: The Making and Marketing of Development Anthropology", American Ethnologist, vol. 18, no. 4, pp. 658–682.

Escobar, A. (1995), Encountering Development: The Making and the Unmaking of the Third World, Princeton University Press, Princeton, 320 pp.

Escobar, A. (1997), "Planning", in W. Sachs (ed), The Development Dictionary, 6th Edn, Zed Books, London, pp. 132–145.

Escobar, A. (2000), "Beyond the Search for a Paradigm? Post-Development and Beyond", Development, vol. 43, no. 4, pp. 11–14.

Esteban, M., Webersik, C., Leary, D. and Thompson-Pomeroy, D. (2008), Innovation in Responding to Climate Change: Nanotechnology, Ocean Energy and Forestry, United Nations University—Institute of Advanced Studies, Yokohama, 46 pp.

Esteva, G. (1985), "Beware of Participation", Development: Seeds of Change, vol. 3, pp. 77–79.

Esteva, G. (1997), "Development", in W. Sachs (ed.), The Development Dictionary, 6th Edn, Zed Books, London, pp. 6–25.

Esteva, G. (2006), "Development — Walking Beyond: From Promotion to Co-Motion", in S. Pimparé and C. Salzano (eds), Emerging and Re-Emerging Learning Communities: Old Wisdoms and New Initiatives from around the World, UNESCO, Paris, pp. 73–86.

ETC Group (2002), "Patenting Elements of Nature: No Patents on Non-Life Either!" Geno-type, vol. 2002, no. March, pp. 1–4, Available: www.etcgroup.org/en/materials/publications.html?pub_id=220.

ETC Group (2003a), The Big Down: From Genomes to Atoms, ETC Group, Ottawa, 80 pp.

ETC Group (2003b), Nanotech Un-Gooed! Is the Grey/Green Goo Brouhaha the Industry's Second Blunder?, Communiqué No. 80, ETC Group, Ottawa, 8 pp.

ETC Group (2004a), "26 Governments Tiptoe Toward Global Nano Governance: Grey Governance", News Release, Available: http://www.etcgroup.org/en/node/98.

ETC Group (2004b), Down on the Farm: The Impact of Nano-Scale Technologies on Food and Agriculture, ETC Group, Ottawa, 68 pp.

ETC Group (2004c), "Itty-Bitty Ethics: Bioethicists See Quantum Plots in Nanotech Concern...And Quantum Bucks in Buckyball Brouhaha?" Geno-type, vol. 2004, no. February, Available: http://www.etcgroup.org/en/materials/publications.html?pub_id=124.

ETC Group (2004d), "The Precautionary Prince II", News Release, Available: http://www.etcgroup.org/en/materials/publications.html?pub_id=124.

ETC Group (2004e), "Scientists Prepare to Use Nanotechnology to Poison Us All? — Jazzing up Jasmine: Atomically Modified Rice in Asia?" News Release, Available: www.etcgroup.org/en/materials/publications.html?pub_id=117.

ETC Group (2005a), Nanogeopolitics: ETC Group Surveys the Political Landscape, Communiqué No. 89, ETC Group, Ottawa, 48 pp.

ETC Group (2005b), Nanotech's 'Second Nature' Patents: Implications for the Global South, Communiqué No. 87, ETC Group, Ottawa, 36 pp.

ETC Group (2005c), The Potential Impacts of Nano-Scale Technologies on Commodity Markets: The Implications for Commodity Dependent Developing Countries, South Centre, Geneva, 59 pp.

ETC Group (2008), Downsizing Development: An Introduction to Nanoscale Technologies and the Implications for the Global South, UN Non-Governmental Liaison Service, Geneva, 102 pp.

Etkind, J. (2006), "Continent's Nano Revolution", AllAfrica.com, Available: http://allafrica.com/stories/printable/200608180122.html.

European Commission (2004), Opening to the World: International Co-Operation, Available: http://www.cordis.lu/nanotechnology/src/intlcoop.htm.

European Patent Office (2005), European Classification (ECLA), Available: http://ep.espacenet.com/ep/en/helpv3/ecla.html.

Evenson, R.E. and Gollin, D. (2003), "Assessing the Impact of the Green Revolution, 1960 to 2000", Science, vol. 300, no. 5620, pp. 758–762.

Everts, S. (ed) (1998), Gender & Technology: Empowering Women, Engendering Development, Zed Books, London, 171 pp.

Express News Service (2005), "Old Indian Medicine System an Offshoot of Nanotechnology", Lucknow Newsline, Available: http://cities. expressindia.com/fullstory.php?newsid=157803.

Feder, B.J. (2003), "As Uses Grow, Tiny Materials' Safety Is Hard to Pin Down", The New York Times, November 3, Business/Financial Desk, p. C1.

Feenberg, A. (1991), Critical Theory of Technology, Oxford University Press, New York, 256 pp.

Feenberg, A. (2002), Transforming Technology: A Critical Theory Revisited, Oxford University Press, New York, 218 pp.

Feenberg, A. (2004), "Modernity Theory and Technology Studies: Reflections on Bridging the Gap", in T.J. Misa, P. Brey and A. Feenberg (eds), Modernity and Technology, MIT Press, Cambridge, Massachusetts, pp. 73–104.

Feynman, R.P. (1960), "There's Plenty of Room at the Bottom", Engineering and Science, vol. 23, no. 5, pp. 22–36.

Financial Times (2008), "Nanotechnology Institute to Undertake R&D", The Sunday Times, Available: http://sundaytimes.lk/080622/ FinancialTimes/ft323.html.

Finkel, E. (2008), "Australia's New Era for GM Crops", Science, vol. 321, no. 5896, p. 1629.

Finnemore, M. (1997), "Redefining Development at the World Bank", in F. Cooper and R.M. Packard (eds), International Development and the Social Sciences: Essays on the History and Politics of Knowledge, University of California Press, Berkeley, pp. 203–227.

Fishburn, D. and Green, S. (2002), "The World in Figures: Industries", The World in 2003 (The Economist — Special Edition), pp. 101–102.

Fisher, William W. III and Rigamonti, C.P. (2005), The South Africa AIDS Controversy: A Case Study in Patent Law and Policy, Harvard Law School, Boston, 56 pp.

Fisher, D.R. (2004), National Governance and the Global Climate Change Regime, Rowman & Littlefield, Plymouth, 191 pp.

Foladori, G. (2006), "Nanoscience and Nanotechnology in Latin America", Nanowerk Spotlight, Available: www.nanowerk.com/spotlight/spotid=767.php.

Foladori, G. and Invernizzi, N. (2007), "Agriculture and Food Workers Challenge Nanotechnologies", Scitizen, Available: http://scitizen.com/stories/NanoSciences/2007/08/Agriculture-and-Food-Workers-Challenge-Nanotechnologies/.

Foladori, G. and Invernizzi, N. (eds) (2008), Nanotechnologies in Latin America, Rosa Luxembourg Foundation, Berlin, 128 pp.

Foladori, G., Rushton, M. and Zayago Lau, E. (2008), "Nanotechnology for Development or Knowledge Enclaves? The World Bank Case for Latin America", Applications Of Nanotechnology, Available: http://estudiosdeldesarrollo.net/relans/documentos/CJDS-Nano-development.pdf.

Foladori, G. and Zayago Lau, E. (2007), "Tracking Nanotechnology in Mexico", Nanotechnology Law and Business Journal, vol. 4, no. 2, pp. 213–224.

Forbes, K.J. (2000), "A Reassessment of the Relationship between Inequality and Growth", The American Economic Review, vol. 90, no. 4, pp. 869–887.

Ford, N., Wilson, D., Costa Chaves, G., Lotrowska, M. and Kijtiwatchakul, K. (2007), "Sustaining Access to Antiretroviral Therapy in the Less-Developed World: Lessons from Brazil and Thailand", AIDS, vol. 21, no. Suppl 4, pp. S21–S29.

Foster, L. (2002), "The State of the Nano Republic", Nano News, Available: http://www.larta.org/NanoNews/020325_Article_one.asp.

Foucault, M. (1972), The Archaeology of Knowledge, Tavistock Publications, London, 218 pp.

Fountain, J.E. (2000), "Constructing the Information Society: Women, Information Technology, and Design", Technology in Society, vol. 22, no. 1, pp. 45–62.

Frank, A.G. (1966), "The Development of Underdevelopment", Monthly Review, vol. 18, no. 4, pp. 17–31.

Frank, A.G. (1969), Capitalism and Underdevelopment in Latin America, Monthly Review Press, New York, 344 pp.

Frank, A.G. (1972), "The Development of Underdevelopment", in J.D. Cockcroft, A.G. Frank and D. Johnson (eds), Dependence and Underdevelopment, Anchor Books, New York, pp. 1–17.

Fransman, M. (1994), "Biotechnology: Generation, Diffusion and Policy", in C. Cooper (ed), Technology and Innovation in the International Economy, Edward Elgar, Aldershot, pp. 51–88.

Frost, R. (2005), "ISO Launches Work on Nanotechnology Standards", International Organisation for Standardisation News and Media, Available: http://www.iso.org/iso/pressrelease.htm?refid=Ref980.

Fukuda-Parr, S. (ed) (2007), The Gene Revolution: GM Crops and Unequal Development, Earthscan, London, 248 pp.

Fukuda-Parr, S., Lopes, C. and Malik, K. (2002), Capacity for Development: New Solutions to Old Problems, Earthscan, New York, 286 pp.

Fukuyama, F. (1992), The End of History and the Last Man, Hamish Hamilton, London, 418 pp.

Gabriele, A. (2001), Science and Technology Policies, Industrial Reform and Technical Progress in China. Can Socialist Property Rights be Compatible with Technological Catching Up?, UNCTAD, New York, 55 pp.

Galembeck, F. (2003), "Ethical Issues of Nanotechnology", in UNESCO (ed), The United Nations Educational, Scientific and Cultural Organisation World Commission on the Ethics of Scientific Knowledge and Technology (COMEST) 3rd Session (proceedings), UNESCO, Paris, pp. 128–132, Available: http://www.fgq.iqm.unicamp.br/textos.html.

Galli, R.E. (1992), "Winners and Losers in Development and Antidevelopment Theory", in R.E. Galli, L. Rudebeck, K.P. Moseley, F.S. Weaver and L. Bloom (eds), Rethinking the Third World: Contributions toward a New Conceptualization, Crane Russak, New York, pp. 1–27.

Gans, J. and Stern, S. (2003), Assessing Australia's Innovative Capacity in the 21st Century, Intellectual Property Research Institute of Australia, Melbourne, 64 pp.

Garrett, M.J. (1999), Health Futures: A Handbook for Health Professionals, WHO, Geneva, 320 pp.

Gavelin, K., Wilson, R. and Doubleday, R. (2007), Democratic Technologies? The Final Report of the Nanotechnology Engagement Group (NEG), Involve, London, 159 pp.

George, S. (2004), Another World Is Possible If... Verso, London, 268 pp.

Georgescu-Roegen, N. (1971), The Entropy Law and the Economic Process, Harvard University Press, Cambridge, Massachussetts, 457 pp.

Georgescu-Roegen, N. (1977), "The Steady State and Ecological Salvation: A Thermodynamic Analysis", BioScience, vol. 27, no. 4, pp. 266–270.

Gerschenkron, A. (1962), Economic Backwardness in Historical Perspective: A Book of Essays, Belknap Press of Harvard University Press, Cambridge, Massachussetts, 456 pp.

Ghose, T.K. and Ghosh, P. (eds) (2003), Biotechnology in India I, Springer-Verlag, Berlin, 290 pp.

Gibbs, G.R. (2002), Qualitative Data Analysis: Explorations with NVivo, Open University Press, Buckingham, 224 pp.

Gibson, C.C., Andersson, K., Ostrom, E. and Shivakumar, S. (2005), The Samaritan's Dilemma: The Political Economy of Development Aid, Oxford University Press, New York, 264 pp.

Giddens, A. (2006), Sociology, 5th Edn, Polity Press, Cambridge, U.K., 1094 pp.

Gillis, M. (2002), "Overview of Nanotechnology at Rice", paper presented at Symposium on Nanotechnology, Houston, March 21, 2002, Available: http://www.professor.rice.edu/professor/03212002.asp?SnID=2.

Glaser, B.G. and Strauss, A.L. (1967), The Discovery of Grounded Theory: Strategies for Qualitative Research, Aldine Transaction, Chicago, 271 pp.

Global Forum for Health Research (2002), The 10/90 Report on Health Research 2001–2002, Global Forum for Health Research, Geneva, 224 pp.

Goldman, L. and Coussens, C. (2005), Implications of Nanotechnology for Environmental Health Research, Institute of Medicine of the National Academies, The National Academies Press, Washington, 70 pp.

Goldstein, D.J. (1992), "The Impact of Food Production Biotechnology Ownership on Developing Nations", in United Nations Department of Economic and Social Development (ed.), Biotechnology and Development: Expanding the Capacity to Produce Food, U.N., New York, pp. 279–286.

Goldthorpe, J.E. (1996), The Sociology of Post-Colonial Societies: Economic Disparity, Cultural Diversity, and Development, Cambridge University Press, Cambridge, U.K., 279 pp.

Goodman, J. (2003), "Nationalism and Globalism: Social Movement Responses", The International Scope Review, vol. 4, no. 8, pp. 1–17.

Gordon, T.J. (2003), "Is the Future Getting Better?" The Futurist, July-August, pp. 28–30.

Gould, K. (2005), "The Treadmill of Production: The Case of Nanotechnology", paper presented at Development, Governance and Nature Panel Discussion, Cornell University, New York, April 4, 2005, Available: http://www.einaudi.cornell.edu/files/calendar/4951/GouldCornellDGNnanotech.pdf.

Graham, S. (2003), "Nanocontainers Deliver Drugs Directly to Cells", Scientific American, vol. Special Edition, Available: http://www.sciam.

com/article.cfm?chanID=sa002&articleID=0001D485-BEC8-1EA9-BDC0809EC588EEDF&catID=7.

Greene, J.C., Caracelli, V.J. and Graham, W.F. (1989), "Toward a Conceptual Framework for Mixed-Method Evaluation Designs", Educational Evaluation and Policy Analysis, vol. 11, no. 3, pp. 255–274.

Grimshaw, D.J. (2004), Small Is Beautiful: A New Dimension? An Assessment of the Implications of Nanotechnology for Development, Intermediate Technology Development Group, Rugby, 22 pp.

Grimshaw, D.J. (2007), "Is Very Small Still Beautiful? An NGO's View of Nanotechnologies", ISO Focus, vol. April 2007, pp. 34–36.

Grimshaw, D.J. (2008), The Role and Potential of New Technologies in International Development, Practical Action, Rugby, 33 pp.

Grimshaw, D.J., Stilgoe, J. and Gudza, L.D. (2006), Globalisation and the Diffusion of Nanotechnologies to Help the Poor. The Role of New Technologies in Potable Water Provision: A Stakeholder Workshop Approach — Can Nanotechnologies Help Achieve the Millennium Development Target of Halving the Number of People without Access to Clean Water by 2015? Report on the Nano-Dialogues Held in Harare, Zimbabwe, 15, 16 and 22 July 2006, Practical Action, Rugby, 25 pp.

Gross, M. (2003), "The Incredible Shrinking Technology", The Australian Financial Review, January 2, Summer Review, pp. 42–43.

Grossman, J. (2008), "Nanotechnology: Risks, Ethics and Law. Edited by Geoffrey Hunt and Michael Mehta. London: Earthscan, 2006. 296pp. [a Review]", Nanoethics, vol. 2, no. 1, pp. 99–100.

Gryskiewicz, S.S. (1999), Positive Turbulence: Developing Climates for Creativity, Innovation and Renewal, Jossey-Bass, San Francisco, 224 pp.

Gutierrez, O. (1989), "Experimental Techniques for Information Requirements Analysis", Information and Management, vol. 16, no. 1, pp. 31–43.

Haberzettl, C.A. (2002), "Nanomedicine: Destination or Journey?" Nanotechnology, vol. 13, no. 4, pp. R9-R13.

Hallen, P. (1991), "Genetic Engineering: 'Miracle of Deliverance' or 'Destroyer of Worlds'?" in R. Haynes (ed), High Tech: High Co$T? Technology, Society and the Environment, Pan MacMillan, Sydney, pp. 35–46.

Hamdan, H. (2005), "Nanotechnology in Malaysia", paper presented at Expert Group Meeting North-South Dialogue on Nanotechnology: Challenges and Opportunities, Trieste, February 10–12, 2005, Available: http://www.ics.trieste.it/Documents/Downloads/df2676.pdf.

Hancock, G. (1989), Lords of Poverty: The Free-Wheeling Lifestyles, Power, Prestige and Corruption of the Multi-Billion Dollar Aid Business, MacMillan, London, 234 pp.

Hannum, H. (1988), "New Developments in Indigenous Rights", Virginia Journal of International Law, vol. 28, pp. 649–678.

Hanson, J. (2006), "Carpe Diem: Now Is the Time to Consider Nanotechnology's Impact on the Human Condition", Treehugger, vol. 2, no. 3, pp. 1 and 3.

Harper, T. (2002), "How Nanotechnology Can Improve Quality of Life", Nanotechweb.org, Available: http://www.nanotechweb.org/articles/column/1/12/1/1.

Harper, T. (2003a), "Nanotechnology in Kabul? Taking the First Steps", Nanotechweb.org, Available: http://www.nanotechweb.org/articles/column/2/8/2/1.

Harper, T. (2003b), "What Is Nanotechnology?" Nanotechnology, vol. 14, no. 1, p. introduction.

Harrod, R.F. (1948), Toward a Dynamic Economics: Some Recent Developments of Economic Theory and Their Application to Policy, Macmillan, London, 172 pp.

Hassan, M.H.A. (2005), "Nanotechnology: Small Things and Big Changes in the Developing World", Science, vol. 309, no. 5731, pp. 65–66.

Haum, R., Petschow, U. and Steinfeldt, M. (2004), Nanotechnology and Regulation within the Framework of the Precautionary Principle, Institut für ökologische Wirtschaftsforschung (IÖW) gGmbH, Berlin, 44 pp.

Healy, S. (2001), "World Economy: No Technofix for Third World Poor", Green Left Weekly, vol. 460, Available: http://www.greenleft.org.au/node/24427.

Heines, H. (2003), "Patent Trends in Nanotechnology", FindLaw, Available: http://library.findlaw.com/2003/Nov/4/133136.html.

Henderson, A. and Kay Weaver, C. (2007), "Talking 'Facts': Identity and Rationality in Industry Perspectives on Genetic Modification", Discourse Studies, vol. 9, no. 1, pp. 9–41.

Henderson, H. (1991), Paradigms in Progress: Life Beyond Economics, Knowledge Systems, Indianapolis, 293 pp.

Henderson, R. (2002), The Next Technological Revolution: Predicting the Technical Future and Its Impacts on Firms, Organisations and Ourselves, Available: http://mitsloan.mit.edu/50th/tech.pdf.

Hettne, B. (1995), Development Theory and the Three Worlds: Towards an International Political Economy of Development, 2nd Edn, Longman Publishing Group, Harlow, 336 pp.

Hettne, B. (2009), Thinking About Development, Zed Books, London, 152 pp.

Hicks, D.A. (1997), "The Inequality-Adjusted Human Development Index: A Constructive Proposal", World Development, vol. 25, no. 8, pp. 1283–1298.

Hillie, T. and Hlophe, M. (2007), "Nanotechnology and the Challenge of Clean Water", Nature Nanotechnology, vol. 2, no. 11, pp. 663–664.

Hines, C. (2000), Localization: A Global Manifesto, Earthscan, London, 290 pp.

Hodge, G. and Bowman, D. (2004), "Governing Nanotechnology: Setting the Regulatory Agenda", The Journal of Contemporary Issues in Business and Government, vol. 10, no. 2, pp. 18–33.

Hodge, G., Bowman, D. and Ludlow, K. (eds) (2007), New Global Frontiers in Regulation: The Age of Nanotechnology, Edward Elgar, Cheltenham, 422 pp.

Hoet, P.H.M., Brüske-Hohlfeld, I. and Salata, O.V. (2004), "Nanoparticles — Known and Unknown Health Risks", Journal of Nanobiotechnology, vol. 2, no. 12, pp. 12–26.

Hong, L.T. (2006), "Nanotechnology Lab Set up in Vietnam Metro", Thanh Nien News, Available: www.thanhniennews.com/education/?catid=4&newsid=23367.

Hood, E. (2004), "Nanotechnology: Looking as We Leap", Environmental Health Perspectives, vol. 112, no. 13, pp. A741-A749.

Hoogvelt, A.M.M. (1997), Globalisation and the Postcolonial World: The New Political Economy of Development, Macmillan, Basingstoke, 291 pp.

Hook, C.C. (2002), "In Whose Image? The Remaking of Man with Cybernetics and Nanotechnology", Commentary from the Center for Bioethics and Human Dignity, vol. Winter 2002, Available: www.cbhd.org/resources/biotech/hook_2002-winter.htm.

Hopkins, R. (2008), The Transition Handbook: From Oil Dependency to Local Resilience, Green Books, Totnes, 240 pp.

Hornig, S. (1992), "Gender Differences in Responses to News About Science and Technology", Science, Technology and Human Values, vol. 17, no. 4, pp. 532–542.

Hsiao, J.C. and Fong, K. (2004), "Making Big Money from Small Technology", Nature, vol. 428, no. 6979, pp. 218–220.

Huaizhi, Z. and Yuantao, N. (2001), "China's Ancient Gold Drugs", Gold Bulletin, vol. 34, no. 1, pp. 24–29.

Huang, Z., Chen, H., Chen, Z.-K. and Roco, M.C. (2004), "International Nanotechnology Development in 2003: Country, Institution, and Technology Field Analysis Based on USPTO Patent Database", Journal of Nanoparticle Research, vol. 6, no. 4, pp. 325–354.

Hulme, D. and Shepherd, A. (2003), "Conceptualizing Chronic Poverty", World Development, vol. 31, no. 3, pp. 403–423.

Hussey, J. and Roger, H. (1997), Business Research: A Practical Guide for Undergraduate and Postgraduate Students, MacMillan Press, London, 357 pp.

Iijima, S. (1991), "Helical Microtubules of Graphitic Carbon", Nature, vol. 354, no. 6348, pp. 56–58.

Intarakamnerd, P., Chairatana, P.-A. and Tangchitpiboon, T. (2002), "National Innovation Systems in Less Successful Developing Countries: The Case of Thailand", Research Policy, vol. 31, nos. 8–9, pp. 1445–1457.

InterAcademy Council (2004), Inventing a Better Future: A Strategy for Building Worldwide Capacities in Science and Technology, InterAcademy Council, Amsterdam, 144 pp.

Interagency Working Group on Nano Science, Engineering, and Technology (2000), National Nanotechnology Initiative — Leading to the Next Industrial Revolution, Available: http://nano.gov/.

Intermediate Technology Development Group (2002), "How to Make Technology Transfer Work for Human Development", paper presented at The People's Global Forum, Johannesburg, South Africa, August 29, 2002, Available: http://practicalaction.org/docs/advocacy/technology_transfer.pdf.

International Labour Organisation (1976), Employment, Growth and Basic Needs: A One-World Problem — Report of the Director-General of the International Labour Office, International Labour Office, Geneva, 177 pp.

Invernizzi, N. and Foladori, G. (2005), "Nanotechnology and the Developing World: Will Nanotechnology Overcome Poverty or Widen Disparities?" Nanotechnology Law and Business Journal, vol. 2, no. 3, pp. 101–110.

Invernizzi, N. and Foladori, G. (2006), "Nanomedicine, Poverty and Development", Development, vol. 49, no. 4, pp. 114–118.

Invernizzi, N. and Foladori, G. (2007), "Nanotechnology for Developing Countries. Asking the Wrong Question", in G. Banse, A. Grunwald,

I. Hronszky and G. Nelson (eds), Assessing Societal Implications of Converging Technological Development, Sigma, Berlin, pp. 229–238.

Invernizzi, N., Foladori, G. and Maclurcan, D.C. (2008), "Nanotechnology's Controversial Role for the South", Science, Technology and Society, vol. 13, no. 1, pp. 123–148.

Invest Australia (2007), Nanotechnology Australian Capability Report, Commonwealth of Australia, Canberra, 96 pp.

Islamic Republic News Agency (2006), "President Calls for Setting up of a National Nanotechnology Organ", Islamic Republic News Agency, Available: http://www4.irna.ir/en/news/view/menu-236/0607158657171656.htm.

ITAR-TASS News Agency (2007), "Russia to Design Nanotechnology Weapons — Commander", ITAR-TASS News, Available: www.itar-tass.com/eng/level2.html?NewsID=12063058&PageNum=0.

Jackson, T. (2009), "Beyond the Growth Economy", Journal of Industrial Ecology, vol. 13, no. 4, pp. 487–490.

Jalali, A. (2000), "Science, Development and Globalisation", in A.M. Cetto, S. Schneegans and H. Moore (eds), World Conference On Science: Science For The Twenty-first Century. A New Commitment (proceedings), Banson for the UNESCO, London, pp. 303–306.

James, P. (1997), "Postdependency? The Third World in an Era of Globalism and Late-Capitalism", Alternatives, vol. 22, no. 2, pp. 205–226.

Jamison, A. (2009), "Can Nanotechnology Be Just? On Nanotechnology and the Emerging Movement for Global Justice", Nanoethics, vol. 3, no. 2, pp. 129–136.

Jihui, Q., Tisue, T. and Volkoff, A. (2000), "Atoms for Peace: Targeting Technical Cooperation for Results", IAEA Bulletin, vol. 42, no. 1, pp. 2–7.

Johnson, K.L., Raybould, A.F., Hudson, M.D. and Poppy, G.M. (2007), "How Does Scientific Risk Assessment of GM Crops Fit within the Wider Risk Analysis?" Trends in Plant Science, vol. 12, no. 1, pp. 1–5.

Johnson, R.B. and Onwuegbuzie, A.J. (2004), "Mixed Methods Research: A Research Paradigm Whose Time Has Come ", Educational Researcher, vol. 33, no. 7, pp. 14–26.

Juma, C. (ed) (2005), Going for Growth: Science, Technology and Innovation in Africa, Syngenta, London, 130 pp.

Juma, C. and Yee-Cheong, L. (2005), Innovation: Applying Knowledge in Development, UN Millennium Project Task Force on Science, Technology and Innovation, Earthscan, London, 194 pp.

Kachonpadungkitti, C. and Macer, D.R.J. (2004), "Attitudes to Bioethics and Biotechnology in Thailand (1993–2000), and Impacts on Employment", Eubios Journal of Asian and International Bioethics, vol. 14, no. 2004, pp. 118–134.

Kalam, A.A.P.J. (2004), "Our Future Lies in Nanotechnology", Hindustan Times, Available: http://www.hindustantimes.com/news/181_921748,00300006.htm.

Kalam, A.A.P.J. (2006), "President's Address at the Inauguration of the Indo-US Nanotechnology Conclave", Press Information Bureau Press Release, Available: http://pib.nic.in/release/rel_print_page1.asp?relid=15766.

Kapoor, I. (2004), "Hyper-Self-Reflexive Development? Spivak on Representing the Third World 'Other'", Third World Quarterly, vol. 25, no. 4, pp. 627–647.

Kearnes, M., Macnaghten, P. and Wynne, B. (2005), "Nanotechnology, Governance and Public Deliberation. What Role for the Social Sciences?" Science Communication, vol. 27, no. 2, pp. 268–291.

Kerr, W.A., Hobbs, J.E. and Yampoin, R. (1991), "Intellectual Property Protection, Biotechnology, and Developing Countries: Will the TRIPs Be Effective?" AgBioForum, vol. 2, nos. 3 and 4, pp. 203–211.

Kerschner, C. (2010), "Economic De-Growth vs. Steady-State Economy", Journal of Cleaner Production, vol. 18, no. 6, pp. 544–551.

Khan, H. (1979), World Economic Development: 1979 and Beyond, Croom Helm, London, 519 pp.

Kilby, P. (2007), "The Australian Aid Program: Dealing with Poverty?" Australian Journal of International Affairs, vol. 61, no. 1, pp. 114–129.

Kim, P., Eng, T.R., Deering, M.J. and Maxfield, A. (1999), "Published Criteria for Evaluating Health Related Web Sites: Review", British Medical Journal, vol. 318, no. 7184, pp. 647–649.

Kimbrell, A. (ed) (2002), Fatal Harvest: The Tragedy of Industrial Agriculture, Foundation for Deep Ecology and Island Press, Sausalito, 396 pp.

Klein, N. (2007), The Shock Doctrine: The Rise of Disaster Capitalism, Henry Holt & Company, New York, 558 pp.

Klotzko, A. (2003), "The Great Leap Forward", Pathways: The Novartis Journal, vol. July/September 2003, pp. 5–6.

Knowledge Ecology International (2008), Thailand's Compulsory Licensing Controversy, Available: http://www.keionline.org/index.php?option=com_content&task=view&id=90.

Kolm, J.E. (1988), "Regional and National Consequences of Globalizing Industries of the Pacific Rim", in J.H. Muroyama and H.G. Stever (eds),

Globalization of Technology: International Perspectives: Proceedings of the Sixth Convocation of the Council of Academies of Engineering and Technological Sciences, National Academy Press, Washington, pp. 106–140.

Korten, D.C. (1990), Getting to the 21st Century: Voluntary Action and the Global Agenda, Kumarian Press, West Hartford, 270 pp.

Korten, D.C. (2006), The Great Turning: From Empire to Earth Community, Berrett-Koehler, San Francisco, 402 pp.

Kroto, H., Heath, J., O'Brien, S., Curl, R. and Smalley, R. (1985), "C60: Buckminsterfullerene", Nature, vol. 318, no. 6042, pp. 162–163.

Kulshrestha, T. (2006), "Nano World: India Goes the Nano Tech Way..." The Financial Express, Available: www.financialexpress.com/news/India-goes-the-nano-tech-way.

Kumar, A., Nair, A.G.C., Reddy, A.V.R. and Garg, A.N. (2006), "Availability of Essential Elements in Bhasmas: Analysis of Ayurvedic Metallic Preparations by INAA", Journal of Radioanalytical and Nuclear Chemistry, vol. 270, no. 1, pp. 173–180.

Kumar, N. and Siddharthan, N.S. (1997), Technology, Market Structure, and Internationalization: Issues and Policies for Developing Countries, Routledge published in association with the United Nations University Press, New York, 165 pp.

Kumaraswamy, M. and Shrestha, G. (2002), "Targeting 'Technology Exchange' for Faster Organisational and Industry Development", Building Research & Information, vol. 30, no. 3, pp. 183–195.

la Prairie, H. (2005), "Is Organic Agriculture a Possible Solution to World Hunger/FAO Proposal", in H.M. Gupta (ed), Organic Farming and Sustainable Agriculture, ABD Publishers, Jaipur, pp. 110–112.

Lalor, G.C. (2000), "Science for Development: The Approach of a Small Island State", in A.M. Cetto, S. Schneegans and H. Moore (eds), World Conference On Science: Science For The Twenty-first Century. A New Commitment (proceedings), Banson for the UNESCO, London, pp. 307–310.

Latham-Koenig, A. (1974), "Intermediate Technologies for True Development", in OECD (ed), Choice and Adaption of Technology in Developing Countries: An Overview of Major Policy Issues, Development Centre of the OECD, Paris, pp. 171–172.

Latouche, S. (1993), In the Wake of the Affluent Society: An Exploration of Post-Development, Zed Books, London, 256 pp.

Latouche, S. (2004), "Why Less Should Be So Much More: Degrowth Economics", Le Monde Diplomatique, Available: http://mondediplo. com/2004/11/14latouche.

Latouche, S. (2009), Farewell to Growth, English Edn, Polity Press, Cambridge, U.K., 124 pp.

LaVan, D.A. and Langer, R. (2001), "Implications of Nanotechnology in the Pharmaceutics and Medical Fields", in M.C. Roco and W.S. Bainbridge (eds), Societal Implications of Nanoscience and Nanotechnology: NSET Workshop Report, Edited Workshop Report, National Science Foundation, Arlington, pp. 79–83.

le Roux, A. (2004), Search Engine Yearbook, Pandecta Magazine, San Francisco, 300 pp.

Leach, M. and Scoones, I. (2006), The Slow Race: Making Technology Work for the Poor, Demos, Upstream, London, 81 pp.

Leahy, S. (2004), "Development: 'Nano Divide' No Small Matter", IPS News, Available: http://www.ipsnews.net/interna.asp?idnews=22193.

Lee-Chua, Q.N. (2003), "Nanotechnology", Inquirer News Service, Available: http://www.inq7.net/inf/2003/jun/25/inf_24-1.htm.

Lee, K., Fustukian, S. and Buse, K. (2002), "An Introduction to Global Health Policy", in K. Lee, K. Buse and S. Fustukian (eds), Health Policy in a Globalising World, Cambridge University Press, Cambridge, U.K., pp. 3–17.

Lee, K. and McInness, C. (2003), Health, Foreign Policy and Security: A Discussion Paper, The Nuffield Trust, London, 57 pp.

Leitch, S. and Davenport, S. (2007), "Strategic Ambiguity as a Discourse Practice: The Role of Keywords in the Discourse on 'Sustainable' Biotechnology", Discourse Studies, vol. 9, no. 1, pp. 43–61.

Leite, J.R. (2004), Questionnaire Response for the International Dialogue on Responsible Research and Development of Nanotechnology, Available: http://www.nanodialogues.org/international.php.

Lemie, M. (2005), "Brazil to Invest US$30 Million in Nanotech by 2006", SciDevNet, Available: www.scidev.net/en/news/brazil-to-invest-us30-million-in-nanotech-by-2006.html.

Leonard-Barton, D. (1995), Wellsprings of Knowledge: Building and Sustaining the Sources of Innovation, Harvard University Press, Boston, 334 pp.

Lewis, W.A. (1954), "Economic Development with Unlimited Supplies of Labour", The Manchester School, vol. 22, no. 2, pp. 139–191.

Li Lin, L. (2000), "Capacity Building in Developing Countries to Facilitate the Implementation of the Cartagena Protocol on Biosafety", paper presented at Fifth Conference of the Parties of the UN Convention on Biological Diversity, Nairobi, May 15–26, 2000, Available: http://www.biosafety-info.net/article.php?aid=8.

Liao, N. (2009), "Combining Instrumental and Contextual Approaches: Nanotechnology and Sustainable Development", Journal of Law, Medicine and Ethics, vol. 37, no. 4, pp. 781–789.

Lin-Liu, J. (2003), "Thailand's Leader Plants the Seeds for a Future in Nanobiotech", Small Times, Available: www.smalltimes.com/document_display.cfm?document_id=5588.

Liu, L. (2003), "Current Status of Nanotech in Thailand", Asia Pacific Nanotechnology Weekly, vol. 1, no. 19, pp. 1–4.

Liu, L. (2004), "Asia Pacific Nanotechnology R&D and Commercialization Efforts", Nanotechnology Law and Business Journal, vol. 1, no. 1, pp. 104–114.

Lloyd, S.M., Lave, L.B. and Matthews, H.S. (2005), "Life Cycle Benefits of Using Nanotechnology to Stabilize Platinum-Group Metal Particles in Automotive Catalysts", Environmental Science and Technology, vol. 39, no. 5, pp. 1384–1392.

Lopes, C. (2002), "Ownership", in S. Fukuda-Parr, C. Lopes and K. Malik (eds), Capacity for Development: New Solutions to Old Problems, Earthscan, London, pp. 121–146.

Lovgren, S. (2003), "Scientists Crack SARS Genetic Sequence", National Geographic News, Available: http://news.nationalgeographic.com/news/2003/04/0415_030415_sarsdna.html.

Lovy, H. (2003), "Do They Know It's Nanotime at All?" NanoBot, Available: http://nanobot.blogspot.com/2003/08/do-they-know-its-nanotime-at-all.html.

Lowe, I. (2009), A Big Fix, 2nd Edn, Black Inc., Melbourne, 116 pp.

Lucas Jr, Henry C., and Sylla, R. (2003), "The Global Impact of the Internet: Widening the Economic Gap between Wealthy and Poor Nations?" Prometheus, vol. 21, no. 1, pp. 3–22.

Lummis, C.D. (1997), "Equality", in W. Sachs (ed), The Development Dictionary, 6th Edn, Zed Books, London, pp. 38–52.

Lundvall, B.-A. (ed) (1992), National Systems of Innovation: Towards a Theory of Innovation and Interactive Learning, Pinter, London, 342 pp.

Lux Research (2003), The Nanotech Report 2003, Lux Research Inc., New York.

Lux Research (2005), Why Big Pharma Is Missing the Nanotech Opportunity, Available: http://biz.yahoo.com/prnews/050215/ntu142_1.html.

Maclurcan, D. (2006), "Molecular Manufacturing and the Developing World: Looking to Nanotechnology for Answers", Nanotechnology Perceptions, vol. 2, no. 1b, pp. 137–141.

Majoni, T. (2006), "Nanotech Debate 'Must Involve Poor Communities'", SciDevNet, Available: www.scidev.net/en/news/nanotech-debate-must-involve-poor-communities.html.

Malaysian National News Agency (2006a), "M'sia to Build Nanotechnology Centre This Year", Bernama.com, Available: www.bernama.com.my/bernama/v3/news_business.php?id=198977.

Malaysian National News Agency (2006b), "Malaysia to Invest in Nanotechnology", Bernama.com, Available: www.bernama.com.my/bernama/state_news/news.php?id=194024&cat=ct.

Malaysian National News Agency (2006c), "Malaysia to Upgrade Six Nanotechnology Centres, Says Najib", Bernama.com, Available: www.bernama.com.my/bernama/v3/news.php?id=220728.

Malaysian National News Agency (2006d), "National Nanotechnology Centre to Be Set up under 9MP", Bernama.com, Available: www.bernama.com.my/bernama/v3/printable.php?id=185498.

Malaysian National News Agency (2007), "Nano Particles Not Harmful to Users", Bernama.com, Available: http://www.bernama.com.my/bernama/v3/news.php?id=284449.

Malsch, I. (2002a), "Biomedical Applications of Nanotechnology", The Industrial Physicist, vol. June/July, pp. 15–17.

Malsch, I. (2002), "Constructive Nanotechnology Assessment", in M.C. Roco and R. Tomellini (eds), 3rd Joint EC-NSF Workshop on Nanotechnology (proceedings), European Communities, Luxembourg, pp. 176–179.

Malsch, I. (2008), "Nanotechnology Solutions for Development Problems. Small Is Beautiful?" The Broker, vol. 1, no. 6, pp. 20–23, Available: www.thebrokeronline.eu/en/articles/small_is_beautiful.

Mander, J. (ed) (2007), Manifesto on Global Economic Transitions, International Forum on Globalization and the Institute for Policy Studies, San Francisco, 36 pp.

Mani, S. (2004), "Government Innovation and Technology Policy: An International Comparative Analysis", International Journal of Technology and Globalisation, vol. 1, no. 1, pp. 29–44.

Mantell, K. (2003), "Developing Nations 'Must Wise up to Nanotechnology'", SciDevNet, Available: http://www.scidevnet/News/index.cfm?fuseaction=readNews&itemid=992&language.

Manzo, K. (1991), "Modernist Discourse and the Crisis of Development Theory", Studies in Comparative International Development, vol. 26, no. 2, pp. 3–36.

Marinova, D. and McAleer, M. (2002), "Nano-Technology Patenting in the USA", in A.E. Rizzoli and A.J. Jakeman (eds), Integrated Assessment and Decision Support: Proceedings of the First Biennial Meeting of the International Environmental Modelling and Software Society (proceedings), iEMSs, Manno, pp. 574–579, Available: http://www.iemss.org/iemss2002/proceedings/.

Marinova, D. and McAleer, M. (2003), "Nanotechnology Strength Indicators: International Rankings Based on US Patents", Nanotechnology, vol. 14, no. 1, pp. R1-R7.

Markert, L.R. (1996), "Gender Related to Success in Science and Technology", Journal of Technology Studies, vol. 22, no. 2, pp. 21–29.

Martens, B. and Saretzki, T. (1993), "Conferences and Courses on Biotechnology. Describing Scientific Communication by Exploratory Methods", Scientometrics, vol. 27, no. 3, pp. 237–260.

Masinda, M.T. (1998), National Systems of Innovation: Implications on Science and Technology Policies in Sub-Saharan Africa, Centre for Policy Research on Science and Technology, Simon Fraser University, Vancouver, pp. 1–8.

Mason, J. (2003), "Nanoelectronics Starts Unfolding a Long and Winding Road-Map", Small Times, Available: www.smalltimes.com/articles/stm_print_screen.cfm?ARTICLE_ID=269094.

Maugh II, T.H. (1996), "Worldwide Study Finds Big Shift in Causes of Death", Los Angeles Times, Available: http://www.aegis.com/news/lt/1996/LT960902.html.

Mauro, F. and Hardison, P.D. (2000), "Traditional Knowledge of Indigenous and Local Communities: International Debate and Policy Initiatives", Ecological Applications, vol. 10, no. 5, pp. 1263–1269.

Max-Neef, M.A. (1991), Human Scale Development: Conception, Application and Further Reflection, The Apex Press, New York, 114 pp.

Maynard, A.D. (2007), "Nanotechnologies: Overview and Issues", in P.P. Simeonova, N. Opopol and M.I. Luster (eds), Nanotechnology — Toxicological Issues and Environmental Safety and Environmental Safety, Springer-Verlag, New York, pp. 1–14.

Maynard, A.D., Aitken, R.J., Butz, T., Colvin, V., Donaldson, K., Oberdörster, G., Philbert, M.A., Ryan, J., Seaton, A., Stone, V., Tinkle, S.S., Tran, L., Walker, N.J. and Warheit, D.B. (2006), "Safe Handling of Nanotechnology", Nature, vol. 444, no. 7117, pp. 267269.

McArthur, J.W. and Sachs, J.D. (2002), "The Growth Competitiveness Index: Measuring Technological Advancement and the Stages of Development", in M.E. Porter, J.D. Sachs, K. Schwab, J. McArthur and P. Cornelius (eds), The Global Competitiveness Report 2001–2002, World Economic Forum, New York, pp. 28–51.

McKibben, B. (2007), Deep Economy: The Wealth of Communities and the Durable Future, Revised Edn, Times Books, New York, 272 pp.

McRobie, G. (1981), Small Is Possible, Jonathon Cape, London, 352 pp.

Meadows, D.H., Meadows, D.I., Randers, J. and Behrens, W. W. III, (1972), The Limits to Growth: A Report for the Club of Rome's Project on the Predicament of Mankind, Earth Island Limited, London, 205 pp.

Mee, W., Lovel, R., Solomon, F., Kearnes, A. and Cameron, F. (2004), Nanotechnology: The Bendigo Workshop Report, CSIRO Minerals, Melbourne, 30 pp.

Mehta, M.D. (2004), "From Biotechnology to Nanotechnology: What Can We Learn from Earlier Technologies?" Bulletin of Science, Technology & Society, vol. 24, no. 1, pp. 34–39.

Meier, G.M. and Seers, D. (eds) (1984), Pioneers in Development, Oxford University Press for the World Bank, New York, 372 pp.

Meléndez-Ortiz, R. and Sánchez, V. (eds) (2005), Trading in Genes: Development Perspectives on Biotechnology, Trade, and Sustainability, Earthscan, London, 294 pp.

Meridian Institute (2004), Report of the International Dialogue on Responsible Research and Development of Nanotechnology, National Science Foundation, Arlington, 50 pp.

Meridian Institute (2005), Nanotechnology and the Poor: Opportunities and Risks — Meeting Summary of the Steering Group Meeting, Meridian Institute, Washington, 37 pp.

Meridian Institute (2006), Workshop on Nanotechnology, Water and Development: Workshop Summary, Meridian Institute, Washington, 21 pp.

Meridian Institute (2007), Nanotechnology, Commodities and Development: Background Paper for the International Workshop on Nanotechnology, Commodities and Development, Meridian Institute, Washington, 41 pp.

Mill, J.S. ([1848] 2001), Principles of Political Economy with Some of Their Applications to Social Philosophy, Batoche, Kitchener, 1136 pp.

Miller, G. (2008), "Nanotechnology and the Public Interest: Repeating the Mistakes of GM Foods?" International Journal of Technology Transfer and Commercialisation, vol. 7, nos. 2–3, pp. 274–280.

Miller, G. and Senjen, R. (2008), Out of the Laboratory and on to Our Plates: Nanotechnology in Food & Agriculture, Friends of the Earth Australia, Melbourne, 62 pp.

Miller, H.I. and Conko, G. (2000), "The Science of Biotechnology Meets the Politics of Global Regulation", Issues in Science and Technology, vol. 17, no. 1, pp. 47–54.

Ministério Das Relaçõs Exteriores (2003), Titulo: Nanotechnology R&D: Sweating the Small Stuff, Available: http://www.mre.gov.br/portugues/noticiario/internacional/selecao_detalhe.asp?ID_RESENHA=4139.

Mittal, A. (2006), "Food Security: Empty Promises of Technological Solutions", Development, vol. 49, no. 4, pp. 33–38.

Mnyusiwalla, A., Daar, A.S. and Singer, P.A. (2003), "'Mind the Gap': Science and Ethics in Nanotechnology", Nanotechnology, vol. 14, no. 3, pp. R9-R13.

Mol, A.P.J. (2001), Globalization and Environmental Reform: The Ecological Modernization of the Global Economy, MIT Press, Cambridge, Massachussetts, 273 pp.

Monceri, F. (2004), "Western Supremacy: The Triumph of an Idea? by Sophie Bessis [Translated by Patrick Camiller] (Book Review)", Political Studies Review, vol. 2, no. 1, p. 130.

Moniruzzaman, M. and Winey, K.I. (2006), "Polymer Nanocomposites Containing Carbon Nanotubes", Macromolecules, vol. 39, no. 16, pp. 5194–5205.

Montague, P. (2004), "Welcome to Nanoworld: Nanotechnology and the Precautionary Principle Imperative", Multinational Monitor, vol. 25, no. 9, pp. 16–19.

Mooney, P. (1999), "The ETC Century: Erosion, Technological Transformation and Corporate Concentration in the 21st Century", Development Dialogue, vol. 1999, nos. 1–2, pp. 1–128.

Mooney, P. (2003a), "Science out of Context? No, a Moratorium Would Put Nano in Proper Context", Smalltimes, vol. 3, no. 2, pp. 8 and 55.

Mooney, P. (2003b), "Small is Dangerous: The Threat of Nano-Technology", paper presented at Is Small Beautiful? A One-Day Conference on the 30th Anniversary of Small is Beautiful by E.F. Schumacher, London, September 3, 2003, Available: http://practicalaction.org/?id=is_small_beautiful_mooney.

Moore Jr, B. (1966), Social Origins of Dictatorship and Democracy: Lord and Peasant in the Making of the Modern World, Beacon Press, Boston, 559 pp.

Morgan, D.L. (1998), "Practical Strategies for Combining Qualitative and Quantitative Method: Applications to Health Research", Qualitative Health Research, vol. 8, no. 3, pp. 362–376.

Morrison, S. (2003), "The Emerging Nanotech Industry; Lessons from Biotech Experience", Analyst Showdown — News and Views on Nanobiotech, Available: http://www.nanobioconvergence.org/files/sMorrison.pdf.

Morse, J.M. (1991), "Approaches to Qualitative-Quantitative Methodological Triangulation", Nursing Research, vol. 40, no. 1, pp. 120–123.

Mowshowitz, A. (1976), The Conquest of Will: Information Processing in Human Affairs, Addison Wesley Longman, Reading, 365 pp.

Moyo, D. (2009), Dead Aid, Farrar, Straus and Giroux, New York, 208 pp.

Munasinghe, M., "Development, Equity and Sustainability (DES) in the Context of Climate Change", in M. Munasinghe and R. Swart (eds), Climate Change and its Linkages with Development, Equity and Sustainability: Proceedings of the Intergovernmental Panel on Climate Change Expert Meeting (proceedings), Intergovernmental Panel on Climate Change, Geneva, pp. 13–66.

Munshi, D., Kurian, P., Bartlett, R.V. and Lakhtakia, A. (2007), "A Map of the Nanoworld: Sizing up the Science, Politics, and Business of the Infinitesimal", Futures, vol. 39, no. 4, pp. 432–452.

Murphy-Medley, D. (2001), "Exportation of Risk: The Case of Bhopal", Onlineethics.org, Available: www.onlineethics.org/CMS/enviro/EECS/Bhopal.aspx.

Najera, J.A. (1989), "Malaria and the Work of the WHO", Bulletin of the World Health Organization, vol. 67, no. 3, pp. 229–243.

National Cancer Institute, U.S. National Institutes of Health (2004), National Cancer Institute Announces Major Commitment to Nanotechnology for Cancer Research, Available: http://www.nci.nih.gov/newscenter/pressreleases/nanotechPressRelease.

National Nanotechnology Initiative (2003), What Is Nanotechnology? Available: http://www.nano.gov/html/facts/whatIsNano.html.

National Science and Technology Council (2000), Newly Released NSTC Reports Provide a Coherent Vision for the Future of Nanotechnology Research & Development, Available: http://www.ostp.gov/html/00_124.html.

National Science and Technology Council, Committee on Technology and Subcommittee on Nanoscale Science, Engineering and Technology (2002), National Nanotechnology Initiative: The Initiative and Its

Implementation Plan, Office of Science and Technology Policy, U.S. Government, Washington, 142 pp.

National Science Foundation Sri Lanka (2002), "Cutting-Edge Technology and Developing Countries", Techwatch Lanka, vol. 2, no. 2, p. 1.

Nature Nanotechnology (2006), "Nan´O·Tech·Nol´O·Gy N." Nature Nanotechnology, vol. 1, no. 1, pp. 8–10.

NDCHealth Corporation (2005), "NDCHealth Announces 2004 U.S. Pharmaceutical Industry Top 10 Rankings", PRNewswire, Available: http://www.prnewswire.com/cgi-bin/micro_stories. pl?ACCT=603950&TICK=NDC&STORY=/www/story/03-07-2005/00 03154658&EDATE=Mar+7,+2005.

Nemets, A. (2004), "China's Nanotech Revolution", Association for Asia Research, Available: www.asianresearch.org/articles/2260.html.

Neuman, W.L. (1997), Social Research Methods: Qualitative and Quantitative Approaches, 3rd Edn, Allyn and Bacon, Needham Heights, 592 pp.

Norberg-Hodge (2000), Shifting Direction: From Global Dependence to Local Interdependence, International Society for Ecology and Culture, London, 16 pp.

North, D.C. (1981), Structure and Change in Economic History, W. W. Norton & Company, New York, 240 pp.

Nuffield Council on Bioethics (2004), The Use of Genetically-Modified Crops in Developing Countries: A Follow-up Discussion Paper, Nuffield Council on Bioethics, London, 144 pp.

Nussbaum, M.C. (2003), "Capabilities as Fundamental Entitlements: Sen and Social Justice", Feminist Economics, vol. 9, nos. 2–3, pp. 33–59.

Oberdörster, G., Oberdörster, E. and Oberdörster, J. (2005), "Nanotoxicology: An Emerging Discipline Evolving from Studies of Ultrafine Particles", Environmental Health Perspectives, vol. 113, no. 7, pp. 823–839.

Oberdorster, G., Sharp, Z., Atudorei, V., Elder, A., Gelein, R. and Kreyling, W. (2004), "Translocation of Inhaled Ultrafine Particles to the Brain", Inhalation Toxicology, vol. 16, nos. 6–7, pp. 437–445.

OECD (2003), DAC List of Aid Recipients — As at 1 January 2003, Available: www.oecd.org/dataoecd/35/9/2488552.pdf.

Ogundiran, T.O. (2005), "Africa Must Come on Board the Genomics Bandwagon", Genomics, Society and Policy, vol. 1, no. 3, pp. 66–77.

Olssen, M.R. (2003), The Construction of the Meaning and Significance of an 'Author' among Information Behaviour Researchers: A Social Constructivist Approach, PhD thesis, University of Technology Sydney, 314 pp.

Oman, C. (2000), Policy Competition for Foreign Direct Investment: A Study of Competition among Governments to Attract FDI, OECD Publishing, Paris, 130 pp.

Osama, A. (2006), "ICT for Development: Hope or Hype?" SciDevNet, Available: www.scidev.net/en/opinions/ict-for-development-hope-or-hype.html.

Paarlberg, R.L. (2003), "Reinvigorating Genetically Modified Crops", Issues in Science and Technology, vol. 19, no. 3, pp. 86–92.

Packard, R.M. (1997), "Visions of Postwar Health and Development and Their Impact on Public Health Interventions in the Developing World", in F. Cooper and R. Packard (eds), International Development and the Social Sciences: Essays on the History and Politics of Knowledge, University of California Press, Berkeley, pp. 93–118.

Padma, T.V. (2006), "Indian Government Says Science Needs Rural Focus", SciDevNet, Available: www.scidev.net/en/news/indian-government-says-science-needs-rural-focus.html.

Padma, T.V. (2007), "India 'Must Regulate Nanotechnology' Urgently", SciDevNet, Available: www.scidev.net/en/news/india-must-regulate-nanotechnology-urgently.html.

Palmberg, C. (2007), Modes, Challenges and Outcomes of Nanotechnology Transfer — A Comparative Analysis of University and Company Researchers, The Research Institute of the Finnish Economy, Helsinki, 33 pp.

Pan American Health Organisation (1999), "Methodological Summaries: Measuring Inequity in Health", Epidemiological Bulletin, vol. 20, no. 1, p. 11.

Pandey, S. (2005), Patenting Nanotechnology in India, National Law University, Jodhpur, 4 pp.

Pang, T., Lansang, M.A. and Haines, A. (2002), "Brain Drain and Health Professionals", British Medical Journal, vol. 324, no. 7336, pp. 499–500.

Panyakeow, S. and Aungkavattana, P. (2002), "Nanotechnology Status in Thailand", in G. Tegart (ed), Nanotechnology the Technology for the 21st Century: Vol. II the Full Report, APEC Center for Technology Foresight, Bangkok, pp. 163–168.

Patil, R. (2005), "If Tomorrow Comes", The Indian Express, Available: http://www.indianexpress.com/full_story.php?content_id=62323.

Paull, J. and Lyons, K. (2008), "Nanotechnology: The Next Challenge for Organics", Journal of Organic Systems, vol. 3, no. 1, pp. 3–22.

Payne, G. and Payne, J. (2004), Key Concepts in Social Research, Sage, London, 242 pp.

Peet, R. and Hartwick, E. (1999), Theories of Development, 2nd Edn, The Guilford Press, New York, 234 pp.

Perera, S.C. (2006), "Government of Sri Lanka Plans to Introduce Nanotechnology to Industrialists", Asian Tribune, Available: www. asiantribune.com/index.php?q=node/2483.

Perkins, J. (2004), Confessions of an Economic Hit Man, Plume, New York, 303 pp.

Peters, S. and Page, P. (2003), "Building Bonds across the Ocean", EQuad News, vol. 15, no. 3, Available: http://www.princeton.edu/~seasweb/ eqnews/spring03/feature1.html.

Peterson, C. (2003), Molecular Manufacturing: Societal Implications of Advanced Nanotechnology, U.S. House of Representatives Committee on Science (Hearing), U.S. House of Representatives, Washington, April 9, 2003, Available: http://stajano.deis.unibo.it/UP2006/11FP/ Nanotech/Peterson.pdf.

Petras, J.F. and Morley, M.H. (1990), US Hegemony under Siege: Class, Politics, and Development in Latin America, Verso, London, 258 pp.

Phillips, J.R. (1988), "Research Blenders", Nursing Science Quarterly, vol. 1, no. 1, pp. 4–5.

Pieterse, J.N. (1998), "My Paradigm or Yours? Alternative Development, Post-Development, Reflexive Development", Development and Change, vol. 29, no. 2, pp. 343–373.

Pieterse, J.N. (2000), "After Post-Development", Third World Quarterly, vol. 21, no. 2, pp. 175–191.

Pieterse, J.N. (2001), Development Theory: Deconstructions/ Reconstructions, Sage, London, 195 pp.

Pleyers, G. (2009), "World Social Forum 2009: A Generation's Challenge", OpenDemocracy, Available: http://www.opendemocracy.net/article/ world-social-forum-2009-a-generation-s-challenge.

Polimeni, J.M., Mayumi, K., Giampietro, M. and Alcott, B. (2008), The Jevons Paradox and the Myth of Resource Efficiency Improvements, Earthscan, London, 184 pp.

Political Bureau (2004), "'India to Tie up with Israel, US for E-Warfare Systems'", The Financial Express, Available: http://www. financialexpress.com/news/India-To-Tie-Up-With-Israel-US-For- Ewarfare-Systems/114308/.

Potoaçnik, J. and Ezin, J.-P. (2008), "Africa-EU Partnership — Science Is Part of the Solution!" AllAfrica.com, Available: www.allafrica.com/ stories/200810080127.html.

Powell, K. (2004), "Green Groups Baulk at Joining Nanotechnology Talks", Nature, vol. 432, no. 7013, p. 5.

Pratap, R. (2005), "Engaging Private Enterprise in Nanotech Research in India", paper presented at Expert Group Meeting North-South Dialogue on Nanotechnology: Challenges and Opportunities, Trieste, February 10–12, 2005, Available: http://www.ics.trieste.it/Documents/ Downloads/df2684.pdf.

Prebisch, R. (1959), "Commercial Policy in the Underdeveloped Countries", The American Economic Review, vol. 49, no. 2, pp. 251–273.

President's Council of Advisors on Science and Technology (2005), The National Nanotechnology Initiative at Five Years: Assessment and Recommendations of the National Nanotechnology Advisory Panel, Office of Science and Technology Policy, U.S. Government, Washington, 50 pp.

Press Trust of India (2006), "India Cannot Afford to Miss Revolution in Nanotechnology: Rao", National Science, Available: www.outlookindia. com/pti_news.asp?id=346054.

Priestly, B.G., Harford, A.J. and Sim, M.R. (2007), "Nanotechnology: A Promising New Technology — But How Safe?" Medical Journal of Australia, vol. 186, no. 4, pp. 187–188.

Prime, R. (2002), Vedic Ecology: Practical Wisdom for Surviving the 21st Century, Mandala Publishing, Novato, 160 pp.

Rader, R.A. (1990), "Trends in Biotechnology Patenting", paper presented at Division of Chemistry and Law, American Chemical Society National Meeting, Washington, August 26–31, 1990, Available: http://www. bioinfo.com/patrev.html.

Radoševic, S. (1999), International Technology Transfer and Catch-Up in Economic Development, Edward Elgar Publishing, Cheltenham, 284 pp.

Rahnema, M. (1997), "Participation", in W. Sachs (ed), The Development Dictionary, 6th Edn, Zed Books, London, pp. 116–131.

Rajvanshi, A.K., "Rocket Science for Rural Development: Ideas for Implementation", in National Collegiate Inventors and Innovators Alliance (ed), NCIIA 10th Annual Conference (proceedings), National Collegiate Inventors and Innovators Alliance, Hadley, pp. 291–294, Available: http://nariphaltan.virtualave.net/rocketscience.pdf

Ramachandran, R. (2006), "Token Gesture for Nanoscience", Frontline, vol. 23, no. 13, pp. 113–116.

Raskin, P., Banuri, T., Gallopin, G., Gutman, P., Hammond, A., Kates, R. and Swart, R. (2002), Great Transition: The Promise and Lure of the Times Ahead, Stockholm Environment Institute, Boston, 111 pp.

Ratner, M.A. and Ratner, D. (2002), Nanotechnology: A Gentle Introduction to the Next Big Idea, Prentice Hall, Upper Saddle River, 208 pp.

Reed Exhibitions Japan (2008), 7th International Bio Forum and Bio Expo Japan, Available: http://www.bio-expo.jp/english/.

Rees, W. (1990), "The Ecology of Sustainable Development", The Ecologist, vol. 20, no. 1, pp. 18–23.

Reid, A., Wood, L.N., Smith, G.H. and Petocz, P. (2005), "Intention, Approach and Outcome: University Mathematics Student's Conceptions of Learning Mathematics", International Journal of Science and Mathematics Education, vol. 3, no. 4, pp. 567–584.

Reid, A.J. (1996), "What We Want: Qualitative Research — Promising Frontier for Family Medicine", Canadian Family Physician, vol. 42, no. March, pp. 387–389.

Reuters (2005), "China Sets up Nanotech Standards Panel", AlertNet, Available: http://www.alertnet.org/thenews/newsdesk/SP29175.htm.

Revaprasadu, N. (2003), "Nanotechnology — The Next Big Thing Is Very, Very Small", Science in Africa: Africa's First On-Line Science Magazine, Available: www.scienceinafrica.co.za/2003/november/nanotech.htm.

Rezaie, R., Frew, S.E., Sammut, S.M., Maliakkal, M.R., Daar, A.S. and Singer, P.A. (2008), "Brazilian Health Biotech — Fostering Crosstalk between Public and Private Sectors", Nature Biotechnology, vol. 26, no. 6, pp. 627–644.

Rip, A. (2006), "The Tension between Fiction and Precaution in Nanotechnology", in E. Fisher, J. Jones and R. von Schomberg (eds), Implementing the Precautionary Principle: Perspectives and Prospects, Edward Elgar Publishing, Cheltenham, pp. 270–283.

Rip, A. and Laredo, P. (2008), "Knowledge, Research and Innovation Systems and Developing Countries", paper presented at Globelics: 6th International Conference, Mexico City, September 22–24, 2008, Available: http://globelics_conference2008.xoc.uam.mx/abstracts.html.

Rist, G. (2002), The History of Development: From Western Origins to Global Faith, Revised Edn, Zed Books, New York, 320 pp.

Roach, S. (2006), "Nano-Herbicide in the Works", Food: Productiondaily.com, Available: www.foodproductiondaily.com/Processing/Nano-herbicide-in-the-works.

Robertson, I. (1991), "Will Biotechnologies Be a Threat or an Opportunity for the South? A Report on the Current Status and Future Targets for Biotechnology-Aided Development in Africa, in Particular Zimbabwe", in A. Sasson and V. Costarini (eds), Biotechnologies in Perspective: Socio-Economic Implications for Developing Countries, UNESCO, Paris, pp. 123–128.

Roco, M.C. (2001), "International Strategy for Nanotechnology Research and Development", Journal of Nanoparticle Research, vol. 3, nos. 5–6, pp. 353–360.

Roco, M.C. (2002), "National Nanotechnology Initiative and a Global Perspective", paper presented at 'Small Wonders', Exploring the Vast Potential of Nanoscience, Washington, 19 March, 2002, Available: http://www.nano.gov/html/res/smwonder_slide.pdf.

Roco, M.C. (2003), "Government Nanotechnology Funding: An International Outlook", JOM: Journal of the Minerals, Metals and Materials Society, vol. 54, no. 9, pp. 22–23.

Roco, M.C. (2004), "National Nanotechnology Initiative: Planning for the Next Five Years", in H.-J. Fecht and M. Werner (eds), The Nano-Micro Interface: Bridging the Micro and Nano Worlds, Wiley-VCH, Weinheim, pp. 3–9.

Roco, M.C. (2007), "National Nanotechnology Initiative — Past, Present, Future", in W.A. Goddard III, D.W. Brenner, S.E. Lyshevski and G.J. Iafrate (eds), Handbook of Nanoscience, Engineering, and Technology, CRC Press, Boca Raton, pp. (3)1–28.

Roco, M.C. and Bainbridge, W.S. (eds) (2001), Societal Implications of Nanoscience and Nanotechnology, Springer, New York, 384 pp.

Roco, M.C. and Bainbridge, W.S. (2003), Converging Technologies for Improving Human Performance: Nanotechnology, Biotechnology, Information Technology and Cognitive Science, Kluwer Academic Publishers, Dordrecht, 467 pp.

Roco, M.C. and Bainbridge, W.S. (2005a), Nanotechnology: Societal Implications — Maximising Benefits for Humanity (Report of the National Nanotechnology Initiative Workshop, December 2–3, 2003), National Science Foundation, Arlington, 120 pp.

Roco, M.C. and Bainbridge, W.S. (2005b), "Societal Implications of Nanoscience and Nanotechnology: Maximizing Human Benefit", Journal of Nanoparticle Research, vol. 7, no. 1, pp. 1–13.

Rosset, P., Collins, J., Lappé, F.M. and Luis, E. (1998), World Hunger: Twelve Myths, 2nd Edn, Earthscan, New York, 224 pp.

Rostow, W. (1960), "The Five Stages of Growth — A Summary", in W. Rostow (ed), The Stages of Economic Growth: A Non-Communist Manifesto, Cambridge University Press, Cambridge, U.K., pp. 4–16.

Runge, C.F. and Ryan, B. (2004), The Global Diffusion of Plant Biotechnology: International Adoption and Research in 2004, University of Minnesota, Minnesota, 117 pp.

Rutherford, D.G. and O'Fallon, M.J. (2007), Hotel Management and Operations, 4th Edn, Wiley and Sons, Hoboken, 478 pp.

Rye Olsen, G. (1995), "North-South Relations in the Process of Change: The Significance of International Civil Society", The European Journal of Development Research, vol. 7, no. 2, pp. 233–256.

Sachs, J.D. (2005), The End of Poverty: Economic Possibilities for Our Time, Penguin Press, New York, 416 pp.

Sachs, W. (ed) (1997a), The Development Dictionary, 6th Edn, Zed Books, London, 306 pp.

Sachs, W. (1997b), "Environment", in W. Sachs (ed), The Development Dictionary, Zed, London, pp. 26–37.

Sachs, W. (1997c), "Introduction", in W. Sachs (ed), The Development Dictionary, 6th Edn, Zed Books, London, pp. 1–5.

Sagasti, F.R. (1980), "The Two Civilizations and the Process of Development", Prospects: Quarterly Review of Education, vol. 10, no. 2, pp. 123–140.

Sahai, S. (1999), "Biotechnology Capacity of LDCs in the Asian Pacific Rim", AgBioForum, vol. 2, nos. 3 and 4, pp. 189–197.

Salam, A. (1991), "Notes on Science, Technology and Science Education in the Development of the South", Minerva, vol. 29, no. 1, pp. 90–108.

Salamanca-Buentello, F., Persad, D.L., Court, E.B., Martin, D.K., Daar, A.S. and Singer, P.A. (2005), "Nanotechnology and the Developing World", PLoS Medicine, vol. 2, no. 4, pp. 300–303.

Sale, J.E.M., Lohfeld, L.H. and Brazil, K. (2002), "Revisiting the Quantitative-Qualitative Debate: Implications for Mixed-Methods Research", Quality and Quantity, vol. 36, no. 1, pp. 43–53.

Salganik, M.J. and Heckathorn, D. D. (2004), "Sampling and Estimation in Hidden Populations Using Respondent-Driven Sampling", Sociological Methodology, vol. 34, no. 1, pp. 193–240.

Salleh, A. (2006), "'Organised Irresponsibility': Contradictions in the Australian Government's Strategy for GM Regulation", Environmental Politics, vol. 15, no. 3, pp. 399–416.

Salleh, A. (ed) (2009), Eco-Sufficiency and Global Justice, Pluto Press, London, 324 pp.

Salvarezza, R.C. (2003), "Why Is Nanotechnology Important for Developing Countries?" in Third Session of the World Commission on the Ethics of Scientific Knowledge and Technology (COMEST) (proceedings), UNESCO, Paris, pp. 133–136.

Sametband, R. (2005), "Ten-Year Nanotechnology Plan Proposed in Argentina", SciDevNet, Available: www.scidev.net/en/news/tenyear-nanotechnology-plan-proposed-in-argentina.html.

Samson, A.E.S. and Symington, A. (2004), "Gender Equality and New Technologies: Nanotechnology", Association for Women's Rights in Development (AWID) Fact Sheet, Available: www.awid.org/publications/primers/nanotech_en.pdf.

Sandelowski, M., Voils, C.I. and Barroso, J. (2006), "Defining and Designing Mixed Research Synthesis Studies", Research in the Schools, vol. 13, no. 1, pp. 29–40.

Sandhu, A. (2008), "Thailand Resorts to Nanotech", Nature Nanotechnology, vol. 3, no. 8, pp. 450–451.

Sass, J., Simms, P. and Negin, E. (2006), "Nanotechnologies: The Promise and the Peril", Sustainable Development Law and Policy, vol. 6, no. 3, pp. 11–14.

Sawahel, W. (2008), "IP Model Proposed for North-South Nanotechnology Divide", Intellectual Property Watch, Available: www.ip-watch.org/weblog/index.php?p=1318.

Saxl, O. (2003), "Nanotechnology — What it Means for the Life Sciences", in World Markets Research Centre (ed), Business Briefing: Life Sciences Technology, World Markets Research Centre, London, pp. 82–86.

Sbert, J.M. (1997), "Planning", in W. Sachs (ed), The Development Dictionary, 6th Edn, Zed Books, London, pp. 132–145.

Schnaiberg, A. (1980), The Environment: From Surplus to Scarcity, Oxford University Press, New York, 464 pp.

Schumacher, E.F. (1973), Small is Beautiful: A Study of Economics as if People Mattered, Blond & Briggs, London, 255 pp.

Schumacher, E.F. (1981), "E.F. Schumacher on Technology for a Democratic Society", in G. McRobie (ed), Small Is Possible, Jonathon Cape, London, pp. 1–15.

Schummer, J. (2007), "The Impact of Nanotechnologies on Developing Countries", in F. Allhoff, P. Lin, J. Moor and J. Weckert (eds), Nanoethics: The Ethical and Social Implications of Nanotechnology, Wiley and Sons, Hoboken, pp. 291–307.

Scialabba, N.E.-H. (2007), Organic Agriculture and Food Security, Food and Agriculture Organisation of the United Nations, Rome, 22 pp.

Sclove, R.E. (1995), Democracy and Technology, The Guilford Press, New York, 338 pp.

Scott, A. (2003), Nanotechnology and Nanoscience (Response to the Royal Society and Royal Academy of Engineering Report on Nanoscience and Nanotechnologies: Opportunities and Uncertainties), Available: http://www.nanotec.org.uk/evidence/77aAndrewScott.htm.

Scott, J. (2002), "New Technologies 'Central to Sustainable Development'", SciDevNet, Available: http://www.scidev.net/News/index.cfm?fuseac tion=readnews&itemid=163&language=1.

Scrinis, G. (2004), "Mega Fear over Something Nano", The Age, Available: http://www.theage.com.au/news/Opinion/Mega-fear-over-something-nano/2004/12/28/1103996550425.html.

Scrinis, G. and Lyons, K. (2007), "The Emerging Nano-Corporate Paradigm: Nanotechnology and the Transformation of Nature, Food and Agri-Food Systems", International Journal of Sociology of Food and Agriculture, vol. 15, no. 2, pp. 22–44.

Seidman, I. (1998), Interviews as Qualitative Research: A Guide for Researchers in Education and the Social Sciences, Teachers College Press, New York, 143 pp.

Selin, C. (2007), "Expectations and the Emergence of Nanotechnology", Science, Technology and Human Values, vol. 32, no. 2, pp. 196–220.

Sen, A. (1976), "Poverty: An Ordinal Approach to Measurement", Econometrica, vol. 44, no. 2, pp. 219–231.

Sen, A. (1999), Development as Freedom, Oxford University Press, London, 366 pp.

Senjen, R. (2006), "Small Nano + Large Corporations = Giant Profits", Chain Reaction, vol. 97, no. June, pp. 32–34.

Service, R.F. (2004), "Nanotech Forum Aims to Head Off Replay of Past Blunders", Science, vol. 306, no. 5698, p. 955.

Shallis, M. (1984), The Silicon Idol: The Micro Revolution and its Social Implications, Schocken Books, New York, 188 pp.

Shanahan, M. (2004), "Nanotech 'Threatens Markets for Poor Nations' Goods'", SciDevNet, Available: www.scidev.net/en/news/nanotech-threatens-markets-for-poor-nations-good.html.

Shand, H. (2003), "New Enclosures: Why Civil Societies and Governments Need to Look Beyond Life Patenting", The New Centennial Review, vol. 3, no. 2, pp. 187–204.

Shiva, V. (1989), Staying Alive: Women, Ecology and Development, Zed Books, London, 121 pp.

Shiva, V. (1992), The Violence of the Green Revolution: Third World Agriculture, Ecology and Politics, Zed Books, London, 164 pp.

Shiva, V. (1993), "Biodiversity and Intellectual Property Rights", in R. Nader (ed), The Case against 'Free Trade': GATT, NAFTA, and the Globalization of Corporate Power, Earth Island Press, San Francisco, pp. 108–120.

Shiva, V. (1997), Biopiracy: The Plunder of Nature and Knowledge, South End Press, Boston, 148 pp.

Shiva, V. (2001), "Special Report: Golden Rice and Neem: Biopatents and the Appropriation of Women's Environmental Knowledge", Women's Studies Quarterly, vol. 29, nos. 1 and 2, pp. 12–23.

Shiva, V. (2002a), "The 'Golden Rice' Hoax — When Public Relations Replaces Science", in M. Ruse and D. Castle (eds), Genetically Modified Foods: Debating Biotechnology (Contemporary Issues), Prometheus Books, Amherst, pp. 58–62.

Shiva, V. (2002b), Water Wars: Privatization, Pollution and Profit, South End Press, Cambridge, Massachussetts, 156 pp.

Shiva, V. (2003), pers.comm., Sydney, October 20, 2003.

Shiva, V. (2006), Earth Democracy: Justice, Sustainability and Peace, Zed Books, London, 224 pp.

Shiva, V. (2008), "The Food Emergency and Food Myths", Seedling, vol. July 2008, pp. 10–12, Available: www.grain.org/seedling/?id=552.

Shiva, V. and Holla-Bhar, R. (1996), "Piracy by Patent: The Case of the Neem Tree", in J. Mander and E. Goldsmith (eds), The Case of the Global Economy: And for a Turn toward the Local, Sierra Club, San Francisco, pp. 146–159.

Shiva, V., Jafri, A.H., Bedi, G. and Holla-Bhar, R. (eds) (1997), The Enclosure and Recovery of the Commons: Biodiversity, Indigenous Knowledge and Intellectual Property Rights, Research Foundation for Science, Technology and Ecology, New Delhi, 182 pp.

Siegel, R.W., Hu, E. and Roco, M.C. (eds) (1999a), WTEC Panel Report on Nanostructure Science and Technology: R&D Status and Trends in Nanoparticles, Nanostructured Materials, and Nanodevices, Kluwer Academic Publishers, Boston, 335 pp.

Siegel, R.W., Hu, E.H. and Roco, M.C. (1999b), Nanostructure Science and Technology: A Worldwide Study, World Technology Evaluation Center, National Science and Technology Council, U.S. Government, Washington, 363 pp.

Simms, A., Johnson, V. and Edwards, M. (2009), Other Worlds are Possible, New Economics Foundation, London, 64 pp.

Singer, P.A. (2003), Bridging the Gap between Ethics and Science in Nanotechnology, Available: http://www.uottawa.ca/vr-recherche-research/frontier/bridging-e.html.

Singer, P.A., Berndtson, K., Tracy, C.S., Cohen, E.R.M., Masum, H., Lavery, J.V. and Daar, A.S. (2007), "Neglected Diseases: A Tough Transition", Nature, vol. 449, no. 7159, pp. 160–163.

Singer, P.A., Salamanca-Buentello, F. and Daar, A.S. (2005), "Harnessing Nanotechnology to Improve Global Equity", Issues in Science and Technology, vol. 21, no. 4, pp. 57–64.

Singh, R.J. (2005a), "The Nano Science of Kajal", The Indian Express, Available: www.indianexpress.com/res/web/ple/full_story.php?content_id=83723.

Singh, R.J. (2005b), "There's Need to Convert Nanotechnology from Science into Technology", Lucknow Newsline, Available: http://cities.expressindia.com/fullstory.php?newsid=160539.

Singleton, D. (2003), "Engineers Can Help Alleviate Poverty", Engineers Australia, vol. 75, no. 5, pp. 40–43.

Slater, D. (2004), Geopolitics and the Post-Colonial: Rethinking North-South Relations, Wiley-Blackwell, Hoboken, New Jersey, 286 pp.

Small Times (2005), "Taking Nano to the Needy: A Small Times Q&A with Fabio Salamanca-Buentello", Small Times, Available: http://www.electroiq.com/index/display/semiconductors-article-display/270094/articles/small-times/profiles/2005/06/taking-nano-to-the-needy-a-small-times-qa-with-fabio-salamanca-buentello.html.

Smith, J. and Wakeford, T. (2003), "Who's in Control?" The Ecologist, vol. 33, no. 4, pp. 40–43.

So, A. (1990), Social Change and Development: Modernisation, Dependency and World-System Theories, Sage, Beverley Hills, 288 pp.

Solow, R.M. (1956), "A Contribution to the Theory of Economic Growth", The Quarterly Journal of Economics, vol. 70, no. 1, pp. 65–94.

Solow, R.M. (1957), "Technical Change and the Aggregate Production Function", The Review of Economics and Statistics, vol. 39, no. 3, pp. 312–320.

Srivastava, N. and Chowdhury, N. (2008), "Regulation of Health Related Nano Applications in India: Exploring the Limitations of the Current Regulatory Design", Social Science Research Network Working Paper Series, Available: http://papers.ssrn.com/sol3/papers.cfm?abstract_id=1105685.

Staff Reporter (2005), "'India Becoming a Pioneer in Nano Technology'", The Hindu: Online Edition of India's National Newspaper, Available: www. hindu.com/2005/03/25/stories/2005032518320300.htm.

Stallings, B. (ed) (1995), Global Change, Regional Response: The New International Context of Development, Cambridge University Press, Cambridge, U.K., 410 pp.

Steer, A. and Lutz, E. (1994), "Measuring Environmentally Sustainable Development", in I. Serageldin and A.D. Steer (eds), Making Development Sustainable: From Concepts to Action, World Bank Publications, Washington, p. 40.

Sterckx, S. (2004), "Patents and Access to Drugs in Developing Countries: An Ethical Analysis", Developing World Bioethics, vol. 4, no. 1, pp. 58–75.

Stern, N. (2009), A Blueprint for a Safer Planet: How to Manage Climate Change and Create a New Era of Progress and Prosperity, Bodley Head, London, 246 pp.

Strand, R. (2001), ELSA Studies of Nanoscience and Nanotechnology, Available: ftp://ftp.cordis.lu/pub/nanotechnology/docs/nanostag-elsa.pdf.

Streeten, P., Burki, S.J., Ul Haq, M., Hicks, N. and Stewart, F. (eds) (1981), First Things First: Meeting Basic Human Needs in the Developing Countries, Oxford University Press, New York, 206 pp.

Subramanian, A. (2004), "Medicines, Patents and TRIPs: Has the Intellectual Property Pact Opened a Pandora's Box for the Pharmaceuticals Industry?" Finance and Development, vol. 41, no. 1, pp. 22–25.

Sutharoj, P. (2005), "Nanotechnology: Ten-Year Plan for Asean Leadership", The Nation, Available: http://www.nationmultimedia. com/2005/06/27/byteline/index.php?news=byteline_17840267. html.

Swaminathan, M.S. (2002), Message from Prof M S Swaminathan President, Pugwash Conferences on Science and World Affairs on the Occasion of the World Science Day for Peace and Development, Available: http:// www.unesco.org/pao/events/india.htm.

Swan (1956), "Economic Growth and Capital Accumulation", Economic Record, vol. 32, no. 2, pp. 334–361.

Swiss Re (2004), Nanotechnology: Small Matter, Many Unknowns, Swiss Reinsurance Company, Zurich, 57 pp.

Szirmai, A. (2005), Dynamics of Socio-Economic Development: An Introduction, Cambridge University Press, Cambridge, U.K., 711 pp.

TajaNews (2004), "India Intends to Usher New Green Revolution with GM Crops", TajaNews, Available: www.tajanews.com/noqnews/nnqview. php?ArtID=4004.

Taniguchi, N. (1974), "On the Basic Concept of Nanotechnology", in International Conference of Production Engineering Part II (proceedings), Japan Society of Precision Engineering, Tokyo, pp. 18–23.

Tanthapanichakoon, W. (2005), "An Overview of Nanotechnology in Thailand", KONA: Powder and Particle, no. 23, pp. 64–68.

Tashakkori, A. and Teddlie, C. (2003), Handbook of Mixed Methods in Social & Behavioral Research, Sage, Thousand Oaks, 768 pp.

Tegart, G. (2001), Nanotechnology the Technology for the 21st Century: Volume II the Full Report, APEC Center for Technology Foresight, Bangkok, 194 pp.

Tegart, G. (2002), "Nanotechnology: The Challenge for the Educators", ATSE Focus, vol. 124, no. November/December 2002, Available: http://www.atse.org.au/publications/focus/focus-tegart5.htm.

Telfer, D.J. (2002), "The Evolution of Tourism and Development Theory", in R. Sharpley and D.J. Telfer (eds), Tourism and Development: Concepts and Issues, Channel View Publications, London, pp. 35–78.

Thai News Agency (2007), "Thai Researchers Adapt Nanotech to Clean Fish Bowls", MCOT, Available: http://etna.mcot.net/query.php?nid=32270.

Thai Press Reports (2006), "Thailand to Produce Anti-Bacterial Surgical Gowns", Medical Device Link News, Available: http://www.e-topics.com/index.asp?layout=topic_story&UserID=20020903133218091732&topic=162&doc_id=a0516263.8ie.

Thajchayapong, P. and Tanthapanichakoon, W. (2003), "Current Status of Nanotechnology Research in Thailand", paper presented at Nano tech 2003 + Future (International Congress and Exhibition on Nanotechnology), Tokyo, Japan, February 26–28, 2003, Available: www.secretariat.ne.jp/nanofuture/current/pdf/S3-2.pdf.

Thao, T. (2004), "Ho Chi Minh City Thinks Nano Labs", VietNamNet Bridge News, Available: http://english.vietnamnet.vn/tech/2004/09/261046/.

The Associated Press (2007), "Russia to Invest $1b in Nanotechnology", Business Week, Available: www.businessweek.com/ap/financialnews/D8OJ4NS80.htm.

The Ecologist (1992), "Excluding the Commons", The Ecologist, vol. 22, no. 5, p. 1.

The Royal Society (2002), Genetically Modified Plants for Food Use and Human Health — An Update, Conquest Litho, London, 20 pp.

The Royal Society and Royal Academy of Engineering (2004), Nanoscience and Nanotechnologies: Opportunities and Uncertainties, Clyvedon Press, London, 116 pp.

The Tribune (2004), "Kalam Talks Nano to Defence Scientists", The Tribune: Online Edition, Available: http://www.tribuneindia.com/2004/20040702/nation.htm#15.

Third World Academy of Sciences (2004), Building Scientific Capacity: A TWAS Perspective, Third World Academy of Sciences, Trieste, 47 pp.

Third World Network (1997), "Technology Transfer, Intellectual Property Rights and the Environment", Earth Summit Plus Briefing No. 4, Available: http://www.engr.sjsu.edu/sbates/images/tech&civ/tech198_TechnologyTransfer&Environment.pdf.

Thomas, J. (2003), "Promising the World, or Costing the Earth", The Ecologist, vol. 33, no. 4, pp. 28–39.

Thorsteinsdóttir, H., Quach, U., Martin, D.K., Daar, A.S. and Singer, P.A. (2004), "Introduction: Promoting Global Health through Biotechnology", Nature Biotechnology, vol. 22, no. Suppl., pp. DC3-7.

Throne-Holst, H. and Stø, E. (2008), "Who Should Be Precautionary? Governance of Nanotechnology in the Risk Society", Technology Analysis and Strategic Management, vol. 20, no. 1, pp. 99–112.

Todaro, M.P. and Smith, S.C. (2002), Economic Development, 8th Edn, Addison-Wesley Higher Education Group, Upper Saddle River, 784 pp.

Turner, G. (2008), "A Comparison of the Limits to Growth with 30 Years of Reality", Global Environmental Change, vol. 18, no. 3, pp. 397–411.

Uldrich, J. and Newberry, D. (2003), The Next Big Thing Is Really Small, Crown Business, New York, 208 pp.

Ullrich, O. (1997), "Technology", in W. Sachs (ed), The Development Dictionary, 6th Edn, Zed Books, London, pp. 275–287.

UNCTAD-International Centre for Trade and Sustainable Development (2005), Resource Book on TRIPs and Development, Cambridge University Press, Cambridge, U.K., 846 pp.

UNCTAD (1997), Trade and Development Report, 1997, U.N., Geneva, 220 pp.

UNCTAD (1999), World Investment Report 1999: Foreign Direct Investment and the Challenge of Development, United Nations, New York, 577 pp.

UNCTAD (2002a), Key Issues in Biotechnology, United Nations, Geneva, 21 pp.

UNCTAD (2002b), The Least Developed Countries Report 2002, United Nations, New York, 320 pp.

UNCTAD (2002c), Trade and Development Report, 2002, United Nations, Geneva, 198 pp.

UNCTAD (2004), Interactive Dialogue on Harnessing Emerging Technologies to Meet the Millennium Development Goals, Available: http://stdev.unctad.org/unsystem/emerging.htm.

UNCTAD (2006), Bridging the Technology Gap between and within Nations, United Nations, Geneva, 20 pp.

UNDP (1990), Human Development Report 1990: Concept and Measurement of Human Development, Oxford University Press, New York, 189 pp.

UNDP (1996), Human Development Report 1996: Economic Growth and Human Development, U.N., Oxford University Press, New York, 229 pp.

UNDP (1997), Human Development Report 1997: Human Development to Eradicate Poverty, Oxford University Press, New York, 245 pp.

UNDP (1998), Human Development Report 1998: Consumption for Human Development, Oxford University Press, New York, 228 pp.

UNDP (1999), Human Development Report 1999: Globalization with a Human Face, Oxford University Press, New York, 172 pp.

UNDP (2000), Human Development Report 2000: Human Rights and Human Development, Oxford University Press, New York, 290 pp.

UNDP (2001), Human Development Report 2001: Making New Technologies Work for Human Development, Oxford University Press, New York, 264 pp.

UNDP (2003), Human Development Report 2003: Millennium Development Goals: A Compact among Nations to End Human Poverty, Oxford University Press, New York, 367 pp.

UNDP (2005), Human Development Report 2005: International Cooperation at a Crossroads: Aid, Trade and Security in an Unequal World, Oxford University Press, New York, 388 pp.

UNDP (2006), Human Development Report: Beyond Scarcity 2006: Power, Poverty and the Global Water Crisis, Oxford University Press, New York, 440 pp.

UNDP (2007a), Human Development Report 2007/2008: Fighting Climate Change: Human Solidarity in a Divided World, Oxford University Press, New York, 399 pp.

UNDP (2007b), Thailand Human Development Report 2007: Sufficiency Economy and Human Development, UNDP, Bangkok, 152 pp.

UNESCO (2001), Forward-Looking Approaches and Innovative Strategies to Promote the Development of Africa in the Twenty-First Century: Final Report, United Nations, Paris, 54 pp.

UNESCO (2003), "Session on Ethics and Nanotechnology", in Third Session of the World Commission on the Ethics of Scientific Knowledge and Technology (COMEST) (proceedings), UNESCO, Paris, pp. 29–32.

UNESCO (2005), Nanotechnology and Ethics Expert Group: Report of the First Meeting, UNESCO, Paris, 19 pp.

UNESCO (2006), The Ethics and Politics of Nanotechnology, UNESCO, Paris, 25 pp.

Unisearch (2004), Final Report: Survey for Current Situation of Nanotechnology Researchers and R&D in Thailand, Chulalongkorn University, Bangkok, 42 pp.

United Nations Conference on Environment and Development (1993), Report of the United Nations Conference on Environment and Development — Volume 1: Resolutions Adopted by the Conference, United Nations, New York, 486 pp.

United Nations Department of Economic and Social Affairs (2001), Report on the World Social Situation 2001, United Nations, New York, 297 pp.

United Nations Environmental Program (1999), Global Environment Outlook 2000, Earthscan, London, 398 pp.

United Nations Industrial Development Organization (2003), "Aide-Memoire", paper presented at Expert Group Meeting on Nanotechnology, Minsk, November 10–12, 2003, Available: http://200.20.105.7/imaac/News/Nanotechnology_Minsk/ExpertMeeting_AideMemoire_020903.pdf.

van Amerom, M. and Ruivenkamp, M. (2006), "Image Dynamics in Nanotechnology's Risk Debate", paper presented at Second International Seville Seminar on Future-Oriented Technology Analysis: Impact of FTA Approaches on Policy and Decision-Making, Seville, September 28–29, 2006, Available: http://foresight.jrc.ec.europa.eu/documents/eposters/pdf/PapersevilleEind1.pdf.

Vargas, M. (2004), "Costa Rica Opens Region's First Lab for Nanotechnology", SciDevNet, Available: http://www.scidev.net/news/index.cfm?fuseaction=readnews&itemid=1602&language=1.

Vaughan, L. and Thelwall, M. (2004), "Search Engine Coverage Bias: Evidence and Possible Causes ", Information Processing & Management, vol. 40, no. 4, pp. 693–707.

Vermeulen, S., Garside, B. and Weber de Morais, G. (2009), "Shifting the Balance: Equity and Sustainable Consumption", IEED Briefing, Available: http://www.iied.org/pubs/pdfs/17048IIED.pdf.

Vernengo, M. (2006), "Technology, Finance, and Dependency: Latin American Radical Political Economy in Retrospect", Review of Radical Political Economics, vol. 38, no. 4, pp. 551–568.

Viet Nam News Agency (2004), "Viet Nam Initially Penetrates into Nanotechnology", Nanotechwire.com, Available: http://nanotechwire.com/news.asp?nid=655&ntid=116&pg=18.

Vu Long, N.D. (2004), "Vietnam Seeks a High-Tech Strategy", SciDevNet, Available: www.scidev.net/en/new-technologies/nanotechnology/news/vietnam-seeks-a-hightech-strategy.html.

Waddell, J.R.E. (1993), Replanting the Banana Tree: A Study in Ecologically Sustainable Development, APACE and the University of Technology Sydney, Sydney, 95 pp.

Waga, M. (2002), "Emerging Nanotechnology Research in Vietnam", Emerging Technology Report, no. 29, Available: http://www.glocom. org/tech_reviews/geti/20021028_geti_s29/.

Wallerstein, I. (1974), "Dependence in an Interdependent World: The Limited Possibilities of Transformation within the Capitalist World Economy", African Studies Review, vol. 17, no. 1, pp. 1–26.

Warris, C. (2004), Nanotechnology Benchmarking Project, Australian Academy of Science, Canberra, 45 pp.

Waruingi, M. and Njoroge, J. (2008), "Why We Need to Teach Nanotechnology in Kenyan Schools", Business Daily Africa, Available: http://www. bdafrica.com/index.php?option=com_content&task=view&id=7899& Itemid=5847.

Warushamana, G. (2007), "Sri Lanka to Set up Nanotechnology Research Institute", Nanowerk, Available: http://www.nanowerk.com/news/newsid=3816.php.

Watanabe, M. (2003), "Small World, Big Hopes", Nature, vol. 426, no. 6965, pp. 478–479.

Watson, R., Crawford, M. and Farley, S. (2003), Strategic Approaches to Science and Technology in Development, World Bank, New York, 62 pp.

Welland, M. (2003), "Technology at the Limit", paper presented at Is Small Beautiful? A One-Day Conference on the 30th Anniversary of Small is Beautiful by E.F. Schumacher, London, September 3, 2003, Available: http://practicalaction.org/print/small_is_working_transcripts?id=is_small_beautiful_welland.

White, E. (2003), Nano-Robots Not Yet on the Patenting Horizon, Available: http://scientific.thomson.com/knowtrend/ipmatters/nanotech/8238656/.

White, L.A. (1949), The Science of Culture: A Study of Man and Civilisation, Farrar and Strauss, New York, 444 pp.

WHO (1997), The World Health Report 1997: Conquering Suffering, Enriching Humanity, WHO, Geneva, 172 pp.

WHO (2002a), Genomics and World Health: Report of the Advisorary Committee on Health Research, WHO, Geneva, 241 pp.

WHO (2002b), The World Health Report 2002 — Reducing Risks, Promoting Healthy Life, WHO, Geneva, 248 pp.

WHO (2005a), Global Tuberculosis Control — Surveillance, Planning, Financing, WHO, Geneva, 258 pp.

WHO (2005b), Modern Food Biotechnology, Human Health and Development: An Evidence Based Study, WHO, Geneva, 85 pp.

WHO Commission on Health and Environment (1992), Our Planet, Our Health: Report of the WHO Commission on Health and Environment, WHO, Geneva, 282 pp.

Wicklein, R.C. and Kachmar, C.J. (2001), "Philosophical Rationale for Appropriate Technology", in R.C. Wicklein (ed), Appropriate Technology for Sustainable Living: 50th Yearbook of the Council of Technology Teacher Education, Glencoe McGraw-Hill, Peoria, pp. 3–17.

Wickson, F. (2007), "Public Engagement Means Listening as Well as Talking", Nature, vol. 448, no. 7154, p. 644.

Wilkins, G. (2002), Technology Transfer for Renewable Energy: Overcoming Barriers in Developing Countries, Earthscan, London, 237 pp.

Wilkinson, R. and Pickett, K. (2009), The Spirit Level: Why More Equal Societies Almost Always Do Better, Penguin Books, London, 331 pp.

Willis, S. and Tranter, B. (2006), "Beyond the 'Digital Divide': Internet Diffusion and Inequality in Australia ", Journal of Sociology, vol. 42, no. 1, pp. 43–59.

Wiltzius, P. and Klabunde, K. (1999), "Applications: Dispersions, Coatings, and Other Large Surface Area Structures", in P. Alivisatos (ed), Nanotechnology Research Directions: IWGN Workshop Report, International Technology Research Institute, Maryland, pp. 69–70.

Wolbring, G. (2006), "Nanotechnology for Health and Development", Development, vol. 49, no. 4, pp. 6–15.

Wolbring, G. (2007), "Nano-Engagement: Some Critical Issues", Journal of Health and Development, vol. 3, nos. 1 and 2, pp. 9–29.

Wolfensohn, J. (2003), "A Better World Is Possible", The Futurist, July-August, p. 66.

Wolfson, J.R. (2004), "Social and Ethical Issues in Nanotechnology: Lessons from Biotechnology and Other High Technologies", Biotechnology Law Report, vol. 22, no. 4, pp. 376–396.

Women's Environment and Development Organization (1998), Mapping Progress: Assessing Implementation of the Beijing Platform, 1998,

Women's Environment and Development Organization, New York, 214 pp.

Wood, S., Geldart, A. and Jones, R. (2003), The Social and Economic Challenges of Nanotechnology, Economic and Social Research Council, Swindon, 63 pp.

Woodrow Wilson International Centre for Scholars (2009), A Nanotechnology Consumer Products Inventory, Available: http://www.nanotechproject.org/inventories/consumer/analysis_draft/.

Woodward, D. and Simms, A. (2006), Growth Isn't Working: The Unbalanced Distribution of Benefits and Costs from Economic Growth, New Economics Foundation, London, 28 pp.

Wootliff, B. (2003), "Canadian Invites the World to Pool It's Resources on Clean Water", Small Times, Available: www.smalltimes.com/document_display.cfm?document_id=6959.

World Bank (1993), World Development Report 1993: Investing in Health, Oxford University Press, New York, 344 pp.

World Bank (1997), World Development Report: The State in a Changing World, Oxford University Press for the World Bank, New York, 248 pp.

World Bank (1998), Bioengineering of Crops: Report of the World Bank Panel on Transgenic Crops, The World Bank, Washington, 30 pp.

World Bank (2001a), World Development Report 1999/2000: Entering the 21st Century: The Changing Development Landscape, The World Bank, Washington, 312 pp.

World Bank (2001b), World Development Report 2000/2001: Attacking Poverty, Oxford University Press, New York, 544 pp.

World Bank (2003), Aligning Assistance for Development Effectiveness: Promising Country Experience, World Bank, New York, 56 pp.

World Bank (2005), The World Bank Annual Report 2005 — Year in Review, World Bank, Washington, 68 pp.

World Bank (2006), Clean Energy and Development: Towards an Investment Framework, World Bank, Washington, 146 pp.

World Bank (2008a), Global Economic Prospects 2008: Technology Diffusion in the Developing World, World Bank, Washington, 224 pp.

World Bank (2008b), The Growth Report: Strategies for Sustained Growth and Inclusive Development, Commission on Growth and Development, The World Bank, Washington, 198 pp.

World Bank Independent Evaluation Group (2007), Development Results in Middle-Income Countries: An Evaluation of the World Bank's Support, World Bank, Washington, 162 pp.

Xinhua News Agency (2003), "China's Nanotechnology Patent Applications Rank Third in the World", China Daily, Available: http://www.chinadaily.com.cn/en/doc/2003-10/03/content_269182.htm.

Yao, S. (2005), Economic Growth, Income Distribution and Poverty Reduction in Contemporary China, Routledge, New York, 271 pp.

Yergin, D. and Stanislaw, J. (2002), The Commanding Heights: The Battle for the World Economy, Revised and Updated Edn, Simon & Schuster, New York, 512 pp.

Yin, R.K. (1994), Case Study Research: Design and Methods, Sage, Thousand Oaks, 204 pp.

Yonas, G. and Picraux, S.T. (2001), "National Needs Drivers for Nanotechnology", in M.C. Roco and W.S. Bainbridge (eds), Societal Implications of Nanoscience and Nanotechnology, Springer, New York, pp. 37–44.

Zhenzhen, L., Jiuchun, Z., Ke, W., Thorsteinsdóttir, H., Quach, U., Singer, P.A. and Daar, A.S. (2004), "Health Biotechnology in China — Reawakening of a Giant", Nature Biotechnology, vol. 22, no. Supplement, pp. DC13-D18.

Zhou, P. and Leydesdorff, L. (2006), "The Emergence of China as a Leading Nation in Science", Research Policy, vol. 35, no. 1, pp. 83–104.

Index